T0329343

Robust Automatic Speech Recognition

A Bridge to Practical Applications

Robust Automatic Speech Recognition
A Bridge to Practical Applications

Jinyu Li

Li Deng

Reinhold Haeb-Umbach

Yifan Gong

AMSTERDAM • BOSTON • HEIDELBERG • LONDON
NEW YORK • OXFORD • PARIS • SAN DIEGO
SAN FRANCISCO • SINGAPORE • SYDNEY • TOKYO
Academic Press is an imprint of Elsevier

Academic Press is an imprint of Elsevier
225 Wyman Street,Waltham,MA 02451, USA
The Boulevard, Langford Lane, Kidlington, Oxford OX5 1GB, UK

Notices
Knowledge and best practice in this field are constantly changing. As new research and experience
broaden our understanding, changes in research methods, professional practices, or medical treatment
may become necessary.

Practitioners and researchers must always rely on their own experience and knowledge in evaluating and
using any information, methods, compounds, or experiments described herein. In using such information
or methods they should be mindful of their own safety and the safety of others, including parties for
whom they have a professional responsibility.

To the fullest extent of the law, neither the Publisher nor the authors, contributors, or editors, assume any
liability for any injury and/or damage to persons or property as a matter of products liability, negligence
or otherwise, or from any use or operation of any methods, products, instructions, or ideas contained in
the material herein.

Library of Congress Cataloging-in-Publication Data
A catalog record for this book is available from the Library of Congress

British Library Cataloguing in Publication Data
A catalogue record for this book is available from the British Library

ISBN: 978-0-12-802398-3

For information on all Academic Press publications
visit our website at http://store.elsevier.com/

Typeset by SPi Global, India
www.spi-global.com

Printed in USA

Working together
to grow libraries in
developing countries

www.elsevier.com • www.bookaid.org

Contents

About the Authors

Jinyu Li received Ph.D. degree from Georgia Institute of Technology, U.S.A. From 2000 to 2003, he was a Researcher at Intel China Research Center and a Research Manager at iFlytek, China. Currently, he is a Principal Applied Scientist at Microsoft, working as a technical lead to design and to improve speech modeling algorithms and technologies that ensure industry state-of-the-art speech recognition accuracy for Microsoft products. His major research interests cover several topics in speech recognition and machine learning, including noise robustness, deep learning, discriminative training, and feature extraction. He has authored over 60 papers and awarded over 10 patents.

Li Deng received Ph.D. degree from the University of Wisconsin-Madison, U.S.A. He was a professor (1989-1999) at the University of Waterloo, Canada. In 1999, he joined Microsoft Research, where he currently leads R&D of application-focused deep learning as Partner Research Manager of its Deep Learning Technology Center. He is also an Affiliate Professor at University of Washington. He is a Fellow of the Acoustical Society of America, Fellow of the IEEE, and Fellow of the International Speech Communication Association. He served as Editor-in-Chief for the IEEE Signal Processing Magazine and for the IEEE/ACM Transactions on Audio, Speech and Language Processing (2009-2014). His technical work has been focused on deep learning for speech, language, image, and multimodal processing, and for other areas of machine intelligence involving big data. He received numerous awards including the IEEE SPS Best Paper Awards, IEEE Outstanding Engineer Award, and APSIPA Industrial Distinguished Leader Award.

Reinhold Haeb-Umbach is a professor with the University of Paderborn, Germany. His main research interests are in the fields of statistical signal processing and pattern recognition, with applications to speech enhancement, acoustic beamforming and source separation, as well as automatic speech recognition. After having worked in industrial research laboratories for more than 10 years, he joined academia as a full professor of Communications Engineering in 2001. He has published more than 150 papers in peer reviewed journals and conferences. He is the co-editor of the book *Robust Speech Recognition of Uncertain or Missing Data—Theory and Applications* (Springer, 2011).

Yifan Gong received Ph.D. (with highest honors) from the University of Henri Poincaré, France. He served the National Scientific Research Center (CNRS) and INRIA, France, as Research Engineer and then joined CNRS as Senior Research Scientist. He was a Visiting Research Fellow at the Communications Research Center of Canada. As Senior Member of Technical Staff, he worked for Texas Instruments at the Speech Technologies Lab, where he developed speech modeling technologies robust against noisy environments, designed systems, algorithms, and software for speech and speaker recognition, and delivered memory- and CPU-efficient recognizers for mobile devices.

He joined Microsoft in 2004, and is currently a Principal Applied Science Manager in the areas of speech modeling, computing infrastructure, and speech model development for speech products. His research interests include automatic speech recognition/interpretation, signal processing, algorithm development, and engineering process/infrastructure and management. He has authored over 130 publications and awarded over 30 patents. Specific contribution includes stochastic trajectory modeling, source normalization HMM training, joint compensation of additive and convolutional noises, and variable parameter HMM. In these areas, he gave tutorials and other invited presentations in international conferences. He has been serving as member of technical committee and session chair for many international conferences, and with IEEE Signal Processing Spoken Language Technical Committees from 1998 to 2002 and since 2013.

List of Figures

List of Tables

Acronyms

AE	autoencoder
AFE	advanced front-end
AIR	acoustic impulse response
ALSD	average localized synchrony detection
ANN	artificial neural network
ASGD	asynchronous stochastic gradient descent
ASR	automatic speech recognition
ATF	acoustic transfer function
BFE	Bayesian feature enhancement
BLSTM	bidirectional long short-term memory
BM	blocking matrix
BMMI	boosted maximum mutual information
BN	bottle-neck
BPC	Bayesian prediction classification
BPTT	backpropagation through time
CAT	cluster adaptive training
CDF	cumulative distribution function
CHiME	computational hearing in multisource environments
CMN	cepstral mean normalization
CMMSE	cepstral minimum mean square error
CMLLR	constrained maximum likelihood linear regression
CMVN	cepstral mean and variance normalization
CNN	convolutional neural network
COSINE	conversational speech in noisy environments
CSN	cepstral shape normalization
CTF	convolutive transfer function
DAE	denoising autoencoder
DBN	deep belief net
DCT	discrete cosine transform
DMT	discriminative mapping transformation
DNN	deep neural network
DPMC	data-driven parallel model combination
DSB	delay-sum beamformer
DSR	distributed speech recognition
DT	discriminative training
EDA	environment-dependent activation
ELR	early-to-late reverberation ratio
EM	expectation-maximization
ESSEM	ensemble speaker and speaking environment modeling
ETSI	European telecommunications standards institute

FBF	fixed beamformer
FCDCN	fixed codeword-dependent cepstral normalization
FIR	finite impulse response
fMPE	feature space minimum phone error
FT	feature transform
GMM	gaussian mixture model
GSC	generalized sidelobe canceller
HEQ	histogram equalization
HLDA	heteroscedastic linear discriminant analysis
HMM	hidden Markov model
IBM	ideal binary mask
IDCT	inverse discrete cosine transform
IIF	invariant-integration features
IIR	infinite impulse response
IRM	ideal ratio mask
IVN	irrelevant variability normalization
JAC	jointly compensate for additive and convolutive
JAT	joint adaptive training
JUD	joint uncertainty decoding
KLD	Kullback-Leibler divergence
LCMV	linearly constrained minimum variance
LHN	linear hidden network
LHUC	learning hidden unit contribution
LIN	linear input network
LMPSC	logarithmic Mel power spectral coefficient
LMS	least mean square
LP	linear prediction
LON	linear output network
MAP	maximum *a posteriori*
MAPLR	maximum *a posteriori* linear regression
MBR	minimum Bayes risk
MC	Monte-Carlo
MCE	minimum classification error
MFCC	Mel-frequency cepstral coefficient
MFCDCN	multiple fixed codeword-dependent cepstral normalization
MIMO	multiple-input multiple-output
MINT	multiple input/output inverse theorem
MLE	maximum likelihood estimation
MLLR	maximum likelihood linear regression
MLP	multi-layer perceptron
MMIE	maximum mutual information estimation
MMSE	minimum mean square error
MWE	minimum word error
MWF	multi-channel wiener filter

MPDCN	multiple phone-dependent cepstral normalization
MPE	minimum phone error
MTF	multiplicative transfer function
MVDR	minimum variance distortionless response
NAT	noise adaptive training
NC	noise cancellation
NMF	non-negative matrix factorization
LDA	linear discriminant analysis
LRSV	late reverberant spectral variance
LSTM	long short-term memory
PCA	principal component analysis
PDF	probability density function
PCMLLR	predictive constrained maximum likelihood linear regression
PDCN	phone-dependent cepstral normalization
PHEQ	polynomial-fit histogram equalization
PMC	parallel model combination
PLP	perceptually based linear prediction
PMVDR	perceptual minimum variance distortionless response
PNCC	power-normalized cepstral coefficients
PSD	power spectral density
QHEQ	quantile-based histogram equalization
RASTA	relative spectral processing
ReLU	rectified linear units
REVERB	reverberant voice enhancement and recognition benchmark
RNN	recurrent neural network
RTF	relative transfer functions
SAT	speaker adaptive training
SC	sparse classification
SDCN	SNR-dependent cepstral normalization
SDW-MWF	speech distortion weighted multi-channel Wiener filter
SE	speech enhancement
SGD	stochastic gradient descent
SMAP	structural maximum *a posteriori*
SMAPLR	structural maximum *a posteriori* linear regression
SME	soft margin estimation
SLDM	switching linear dynamic model
SNR	signal-to-noise ratio
SNT	source normalization training
SPARK	sparse auditory reproducing kernel
SPDCN	SNR-phone-dependent cepstral normalization
SPINE	speech in noisy environments
SPLICE	stereo-based piecewise linear compensation for environments
SS	spectral subtraction
STDFT	short-time discrete fourier transform

SVD	singular value decomposition
SVM	support vector machine
THEQ	table-based histogram equalization
TRAP	temporal pattern
TSN	temporal structure normalization
UBM	universal background model
ULA	uniform linear array
UT	unscented transform
VAD	voice activity detector
VADNN	variable-activation deep neural network
VCDNN	variable-component deep neural network
VIDNN	variable-input deep neural network
VODNN	variable-output deep neural network
VPDNN	variable-parameter deep neural network
VPHMM	variable-parameter hidden Markov model
VTLN	vocal tract length normalization
VTS	vector Taylor series
VQ	vector quantization
WER	word error rate
WPE	weighted prediction error
WSJ	Wall Street Journal
ZCPA	zero crossing peak amplitude

Notations

Mathematical language is an essential tool in this book. We thus introduce our mathematical notations right from the start in the following table, separated in five general categories. Throughout this book, both matrices and vectors are in bold type, and matrices are capitalized.

Definitions of a Subset of Commonly Used Symbols and Notations, Grouped in Five Separate General Categories

General notation

$\approx$	approximately equal to
$\propto$	proportional to
s	scalar quantity (lowercase plain letter)
$\mathbf{v}$	vector quantity (lowercase bold letter)
v_i	the ith element of vector $\mathbf{v}$
$\mathbf{M}$	matrix (uppercase bold letter)
m_{ij}	the (i,j)th element of the matrix $\mathbf{M}$
$\mathbf{M}^T$	transpose of matrix bfM
$\lvert \cdot \rvert$	determinant of a square matrix
$(\cdot)^{-1}$	inverse of a square matrix
$\mathrm{diag}(\mathbf{v})$	diagonal matrix with vector $\mathbf{v}$ as its diagonal elements
$\mathrm{diag}(\mathbf{M})$	diagonal matrix derived from a squared matrix $\mathbf{M}$
$.*$	element-wise production

Functions

$F(\cdot)$	objective function or a mapping function
$Q(\cdot;\cdot)$	auxiliary function at the current estimates of parameters
$\frac{\partial}{\partial \mathbf{x}}f(\mathbf{x})$	derivate of a function
$\frac{\partial^2}{\partial \mathbf{x}\partial \mathbf{x}}f(\mathbf{x})$	Hessian of a function
$f(\mathbf{x})\big|_{\mathbf{x}=\hat{\mathbf{x}}}$	value of function $f(\mathbf{x})$ at $\mathbf{x}=\hat{\mathbf{x}}$
$\arg\max_{\mathbf{x}} f(\mathbf{x})$	value of $\mathbf{x}$ that maximizes $f(\mathbf{x})$
$\arg\min_{\mathbf{x}} f(\mathbf{x})$	value of $\mathbf{x}$ that minimizes $f(\mathbf{x})$

Probability distributions

$p(\cdot)$	probability density function
$p(\cdot\lvert\cdot)$	conditional probability density
$P(\cdot)$	probability mass distribution
$P(\cdot\lvert\cdot)$	conditional probability mass distribution
$\sigma(\cdot)$	sigmoid function

HMM model parameters and speech sequence

T	number of frames in a speech sequence
Λ	acoustic model parameter
Γ	language model parameter
$\hat{\Lambda}$	adapted acoustic model parameter
$\mathbf{W}$	hypothesis or word sequence
D	dimension of feature vector
a_{ij}	discrete state transition probability from state i to state j
$\mathbf{X}$	sequence of clean speech vectors $(\mathbf{x}_1, \mathbf{x}_2, \dots, \mathbf{x}_T)$
$\mathbf{Y}$	sequence of distorted speech vectors $(\mathbf{y}_1, \mathbf{y}_2, \dots, \mathbf{y}_T)$
$\hat{\mathbf{X}}$	estimated sequence of clean speech vectors
$\boldsymbol{\theta}$	sequence of speech states $(\theta_1, \theta_2, \dots, \theta_T)$
θ_t	speech state at time t
$\mathcal{N}(\mathbf{x}; \mu, \Sigma)$	Gaussian multivariate distributions of $\mathbf{x}$
$c(m)$	weight for the mth Gaussian component
$\gamma_t(m)$	posterior probability of component m at time t
$\mathbf{v}^l$	the input at the lth layer in a deep neural network (DNN)
$\mathbf{A}^l$	the weight matrix at the lth layer in a DNN
$\mathbf{b}^l$	the bias vector at the lth layer in a DNN
$\mathbf{e}^l$	the error signal at the lth layer in a DNN

Environment robustness

r	utterance index
$\mathbf{C}$	discrete cosine transform (DCT) matrix
$\mathbf{x}$	clean speech feature
$\mathbf{y}$	distorted speech feature
$\mathbf{n}$	noise feature
$\mathbf{h}$	channel feature
$\mu_{\mathbf{x}}$	clean speech mean
$\mu_{\mathbf{y}}$	distorted speech mean
$\mu_{\mathbf{n}}$	noise mean
$\mathbf{A}$	linear transform matrix
$\mathbf{b}$	bias vector
$\mathbf{W}$	affine transform, $\mathbf{W} = [\mathbf{Ab}]$
r_m	regression class corresponding to mth Gaussian component

Introduction

1

CHAPTER OUTLINE

1.1 AUTOMATIC SPEECH RECOGNITION

Automatic speech recognition (ASR) is the process and the related technology for converting the speech signal into its corresponding sequence of words or other linguistic entities by means of algorithms implemented in a device, a computer, or computer clusters (Deng and O'Shaughnessy, 2003; Huang et al., 2001b). ASR by machine has been a field of research for more than 60 years (Baker et al., 2009a,b; Davis et al., 1952). The industry has developed a broad range of commercial products where speech recognition as user interface has become ever useful and pervasive.

Historically, ASR applications have included voice dialing, call routing, interactive voice response, data entry and dictation, voice command and control, gaming, structured document creation (e.g., medical and legal transcriptions), appliance control by voice, computer-aided language learning, content-based spoken audio search, and robotics. More recently, with the exponential growth of big data and computing power, ASR technology has advanced to the stage where more challenging applications are becoming a reality. Examples are voice search, digital assistance and interactions with mobile devices (e.g., Siri on iPhone, Bing voice search and Cortana on winPhone and Windows 10 OS, and Google Now on Android), voice control in home entertainment systems (e.g., Kinect on xBox), machine translation, home automation, in-vehicle navigation and entertainment, and various speech-centric information processing applications capitalizing on downstream processing of ASR outputs (He and Deng, 2013).

Robust Automatic Speech Recognition. http://dx.doi.org/10.1016/B978-0-12-802398-3.00001-5

1.2 ROBUSTNESS TO NOISY ENVIRONMENTS

New waves of consumer-centric applications increasingly require ASR to be robust to the full range of real-world noise and other acoustic distorting conditions. However, reliably recognizing spoken words in realistic acoustic environments is still a challenge. For such large-scale, real-world applications, noise robustness is becoming an increasingly important core technology since ASR needs to work in much more difficult acoustic environments than in the past (Deng et al., 2002).

Noise refers to any unwanted disturbances superposed upon the intended speech signal. Robustness is the ability of a system to maintain its good performance under varying operating conditions, including those unforeseeable or unavailable at the time of system development.

Speech as observed and digitized is generated by a complex process, from the thoughts to actual speech signals. This process can be described in five stages as shown in Figure 1.1, where a number of variables affect the outcome of each stage. Some major stages in this long chain have been analyzed and modeled mathematically in Deng (1999, 2006).

All of the above could lead to ASR robustness issues. This book addresses challenges mostly in the acoustic channel area where interfering signals lead to ASR performance degradation.

In this area, robustness of ASR to noisy background can be approached from two directions:

- reducing the noise level by exploring hardware utilizing spatial or directional information from microphone technology and transducer principles, such as noise canceling microphones and microphone arrays;
- software algorithmic processing taking advantage of the spectral and temporal separation between speech and interfering signals, which is the major focus of this book.

1.3 EXISTING SURVEYS IN THE AREA

Researchers and practitioners have been trying to improve ASR robustness to operating conditions for many years (Huang et al., 2001a; Huang and Deng, 2010). A survey of the 1970s speech recognition systems has identified (Lea, 1980) that "a primary difficulty with speech recognition is this ability of the input to pick up other sounds in the environment that act as interfering noise." The term "robust speech recognition" emerged in the late 1980s. Survey papers in the 1990s include (Gong, 1995; Juang, 1991; Junqua and Haton, 1995). By 2000, robust speech recognition has gained significant importance in the speech and language processing fields. Actually, it was the most popular area in the International Conference on Acoustics, Speech and Signal Processing, at least during 2001-2003 (Gong, 2004). Since 2010, robust ASR remains one of the most popular areas in the speech processing community, and tremendous and steady progress in noisy speech recognition have been made.

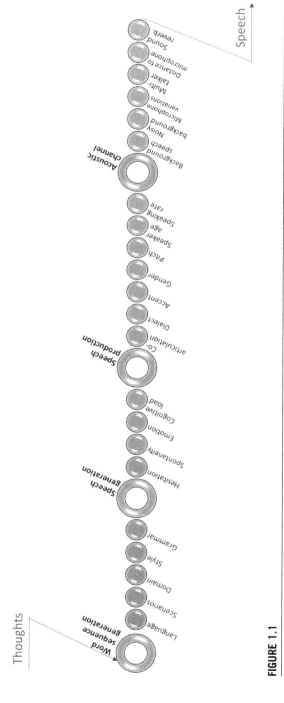

FIGURE 1.1

From thoughts to speech.

A large number of noise-robust ASR methods, in the order of hundreds, have been proposed and published over the past 30 years or so, and many of them have created significant impact on either research or commercial use. Such accumulated knowledge deserves thorough examination not only to define the state of the art in this field from a fresh and unifying perspective, but also to point to potentially fruitful future directions. Nevertheless, a well-organized framework for relating and analyzing these methods is conspicuously missing. The existing survey papers (Acero, 1993; Deng, 2011; Droppo and Acero, 2008; Gales, 2011; Gong, 1995; Haeb-Umbach, 2011; Huo and Lee, 2001; Juang, 1991; Kumatani et al., 2012; Lee, 1998) in noise-robust ASR either do not cover all recent advances in the field or focus only on a specific sub-area. Although there are also few recent books (Kolossa and Haeb-Umbach, 2011; Virtanen et al., 2012), they are collections of topics with each chapter written by different authors and it is hard to provide a unified view across all topics. Given the importance of noise-robust ASR, the time is ripe to analyze and unify the solutions. The most recent overview paper (Li et al., 2014) elaborates on the basic concepts in noise-robust ASR and develops categorization criteria and unifying themes. Specifically, it hierarchically classifies the major and significant noise-robust ASR methods using a consistent and unifying mathematical language. It establishes their interrelations and differentiates among important techniques, and discusses current technical challenges and future research directions. It also identifies relatively promising, short-term new research areas based on a careful analysis of successful methods, which can serve as a reference for future algorithm development in the field. Furthermore, in the literature spanning over 30 years on noise-robust ASR, there is inconsistent use of basic concepts and terminology as adopted by different researchers in the field. This kind of inconsistency is confusing at times, especially for new researchers and students. It is, therefore, important to examine discrepancies in the current literature and re-define a consistent terminology. However, due to the restriction of page length, the overview paper (Li et al., 2014) did not discuss the technologies in depth. More importantly, all the aforementioned books and articles largely assumed that the acoustic models for ASR are based on Gaussian mixture model hidden Markov models (GMM-HMMs).

More recently, a new acoustic modeling technique, referred to as the context-dependent deep neural network hidden Markov model (CD-DNN-HMM) which employs deep learning, has been developed (Deng and Yu, 2014; Yu and Deng, 2011, 2014). This new DNN-based acoustic model has been shown, by many groups, to significantly outperform the conventional state-of-the-art GMM-HMMs in many ASR tasks (Dahl et al., 2012; Hinton et al., 2012). As of the writing of this book, DNN-based ASR has been widely adopted by almost all major speech recognition products and public tools worldwide.

DNNs combine acoustic feature extraction and speech phonetic symbol classification into a single framework. By design, they ensure that both feature extraction and classification are jointly optimized under a discriminative criterion. With their complex non-linear mapping built on top of successive applications of simple non-linear mapping, DNNs force input features distorted by a variety of noise and channels as well as other factors to be mapped to a same output vector of phonetic symbol classes. Such an ability provides the potential for substantial performance improvement in noisy speech recognition.

However, while DNNs dramatically reduce overall word error rate for speech recognition, many new questions are raised: How much are DNNs more robust than GMMs? How should we introduce a physical model of speech, noise, and channel into a DNN model so that a better DNN can be trained given the same data? Will feature cleaning for a DNN add value to the DNN modeling? Can we model speech with a DNN such that complete, expensive retraining can be avoided upon a change in noise? To what extend the noise robustness methods developed for GMMs can enhance the robustness of DNNs? etc. More generally, what the future of noise-robust ASR technologies would hold in the new era of DNNs for ASR is a question not addressed in the existing survey literature on noise-robust ASR. One of the main goals of this book is to survey the recent noise-robust methods developed for DNNs as the acoustic models of speech, and to discuss the future research directions.

1.4 BOOK STRUCTURE OVERVIEW

This book is devoted to providing a summary of the current, fast expanding knowledge and approaches to solving a variety of problems in noise-robust ASR. A more specific purpose is to assist readers in acquiring a structured understanding of the state of the art and to continue to enrich the knowledge.

In this book, we aim to establish a solid, consistent, and common mathematical foundation for noise-robust ASR. We emphasize the methods that are proven to be effective and successful and that are likely to sustain or expand their future applicability. For the methods described in this book, we attempt to present the basic ideas, the assumptions, and the relationships with other methods. We categorize a wide range of noise-robust techniques using different criteria to equip the reader with the insight to choose among techniques and with the awareness of the performance-complexity tradeoffs. The pros and cons of using different noise-robust ASR techniques in practical application scenarios are provided as a guide to interested practitioners. The current challenges and future research directions especially in the era of DNNs and deep learning are carefully analyzed.

This book is organized as follows. We provide the basic concepts and formulations of ASR in Chapter 2. In Chapter 3, we discuss the fundamentals of noise-robust ASR. The impact of noise and channel distortions on clean speech is examined. Then, we build a general framework for noise-robust ASR and define five ways of categorizing and analyzing noise-robust ASR techniques. Chapter 4 is devoted to the first category—feature-domain vs. model-domain techniques. Various feature-domain processing methods are covered in detail, including noise-resistant features, feature moment normalization, and feature compensation, as well as a few most prominent model-domain methods. The second category, detailed in Chapter 5, comprises methods that exploit prior knowledge about the signal distortion.

Examples of such models are mapping functions between the clean and noisy speech features, and environment-specific models combined during online operation of the noise-robust algorithms. Methods that incorporate an explicit distortion model to predict the distorted speech from a clean one define the third category, covered in Chapter 6. The use of uncertainty constitutes the fourth way to categorize a wide range of noise-robust ASR algorithms, and is covered in Chapter 7. Uncertainty in either the model space or feature space may be incorporated within the Bayesian framework to promote noise-robust ASR. The final, fifth way to categorize and analyze noise-robust ASR techniques exploits joint model training, described in Chapter 8. With joint model training, environmental variability in the training data is removed in order to generate canonical models. After the noise-robust techniques for single-microphone non-reverberant ASR are comprehensively discussed above, the book includes two chapters, covering reverberant ASR and multi-channel processing for noise-robust ASR, respectively. We conclude this book in Chapter 11, with discussions on future directions for noise-robust ASR.

REFERENCES

Acero, A., 1993. Acoustical and Environmental Robustness in Automatic Speech Recognition. Cambridge University Press, Cambridge, UK.

Baker, J., Deng, L., Glass, J., Khudanpur, S., Lee, C.H, Morgan, N., et al., 2009a. Research developments and directions in speech recognition and understanding, Part I. IEEE Signal Process. Mag. 26 (3), 75-80.

Baker, J., Deng, L., Glass, J., Khudanpur, S., Lee, C.H, Morgan, N., et al., 2009b. Updated minds report on speech recognition and understanding (research developments and directions in speech recognition and understanding, Part II). IEEE Signal Process. Mag. 26 (4), 78-85.

Dahl, G., Yu, D., Deng, L, Acero, A. 2012. Context-dependent pre-trained deep neural networks for large-vocabulary speech recognition. IEEE Trans. Audio Speech Lang. Process. 20 (1), 30-42.

Davis, K.H., Biddulph, R, Balashek, S., 1952. Automatic recognition of spoken digits. J. Acoust. Soc. Am. 24 (6), 627-642.

Deng, L., 1999. Computational models for speech production. In: Computational Models of Speech Pattern Processing. Springer-Verlag, New York, pp. 199-213.

Deng, L., 2006. Dynamic Speech Models—Theory, Algorithm, and Applications. Morgan and Claypool, San Rafael, CA.

Deng, L., 2011. Front-end, back-end, and hybrid techniques for noise-robust speech recognition. In: Robust Speech Recognition of Uncertain or Missing Data: Theory and Application. Springer, New York, pp. 67-99.

Deng, L., O'Shaughnessy, D., 2003. Speech Processing-A Dynamic and Optimization-Oriented Approach. Marcel Dekker Inc., New York.

Deng, L., Wang, K, A. Acero, H.H., Huang, X., 2002. Distributed speech processing in MiPad's multimodal user interface. IEEE Trans. Audio Speech Lang. Process. 10 (8), 605-619.

Deng, L., Yu, D., 2014. Deep Learning: Methods and Applications. Now Publishers, Hanover, MA.

Droppo, J., Acero, A., 2008. Environmental robustness. In: Benesty, J, Sondhi, M.M., Huang, Y. (Eds.), Handbook of Speech Processing. Springer, New York.

Gales, M.J.F., 2011. Model-based approaches to handling uncertainty. In: Robust Speech Recognition of Uncertain or Missing Data: Theory and Application. Springer, New York, pp. 101-125.

Gong, Y., 1995. Speech recognition in noisy environments: A survey. Speech Commun. 16, 261-291.

Gong, Y., 2004. Speech recognition in noisy environments on mobile devices—a tutorial. In: IEEE International Conference on Acoustics, Speech, and Signal Processing.

Haeb-Umbach, R., 2011. Uncertainty decoding and conditional Bayesian estimation. In: Robust Speech Recognition of Uncertain or Missing Data: Theory and Application. Springer, New York, pp. 9-34.

He, X., Deng, L., 2013. Speech-centric information processing: An optimization-oriented approach. Proc. IEEE 101 (5), 1116-1135.

Hinton, G., Deng, L., Yu, D., Dahl, G.E., Mohamed, A., Jaitly, N., et al., 2012. Deep neural networks for acoustic modeling in speech recognition: The shared views of four research groups. IEEE Signal Process. Mag. 29 (6), 82-97.

Huang, X., Acero, A, Chelba, C., Deng, L, Droppo, J., Duchene, D., et al., 2001a. MiPad: a multimodal interaction prototype. In: Proc. International Conference on Acoustics, Speech and Signal Processing (ICASSP).

Huang, X., Acero, A., Hon, H.W., 2001b. Spoken Language Processing. Prentice-Hall, Upper Saddle River, NJ.

Huang, X., Deng, L., 2010. An overview of modern speech recognition. In: Indurkhya, N, Damerau, F.J. (Eds.), Handbook of Natural Language Processing, 2nd ed. CRC Press, Taylor and Francis Group, Boca Raton, FL.

Huo, Q., Lee, C.H., 2001. Robust speech recognition based on adaptive classification and decision strategies. Speech Commun. 34 (1-2), 175-194.

Juang, B., 1991. Speech recognition in adverse environments. Comput. Speech Lang. 5 (3), 275-294.

Junqua, J.C., Haton, J.P., 1995. Robustness in Automatic Speech Recognition: Fundamentals and Applications. Kluwer Academic Publishers, Boston, MA.

Kolossa, D., Haeb-Umbach, R. (Eds.), 2011. Robust Speech Recognition of Uncertain or Missing Data: Theory and Applications. Springer, New York.

Kumatani, K., McDonough, J.W, Raj, B., 2012. Microphone array processing for distant speech recognition: From close-talking microphones to far-field sensors. IEEE Signal Process. Mag. 29 (6), 127-140.

Lea, W.A., 1980. The value of speech recognition systems. In: Trends in Speech Recognition. Prentice Hall, Upper Saddle River, NJ, pp. 3-18.

Lee, C.H., 1998. On stochastic feature and model compensation approaches to robust speech recognition. Speech Commun. 25, 29-47.

Li, J., Deng, L., Gong, Y, Haeb-Umbach, R., 2014. An overview of noise-robust automatic speech recognition. IEEE/ACM Trans. Audio Speech Lang. Process. 22 (4), 745-777.

Virtanen, T., Singh, R, Raj, B. (Eds.), 2012. Techniques for noise robustness in automatic speech recognition. John Wiley & Sons, West Sussex, UK.

Yu, D., Deng, L., 2011. Deep learning and its applications to signal and information processing. IEEE Signal Processing Mag. 28, pp. 145-154.

Yu, D., Deng, L., 2014. Automatic Speech Recognition—A Deep Learning Approach. Springer, New York.

Fundamentals of speech recognition

CHAPTER OUTLINE

2.1 INTRODUCTION: COMPONENTS OF SPEECH RECOGNITION

Speech recognition has been an active research area for many years. It is not until recently, over the past 2 years or so, the technology has passed the usability bar for many real-world applications under most realistic acoustic environments (Yu and Deng, 2014). Speech recognition technology has started to change the way we live and work and has become one of the primary means for humans to interact with mobile devices (e.g., Siri, Google Now, and Cortana). The arrival of this new trend is attributed to the significant progress made in a number of areas. First, Moore's law continues to dramatically increase computing power, which, through multi-core processors, general purpose graphical processing units, and clusters, is nowadays several orders of magnitude higher than that available only a decade ago (Baker et al., 2009a,b; Yu and Deng, 2014). The high power of computation

Robust Automatic Speech Recognition. http://dx.doi.org/10.1016/B978-0-12-802398-3.00002-7

makes training of powerful deep learning models possible, dramatically reducing the error rates of speech recognition systems (Sak et al., 2014a). Second, much more data are available for training complex models than in the past, due to the continued advances in Internet and cloud computing. Big models trained with big and real-world data allow us to eliminate unrealistic model assumptions (Bridle et al., 1998; Deng, 2003; Juang, 1985), creating more robust ASR systems than in the past (Deng and O'Shaughnessy, 2003; Huang et al., 2001b; Rabiner, 1989). Finally, mobile devices, wearable devices, intelligent living room devices, and in-vehicle infotainment systems have become increasingly popular. On these devices, interaction modalities such as keyboard and mouse are less convenient than in personal computers. As the most natural way of human-human communication, speech is a skill that all people already are equipped with. Speech, thus, naturally becomes a highly desirable interaction modality on these devices.

From the technical point of view, the goal of speech recognition is to predict the optimal word sequence $\mathbf{W}$, given the spoken speech signal $\mathbf{X}$, where optimality refers to maximizing the *a posteriori* probability (maximum *a posteriori*, MAP) :

$$\hat{\mathbf{W}} = \underset{\mathbf{W}}{\mathrm{argmax}}\, P_{\Lambda,\Gamma}(\mathbf{W}|\mathbf{X}), \tag{2.1}$$

where Λ and Γ are the acoustic model and language model parameters. Using Bayes' rule

$$P_{\Lambda,\Gamma}(\mathbf{W}|\mathbf{X}) = \frac{p_\Lambda(\mathbf{X}|\mathbf{W})P_\Gamma(\mathbf{W})}{p(\mathbf{X})}, \tag{2.2}$$

Equation 2.1 can be re-written as:

$$\hat{\mathbf{W}} = \underset{\mathbf{W}}{\mathrm{argmax}}\, p_\Lambda(\mathbf{X}|\mathbf{W})P_\Gamma(\mathbf{W}), \tag{2.3}$$

where $p_\Lambda(\mathbf{X}|\mathbf{W})$ is the AM likelihood and $P_\Gamma(\mathbf{W})$ is the LM probability. When the time sequence is expanded and the observations $\mathbf{x}_t$ are assumed to be generated by hidden Markov models (HMMs) with hidden states θ_t, we have

$$\hat{\mathbf{W}} = \underset{\mathbf{W}}{\mathrm{argmax}}\, P_\Gamma(\mathbf{W}) \sum_\theta \prod_{t=1}^{T} p_\Lambda(\mathbf{x}_t|\theta_t)P_\Lambda(\theta_t|\theta_{t-1}), \tag{2.4}$$

where θ belongs to the set of all possible state sequences for the transcription W. The speech signal is first processed by the feature extraction module to obtain the acoustic feature. The feature extraction module is often referred as the front-end of speech recognition systems. The acoustic features will be passed to the acoustic model and the language model to compute the probability of the word sequence under consideration. The output is a word sequence with the largest probability from acoustic and language models. The combination of acoustic and language models are usually referred as the back-end of speech recognition systems. The focus of this book is on the noise-robustness of front-end and acoustical model, therefore, the robustness of language model is not considered in the book.

Acoustic models are used to determine the likelihood of acoustic feature sequences given hypothesized word sequences. The research in speech recognition has been under a long period of development since the HMM was introduced in 1980s as the acoustic model (Juang, 1985; Rabiner, 1989). The HMM is able to gracefully represent the temporal evolution of speech signals and characterize it as a parametric random process. Using the Gaussian mixture model (GMM) as its output distribution, the HMM is also able to represent the spectral variation of speech signals.

In this chapter, we will first review the GMM, and then review the HMM with the GMM as its output distribution. Finally, the recent development in speech recognition has demonstrated superior performance of the deep neural network (DNN) over the GMM in discriminating speech classes (Dahl et al., 2011; Yu and Deng, 2014). A review of the DNN and related deep models will thus be provided.

2.2 GAUSSIAN MIXTURE MODELS

As part of acoustic modeling in ASR and according to how the acoustic emission probabilities are modeled for the HMMs' state, we can have discrete HMMs (Liporace, 1982), semi-continuous HMMs (Huang and Jack, 1989), and continuous HMMs (Levinson et al., 1983). For the continuous output density, the most popular one is the Gaussian mixture model (GMM), in which the state output density is modeled as:

$$P_\Lambda(o) = \sum_i c(i) \mathcal{N}(o; \mu(i), \sigma^2(i)), \qquad (2.5)$$

where $\mathcal{N}(o; \mu(i), \sigma^2(i))$ is a Gaussian with mean $\mu(i)$ and variance $\sigma^2(i)$, and $c(i)$ is the weight for the ith Gaussian component. Three fundamental problems of HMMs are probability evaluation, determination of the best state sequence, and parameter estimation (Rabiner, 1989). The probability evaluation can be realized easily with the forward algorithm (Rabiner, 1989).

The parameter estimation is solved with maximum likelihood estimation (MLE) (Dempster et al., 1977) using a forward-backward procedure (Rabiner, 1989). The quality of the acoustic model is the most important issue for ASR. MLE is known to be optimal for density estimation, but it often does not lead to minimum recognition error, which is the goal of ASR. As a remedy, several discriminative training (DT) methods have been proposed in recent years to boost ASR system accuracy. Typical methods are maximum mutual information estimation (MMIE) (Bahl et al., 1997), minimum classification error (MCE) (Juang et al., 1997), minimum word/phone error (MWE/MPE) (Povey and Woodland, 2002), minimum Bayes risk (MBR) (Gibson and Hain, 2006), and boosted MMI (BMMI) (Povey et al., 2008). Other related methods can be found in He and Deng (2008), He et al. (2008), and Xiao et al. (2010).

Inspired by the high success of margin-based classifiers, there is a trend towards incorporating the margin concept into hidden Markov modeling for ASR. Several attempts based on margin maximization were proposed, with three major classes of methods: large margin estimation (Jiang et al., 2006; Li and Jiang, 2007), large margin HMMs (Sha, 2007; Sha and Saul, 2006), and soft margin estimation (SME) (Li et al., 2006, 2007b). The basic concept behind all these margin-based methods is that by securing a margin from the decision boundary to the nearest training sample, a correct decision can still be made if the mismatched test sample falls within a tolerance region around the original training samples defined by the margin.

The main motivations of using the GMM as a model for the distribution of speech features are discussed here. When speech waveforms are processed into compressed (e.g., by taking logarithm of) short-time Fourier transform magnitudes or related cepstra, the GMM has been shown to be quite appropriate to fit such speech features when the information about the temporal order is discarded. That is, one can use the GMM as a model to represent frame-based speech features.

Both inside and outside the ASR domain, the GMM is commonly used for modeling the data and for statistical classification. GMMs are well known for their ability to represent arbitrarily complex distributions with multiple modes. GMM-based classifiers are highly effective with widespread use in speech research, primarily for speaker recognition, denoising speech features, and speech recognition. For speaker recognition, the GMM is directly used as a universal background model (UBM) for the speech feature distribution pooled from all speakers. In speech feature denoising or noise tracking applications, the GMM is used in a similar way and as a prior distribution for speech (Deng et al., 2003, 2002a,b; Frey et al., 2001a; Huang et al., 2001a). In ASR applications, the GMM is integrated into the doubly stochastic model of HMM as its output distribution conditioned on a state, which will be discussed later in more detail.

GMMs have several distinct advantages that make them suitable for modeling the distributions over speech feature vectors associated with each state of an HMM. With enough components, they can model distributions to any required level of accuracy, and they are easy to fit to data using the EM algorithm. A huge amount of research has gone into finding ways of constraining GMMs to increase their evaluation speed and to optimize the tradeoff between their flexibility and the amount of training data required to avoid overfitting. This includes the development of parameter- or semi-tied GMMs and subspace GMMs.

Despite all their advantages, GMMs have a serious shortcoming. That is, GMMs are statistically inefficient for modeling data that lie on or near a nonlinear manifold in the data space. For example, modeling the set of points that lie very close to the surface of a sphere only requires a few parameters using an appropriate model class, but it requires a very large number of diagonal Gaussians or a fairly large number of full-covariance Gaussians. It is well known that speech is produced by modulating a relatively small number of parameters of a dynamical system (Deng, 1999, 2006; Lee et al., 2001). This suggests that the true underlying structure of speech is of a much lower dimension than is immediately apparent in a window that contains hundreds of coefficients. Therefore, other types of model that can capture better the properties of speech features are expected to work better than GMMs for acoustic modeling of speech. In particular, the new models should more effectively exploit information embedded in a large window of frames of speech features than GMMs. We will return to this important problem of characterizing speech features after discussing a model, the HMM, for characterizing temporal properties of speech next.

2.3 HIDDEN MARKOV MODELS AND THE VARIANTS

As a highly special or degenerative case of the HMM, we have the Markov chain as an information source capable of generating observational output sequences. Then we can call the Markov chain an observable (non-hidden) Markov model because its output has one-to-one correspondence to a state in the model. That is, each state corresponds to a deterministically observable variable or event. There is no randomness in the output in any given state. This lack of randomness makes the Markov chain too restrictive to describe many real-world informational sources, such as speech feature sequences, in an adequate manner.

The Markov property, which states that the probability of observing a certain value of the random process at time t only depends on the immediately preceding observation at $t - 1$, is rather restrictive in modeling correlations in a random process. Therefore, the Markov chain is extended to give rise to a HMM, where the states, that is, the values of the Markov chain, are "hidden" or non-observable. This extension is accomplished by associating an observation probability distribution with each state in the Markov chain. The HMM thus defined is a doubly embedded random sequence whose underlying Markov chain is not directly observable. The underlying Markov chain in the HMM can be observed only through a separate random function characterized by the observation probability distributions. Note that the observable random process is no longer a Markov process and thus the probability of an observation not only depends on the immediately preceding observations.

2.3.1 HOW TO PARAMETERIZE AN HMM

We can give a formal parametric characterization of an HMM in terms of its model parameters:

1. State transition probabilities, $\mathbf{A} = [a_{ij}]$, $i, j = 1, 2, \ldots, N$, of a homogeneous Markov chain with a total of N states

$$a_{ij} = P(\theta_t = j | \theta_{t-1} = i), \qquad i, j = 1, 2, \ldots, N. \tag{2.6}$$

2. Initial Markov chain state-occupation probabilities: $\pi = [\pi_i]$, $i = 1, 2, \ldots, N$, where $\pi_i = P(\theta_1 = i)$.
3. Observation probability distribution, $P(\mathbf{o}_t | \theta_t = i)$, $i = 1, 2, \ldots, N$. if $\mathbf{o}_t$ is discrete, the distribution associated with each state gives the probabilities of symbolic observations $\{\mathbf{v}_1, \mathbf{v}_2, \ldots, \mathbf{v}_K\}$:

$$b_i(k) = P[\mathbf{o}_t = \mathbf{v}_k | \theta_t = i], \qquad i = 1, 2, \ldots, N. \tag{2.7}$$

If the observation probability distribution is continuous, then the parameters, Λ_i, in the probability density function (PDF) characterize state i in the HMM.

The most common and successful distribution used in ASR for characterizing the continuous observation probability distribution in the HMM is the GMM discussed in the preceding section. The GMM distribution with vector-valued observations ($\mathbf{o}_t \in \mathcal{R}^D$) has the mathematical form:

$$
\begin{aligned}
b_i(\mathbf{o}_t) &= P(\mathbf{o}_t | \theta_t = i) \\
&= \sum_{m=1}^{M} \frac{c(i,m)}{(2\pi)^{D/2} |\mathbf{\Sigma}(i,m)|^{1/2}} \exp\left[-\frac{1}{2}(\mathbf{o}_t - \boldsymbol{\mu}(i,m))^{\mathrm{T}} \mathbf{\Sigma}^{-1}(i,m)(\mathbf{o}_t - \boldsymbol{\mu}(i,m))\right]
\end{aligned}
$$
$$(2.8)$$

In this GMM-HMM, the parameter set Λ_i comprises scalar mixture weights, $c(i,m)$, Gaussian mean vectors, $\boldsymbol{\mu}(i,m) \in \mathcal{R}^D$, and Gaussian covariance matrices, $\mathbf{\Sigma}(i,m) \in \mathcal{R}^{D \times D}$.

When the number of mixture components is reduced to one: $M = 1$, the state-dependent output PDF reverts to a (uni-modal) Gaussian:

$$b_i(\mathbf{o}_t) = \frac{1}{(2\pi)^{D/2} |\mathbf{\Sigma}_i|^{1/2}} \exp\left[-\frac{1}{2}(\mathbf{o}_t - \boldsymbol{\mu}(i))^{\mathrm{T}} \mathbf{\Sigma}^{-1}(i)(\mathbf{o}_t - \boldsymbol{\mu}(i))\right] \tag{2.9}$$

and the corresponding HMM is commonly called a (continuous-density) Gaussian HMM.

2.3.2 EFFICIENT LIKELIHOOD EVALUATION FOR THE HMM

Likelihood evaluation is a basic task needed for speech processing applications involving an HMM that uses a hidden Markov sequence to approximate vectorized speech features.

Let $\theta_1^T = (\theta_1, \ldots, \theta_T)$ be a finite-length sequence of states in a Gaussian-mixture HMM or GMM-HMM, and let $P(\mathbf{o}_1^T, \theta_1^T)$ be the joint likelihood of the observation sequence $\mathbf{o}_1^T = (\mathbf{o}_1, \ldots, \mathbf{o}_T)$ and the state sequence θ_1^T. Let $P(\mathbf{o}_1^T | \theta_1^T)$ denote the likelihood that the observation sequence $\mathbf{o}_1^T$ is generated by the model conditioned on the state sequence θ_1^T.

In the Gaussian-mixture HMM, the conditional likelihood $P(\mathbf{o}_1^T | \theta_1^T)$ is in the form of

$$
\begin{aligned}
P(\mathbf{o}_1^T | \boldsymbol{\theta}_1^T) = \prod_{t=1}^{T} b_i(\mathbf{o}_t) &= \prod_{t=1}^{T} \sum_{m=1}^{M} \frac{c(i,m)}{(2\pi)^{D/2} |\mathbf{\Sigma}_{i,m}|^{1/2}} \\
&\exp\left[-\frac{1}{2}(\mathbf{o}_t - \boldsymbol{\mu}(i,m))^{T} \mathbf{\Sigma}^{-1}(i,m)(\mathbf{o}_t - \boldsymbol{\mu}(i,m))\right]
\end{aligned}
\tag{2.10}
$$

On the other hand, the probability of state sequence θ_1^T is just the product of transition probabilities, that is,

$$P(\boldsymbol{\theta}_1^T) = \pi_{\theta_1} \prod_{t=1}^{T-1} a_{\theta_t \theta_{t+1}}. \tag{2.11}$$

In the remainder of the chapter, for notational simplicity, we consider the case where the initial state distribution has probability of one in the starting state: $\pi_1 = P(\theta_1 = 1) = 1$.

Note that the joint likelihood $P(\mathbf{o}_1^T, \theta_1^T)$ can be obtained by the product of likelihoods in Equations 2.10 and 2.11:

$$P(\mathbf{o}_1^T, \boldsymbol{\theta}_1^T) = P(\mathbf{o}_1^T | \boldsymbol{\theta}_1^T) P(\boldsymbol{\theta}_1^T). \tag{2.12}$$

In principle, the total likelihood for the observation sequence can be computed by summing the joint likelihoods in Equation 2.12 over all possible state sequences θ_1^T:

$$P(\mathbf{o}_1^T) = \sum_{\boldsymbol{\theta}_1^T} P(\mathbf{o}_1^T, \boldsymbol{\theta}_1^T). \tag{2.13}$$

However, the computational effort is exponential in the length of the observation sequence, T, and hence the naive computation of $P(\mathbf{o}_1^T)$ is not tractable. The forward-backward algorithm (Baum and Petrie, 1966) computes $P(\mathbf{o}_1^T)$ for the HMM with complexity linear in T.

To describe this algorithm, we first define the forward probabilities by

$$\alpha_t(i) = P(\theta_t = i, \mathbf{o}_1^t), \quad t = 1, \dots, T, \tag{2.14}$$

and the backward probabilities by

$$\beta_t(i) = P(\mathbf{o}_{t+1}^T | \theta_t = i), \quad t = 1, \dots, T - 1, \tag{2.15}$$

both for each state i in the Markov chain. The forward and backward probabilities can be calculated recursively from

$$\alpha_t(j) = \sum_{i=1}^{N} \alpha_{t-1}(i) a_{ij} b_j(\mathbf{o}_t), \quad t = 2, 3, \dots, T; \quad j = 1, 2, \dots, N \tag{2.16}$$

$$\beta_t(i) = \sum_{j=1}^{N} \beta_{t+1}(j) a_{ij} b_j(\mathbf{o}_{t+1}), \quad t = T - 1, T - 2, \dots, 1; \quad i = 1, 2, \dots, N \tag{2.17}$$

Proofs of these recursions are given in the following section. The starting value for the α recursion is, according to the definition in Equation 2.14,

$$\alpha_1(i) = P(\theta_1 = i, \mathbf{o}_1) = P(\theta_1 = i) P(\mathbf{o}_1 | \theta_1) = \pi_i b_i(\mathbf{o}_1), \quad i = 1, 2, \dots N \tag{2.18}$$

and that for the β recursion is chosen as

$$\beta_T(i) = 1, \quad i = 1, 2, ...N, \tag{2.19}$$

so as to provide the correct values for β_{T-1} according to the definition in Equation 2.15.

To compute the total likelihood $P(\mathbf{o}_1^T)$ in Equation 2.13, we first compute

$$
\begin{aligned}
P(\theta_t = i, \mathbf{o}_1^T) &= P(\theta_t = i, \mathbf{o}_1^t, \mathbf{o}_{t+1}^T) \\
&= P(\theta_t = i, \mathbf{o}_1^t) P(\mathbf{o}_{t+1}^T | \mathbf{o}_1^t, \theta_t = i) \\
&= P(\theta_t = i, \mathbf{o}_1^t) P(\mathbf{o}_{t+1}^T | \theta_t = i) \\
&= \alpha_t(i) \beta_t(i),
\end{aligned}
\tag{2.20}
$$

for each state i and $t = 1, 2, \ldots, T$ using definitions in Equations 2.14 and 2.15. Note that $P(\mathbf{o}_{t+1}^T | \mathbf{o}_1^t, \theta_t = i) = P(\mathbf{o}_{t+1}^T | \theta_t = i)$ because the observations are independent, given the state in the HMM. Given this, $P(\mathbf{o}_1^T)$ can be computed as

$$P(\mathbf{o}_1^T) = \sum_{i=1}^{N} P(\theta_t = i, \mathbf{o}_1^T) = \sum_{i=1}^{N} \alpha_t(i) \beta_t(i). \tag{2.21}$$

With Equation 2.20 we find for the posterior probability of being in state i at time t given the whole sequence of observed data

$$\gamma_t(i) = P(\theta_t = i | \mathbf{o}_1^T) = \frac{\alpha_t(i) \beta_t(i)}{P(\mathbf{o}_1^T)}. \tag{2.22}$$

Further, we can find for the posterior probability of the state-transition probabilities

$$\xi_t(i, j) = P(\theta_t = j, \theta_{t-1} = i | \mathbf{o}_1^T) = \frac{\alpha_{t-1}(i) a_{ij} P(\mathbf{o}_t | \theta_t = j) \beta_t(j)}{P(\mathbf{o}_1^T)} \tag{2.23}$$

These posteriors are needed to learn about the HMM parameters, as will be explained in the following section.

Taking $t = T$ in Equation 2.21 and using Equation 2.19 lead to

$$P(\mathbf{o}_1^T) = \sum_{i=1}^{N} \alpha_T(i). \tag{2.24}$$

Thus, strictly speaking, the backward recursion, Equation 2.17 is not necessary for the forward scoring computation, and hence the algorithm is often called the forward algorithm. However, the β computation is a necessary step for solving the model parameter estimation problem, which will be briefly described in the following section.

2.3.3 EM ALGORITHM TO LEARN THE HMM PARAMETERS

Despite many unrealistic aspects of the HMM as a model for speech feature sequences, one most important reason for its wide-spread use in ASR is the Baum-Welch algorithm developed in 1960s (Baum and Petrie, 1966), which is a prominent instance of the highly popular EM (expectation-maximization) algorithm (Dempster et al., 1977), for efficient training of the HMM parameters from data.

The EM algorithm is a general iterative technique for maximum likelihood estimation, with local optimality, when hidden variables exist. When such hidden variables take the form of a Markov chain, the EM algorithm becomes the Baum-Welch algorithm. Here we use a Gaussian HMM as the example to describe steps involved in deriving E- and M-step computations, where the complete data in the general case of EM above consists of the observation sequence and the hidden Markov-chain state sequence; that is, $[\mathbf{o}_1^T, \theta_1^T]$.

Each iteration in the EM algorithm consists of two steps for any incomplete data problem including the current HMM parameter estimation problem. In the E (expectation) step of the Baum-Welch algorithm, the following conditional expectation, or the auxiliary function $Q(\theta|\theta_0)$, need to be computed:

$$Q(\Lambda; \Lambda_0) = E[\log P(\mathbf{o}_1^T, \boldsymbol{\theta}_1^T|\Lambda)|\mathbf{o}_1^T, \Lambda_0], \tag{2.25}$$

where the expectation is taken over the "hidden" state sequence θ_1^T. For the EM algorithm to be of utility, $Q(\Lambda; \Lambda_0)$ has to be sufficiently simplified so that the M (maximization) step can be carried out easily. Estimates of the model parameters are obtained in the M-step via maximization of $Q(\Lambda; \Lambda_0)$, which is in general much simpler than direct procedures for maximizing $P(\mathbf{o}_1^T|\Lambda)$.

An iteration of the above two steps will lead to maximum likelihood estimates of model parameters with respect to the objective function $P(\mathbf{o}_1^T|\Lambda)$. This is a direct consequence of Baum's inequality (Baum and Petrie, 1966), which asserts that

$$\log\left(\frac{P(\mathbf{o}_1^T|\Lambda)}{P(\mathbf{o}_1^T|\Lambda_0)}\right) \geq Q(\Lambda; \Lambda_0) - Q(\Lambda_0; \Lambda_0).$$

After carrying out the E- and M-steps for the Gaussian HMM, details of which are omitted here but can be found in Rabiner (1989) and Huang et al. (2001b), we can establish the re-estimation formulas for the maximum-likelihood estimates of its parameters:

$$\hat{a}_{ij} = \frac{\sum_{t=1}^{T-1} \xi_t(i,j)}{\sum_{t=1}^{T-1} \gamma_t(i)}, \tag{2.26}$$

where $\xi_t(i,j)$ and $\gamma_t(i)$ are the posterior state-transition and state-occupancy probabilities computed from the E-step.

The re-estimation formula for the covariance matrix in state i of an HMM can be derived to be

$$\hat{\boldsymbol{\Sigma}}_i = \frac{\sum_{t=1}^{T} \gamma_t(i)(\mathbf{o}_t - \hat{\boldsymbol{\mu}}(i))(\mathbf{o}_t - \hat{\boldsymbol{\mu}}(i))^{\mathrm{T}}}{\sum_{t=1}^{T} \gamma_t(i)} \tag{2.27}$$

for each state: $i = 1, 2, \ldots, N$, where $\hat{\boldsymbol{\mu}}(i)$ is the re-estimate of the mean vector in the Gaussian HMM in state i, whose re-estimation formula is also straightforward to derive and has the following easily interpretable form:

$$\hat{\boldsymbol{\mu}}(i) = \frac{\sum_{t=1}^{T} \gamma_t(i)\mathbf{o}_t}{\sum_{t=1}^{T} \gamma_t(i)}. \tag{2.28}$$

The above derivation is for the single-Gaussian HMM. The EM algorithm for the GMM-HMM can be similarly determined by considering the Gaussian component of each frame at each state as another hidden variable. In a later section, we will describe the deep neural network (DNN)-HMM hybrid system in which the observation probability is estimated using a DNN.

2.3.4 HOW THE HMM REPRESENTS TEMPORAL DYNAMICS OF SPEECH

The popularity of the HMM in ASR stems from its ability to serve as a generative sequence model of acoustic features of speech; see excellent reviews of HMMs for selected speech modeling and recognition applications as well as the limitations of HMMs in Rabiner (1989), Jelinek (1976), Baker (1976), and Baker et al. (2009a,b). One most interesting and unique problem in speech modeling and in the related speech recognition application lies in the nature of variable length in acoustic-feature sequences. This unique characteristic of speech rests primarily in its temporal dimension. That is, the actual values of the speech feature are correlated lawfully with the elasticity in the temporal dimension. As a consequence, even if two word sequences are identical, the acoustic data of speech features typically have distinct lengths. For example, different acoustic samples from the same sentence usually contain different data dimensionality, depending on how the speech sounds are produced and in particular how fast the speaking rate is. Further, the discriminative cues among separate speech classes are often distributed over a reasonably long temporal span, which often crosses neighboring speech units. Other special aspects of speech include class-dependent acoustic cues. These cues are often expressed over diverse time spans that would benefit from different lengths of analysis windows in speech analysis and feature extraction.

Conventional wisdom posits that speech is a one-dimensional temporal signal in contrast to image and video as higher dimensional signals. This view is simplistic and does not capture the essence and difficulties of the speech recognition problem. Speech is best viewed as a two-dimensional signal, where the spatial (or frequency or tonotopic) and temporal dimensions have vastly different characteristics, in contrast to images where the two spatial dimensions tend to have similar properties.

The spatial dimension in speech is associated with the frequency distribution and related transformations, capturing a number of variability types including primarily those arising from environments, speakers, accent, and speaking style and rate. The latter induces correlations between spatial and temporal dimensions, and the environment factors include microphone characteristics, speech transmission channel, ambient noise, and room reverberation.

The temporal dimension in speech, and in particular its correlation with the spatial or frequency-domain properties of speech, constitutes one of the unique challenges for speech recognition. The HMM addresses this challenge to a limited extent. In the following two sections, a selected set of advanced generative models, as various extensions of the HMM, will be described that are aimed to address the same challenge, where Bayesian approaches are used to provide temporal constraints as prior knowledge about aspects of the physical process of human speech production.

2.3.5 GMM-HMMs FOR SPEECH MODELING AND RECOGNITION

In speech recognition, one most common generative learning approach is based on the Gaussian-mixture-model based hidden Markov models, or GMM-HMM (Bilmes, 2006; Deng and Erler, 1992; Deng et al., 1991a; Juang et al., 1986; Rabiner, 1989; Rabiner and Juang, 1993). As discussed earlier, a GMM-HMM is a statistical model that describes two dependent random processes, an observable process and a hidden Markov process. The observation sequence is assumed to be *generated* by each hidden state according to a Gaussian mixture distribution. A GMM-HMM is parameterized by a vector of state prior probabilities, the state transition probability matrix, and by a set of state-dependent parameters in Gaussian mixture models. In terms of modeling speech, a state in the GMM-HMM is typically associated with a sub-segment of a phone in speech. One important innovation in the use of HMMs for speech recognition is the introduction of context-dependent states (Deng et al., 1991b; Huang et al., 2001b), motivated by the desire to reduce output variability of speech feature vectors associated with each state, a common strategy for "detailed" generative modeling. A consequence of using context dependency is a vast expansion of the HMM state space, which, fortunately, can be controlled by regularization methods such as state tying. It turns out that such context dependency also plays a critical role in the recent advance of speech recognition in the area of discrimination-based deep learning (Dahl et al., 2011, 2012; Seide et al., 2011; Yu et al., 2010).

The introduction of the HMM and the related statistical methods to speech recognition in mid-1970s (Baker, 1976; Jelinek, 1976) can be regarded as the most significant paradigm shift in the field, as discussed and analyzed in Baker et al. (2009a,b). One major reason for this early success is the highly efficient EM algorithm (Baum and Petrie, 1966), which we described earlier in this chapter. This maximum likelihood method, often called Baum-Welch algorithm, had been a principal way of training the HMM-based speech recognition systems until 2002, and is still one major step (among many) in training these systems nowadays. It is interesting to note that the Baum-Welch algorithm serves as one major motivating example for the later development of the more general EM algorithm (Dempster et al., 1977). The goal of maximum likelihood or EM method in training GMM-HMM speech recognizers is to minimize the empirical risk with respect to the joint likelihood loss involving a sequence of linguistic labels and a sequence of acoustic data of speech, often extracted at the frame level. In large-vocabulary speech recognition systems, it is normally the case that word-level labels are provided, while state-level labels are latent. Moreover, in training GMM-HMM-based speech recognition systems, parameter tying is often used as a type of regularization. For example, similar acoustic states of the triphones can share the same Gaussian mixture model.

The use of the generative model of HMMs for representing the (piecewise stationary) dynamic speech pattern and the use of EM algorithm for training the tied HMM parameters constitute one of the most prominent and successful examples of generative learning in speech recognition. This success has been firmly established by the speech community, and has been spread widely to machine learning and related communities. In fact, the HMM has become a standard tool not only in speech recognition but also in machine learning as well as their related fields such as bioinformatics and natural language processing. For many machine learning as well as speech recognition researchers, the success of HMMs in speech recognition is a bit surprising due to the well-known weaknesses of the HMM in modeling speech dynamics. The following section is aimed to address ways of using more advanced dynamic generative models and related techniques for speech modeling and recognition.

2.3.6 HIDDEN DYNAMIC MODELS FOR SPEECH MODELING AND RECOGNITION

Despite great successes of GMM-HMMs in speech modeling and recognition, their weaknesses, such as the conditional independence and piecewise stationary assumptions, have been well known for speech modeling and recognition applications since early days (Bridle et al., 1998; Deng, 1992, 1993; Deng et al., 1994a; Deng and Sameti, 1996; Deng et al., 2006a; Ostendorf et al., 1996, 1992). Conditional independence refers to the fact the observation probability at time t only depends on the state θ_t and is independent of the preceding states or observations, if θ_t is given.

Since early 1990s, speech recognition researchers have begun the development of statistical models that capture more realistically the dynamic properties of speech in the temporal dimension than HMMs do. This class of extended HMM models have been variably called stochastic segment model (Ostendorf et al., 1996, 1992), trended or nonstationary-state HMM (Chengalvarayan and Deng, 1998; Deng, 1992; Deng et al., 1994a), trajectory segmental model (Holmes and Russell, 1999; Ostendorf et al., 1996), trajectory HMM (Zen et al., 2004; Zhang and Renals, 2008), stochastic trajectory model (Gong et al., 1996), hidden dynamic model (Bridle et al., 1998; Deng, 1998, 2006; Deng et al., 1997; Ma and Deng, 2000, 2003, 2004; Picone et al., 1999; Russell and Jackson, 2005), buried Markov model (Bilmes, 2003, 2010; Bilmes and Bartels, 2005), structured speech model, and hidden trajectory model (Deng, 2006; Deng and Yu, 2007; Deng et al., 2006a,b; Yu and Deng, 2007; Yu et al., 2006; Zhou et al., 2003), depending on different "prior knowledge" applied to the temporal structure of speech and on various simplifying assumptions to facilitate the model implementation. Common to all these beyond-HMM model variants is some temporal dynamic structure built into the models. Based on the nature of such structure, we can classify these models into two main categories. In the first category are the models focusing on the temporal correlation structure at the "surface" acoustic level. The second category consists of deep hidden or latent dynamics, where the underlying speech production mechanisms are exploited as a prior to represent the temporal structure that accounts for the visible speech pattern. When the mapping from the hidden dynamic layer to the visible layer is limited to be linear and deterministic, then the generative hidden dynamic models in the second category reduce to the first category.

The temporal span in many of the generative dynamic/trajectory models above is often controlled by a sequence of linguistic labels, which segment the full sentence into multiple regions from left to right; hence the name segment models.

2.4 DEEP LEARNING AND DEEP NEURAL NETWORKS
2.4.1 INTRODUCTION

Deep learning is a set of algorithms in machine learning that attempt to model high-level abstractions in data by using model architectures composed of multiple non-linear transformations. It is part of a broader family of machine learning methods based on learning representations of data. The Deep Neural Network (DNN) is the most important and popular deep learning model, especially for the applications in speech recognition (Deng and Yu, 2014; Yu and Deng, 2014).

In the long history of speech recognition, both shallow forms and deep forms (e.g., recurrent nets) of artificial neural networks had been explored for many years during 1980s, 1990s and a few years into 2000 (Boulard and Morgan, 1993; Morgan and Bourlard, 1990; Neto et al., 1995; Renals et al., 1994; Waibel et al., 1989). But these methods never won over the GMM-HMM technology based on generative models of speech acoustics that are trained discriminatively (Baker et al., 2009a,b). A number of key difficulties had been methodologically analyzed in 1990s, including gradient diminishing and weak temporal correlation structure in the neural predictive models (Bengio, 1991; Deng et al., 1994b). All these difficulties were in addition to the lack of big training data and big computing power in these early days. Most speech recognition researchers who understood such barriers hence subsequently moved away from neural nets to pursue generative modeling approaches until the recent resurgence of deep learning starting around 2009-2010 that had overcome all these difficulties.

The use of deep learning for acoustic modeling was introduced during the later part of 2009 by the collaborative work between Microsoft and the University of Toronto, which was subsequently expanded to include IBM and Google

(Hinton et al., 2012; Yu and Deng, 2014). Microsoft and University of Toronto co-organized the 2009 NIPS Workshop on Deep Learning for Speech Recognition (Deng et al., 2009), motivated by the urgency that many versions of deep and dynamic generative models of speech could not deliver what speech industry wanted. It is also motivated by the arrival of a big-compute and big-data era, which would warrant a serious try of the DNN approach. It was then (incorrectly) believed that pre-training of DNNs using generative models of deep belief net (DBN) would be the cure for the main difficulties of neural nets encountered during 1990s. However, soon after the research along this direction started at Microsoft Research, it was discovered that when large amounts of training data are used and especially when DNNs are designed correspondingly with large, context-dependent output layers, dramatic error reduction occurred over the then state-of-the-art GMM-HMM and more advanced generative model-based speech recognition systems without the need for generative DBN pre-training. The finding was verified subsequently by several other major speech recognition research groups. Further, the nature of recognition errors produced by the two types of systems was found to be characteristically different, offering technical insights into how to artfully integrate deep learning into the existing highly efficient, run-time speech decoding system deployed by all major players in speech recognition industry.

One fundamental principle of deep learning is to do away with hand-crafted feature engineering and to use raw features. This principle was first explored successfully in the architecture of deep autoencoder on the "raw" spectrogram or linear filter-bank features (Deng et al., 2010), showing its superiority over the Mel-Cepstral features which contain a few stages of fixed transformation from spectrograms. The true "raw" features of speech, waveforms, have more recently been shown to produce excellent larger-scale speech recognition results (Tuske et al., 2014).

Large-scale automatic speech recognition is the first and the most convincing successful case of deep learning in the recent history, embraced by both industry and academic across the board. Between 2010 and 2014, the two major conferences on signal processing and speech recognition, IEEE-ICASSP and Interspeech, have seen near exponential growth in the numbers of accepted papers in their respective annual conferences on the topic of deep learning for speech recognition. More importantly, all major commercial speech recognition systems (e.g., Microsoft Cortana, Xbox, Skype Translator, Google Now, Apple Siri, Baidu and iFlyTek voice search, and a range of Nuance speech products, etc.) nowadays are based on deep learning methods.

Since the initial successful debut of DNNs for speech recognition around 2009-2011, there has been huge progress made. This progress (as well as future directions) has been summarized into the following eight major areas in Deng and Yu (2014) and Yu and Deng (2014): (1) scaling up/out and speedup DNN training and decoding; (2) sequence discriminative training of DNNs; (3) feature processing by deep models with solid understanding of the underlying mechanisms; (4) adaptation of DNNs and of related deep models; (5) multi-task and transfer learning by DNNs and related deep models; (6) convolution neural networks and how to design them to best exploit domain knowledge of speech; (7) recurrent neural network and its rich long short-term memory (LSTM) variants; (8) other types of deep models including tensor-based models and integrated deep generative/discriminative models.

2.4.2 A BRIEF HISTORICAL PERSPECTIVE

For many years and until the recent rise of deep learning technology as discussed earlier, speech recognition technology had been dominated by a "shallow" architecture—HMMs with each state characterized by a GMM. While significant technological successes had been achieved using complex and carefully engineered variants of GMM-HMMs and acoustic features suitable for them, researchers had for long anticipated that the next generation of speech recognition would require solutions to many new technical challenges under diversified deployment environments and that overcoming these challenges would likely require *deep* architectures that can at least functionally emulate the human speech recognition system known to have dynamic and hierarchical structure in both speech production and speech perception (Deng, 2006; Deng and O'Shaughnessy, 2003; Divenyi et al., 2006; Stevens, 2000). An attempt to incorporate a primitive level of understanding of this deep speech structure, initiated at the 2009 NIPS Workshop on Deep Learning for Speech Recognition (Deng et al., 2009; Mohamed et al., 2009) has helped create an impetus in the speech recognition community to pursue a deep representation learning approach based on the DNN architecture, which was pioneered by the machine learning community only a few years earlier (Hinton et al., 2006; Hinton and Salakhutdinov, 2006) but rapidly evolved into the new state of the art in speech recognition with industry-wide adoption (Deng et al., 2013b; Hannun et al., 2014; Hinton et al., 2012; Kingsbury et al., 2012; Sainath et al., 2013a; Seide et al., 2011, 2014; Vanhoucke et al., 2011, 2013; Yu and Deng, 2011; Yu et al., 2010).

In the remainder of this section, we will describe the DNN and related methods with some technical detail.

2.4.3 THE BASICS OF DEEP NEURAL NETWORKS

The most successful version of the DNN in speech recognition is the context-dependent deep neural network hidden Markov model (CD-DNN-HMM) , where the HMM is interfaced with the DNN to handle the dynamic process of speech feature sequences and context-dependent phone units, also known as the senones, are used as the output layer of the DNN. It has been shown by many groups (Dahl et al., 2011, 2012; Deng et al., 2013b; Hinton et al., 2012; Mohamed et al., 2012; Sainath et al., 2011, 2013b; Tuske et al., 2014; Yu et al., 2010), to outperform the conventional GMM-HMMs in many ASR tasks.

The CD-DNN-HMM is a hybrid system. Three key components of this system are shown in Figure 2.1, which is based on Dahl et al. (2012). First, the CD-DNN-HMM models senones (tied states) directly, which can be as many as tens of thousands

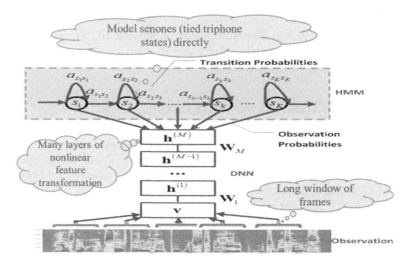

FIGURE 2.1

Illustration of the CD-DNN-HMM and its three core components.

of senones in English, making the output layer of the DNN unprecedentedly large. Second, a deep instead of a shallow multi-layer perceptrons are used. Third, the system takes a long and fixed contextual window of frames as the input. All these three elements of the CD-DNN-HMM have been shown to be critical for achieving the huge accuracy improvement in speech recognition (Dahl et al., 2012; Deng et al., 2013c; Sainath et al., 2011; Yu et al., 2010). Although some conventional shallow neural nets also took a long contextual window as the input, the key to the success of the CD-DNN-HMM is due to a combination of these components. In particular, the deep structure in the DNN allows the system to perform transfer or multi-task learning (Ghoshal et al., 2013; Heigold et al., 2013; Huang et al., 2013), outperforming the shallow models that are unable to carry out transfer learning (Lin et al., 2009; Plahl et al., 2011; Schultz and Waibel, 1998; Yu et al., 2009).

Further, it is shown in Seltzer et al. (2013) and many other research groups that with the excellent modeling power of the DNN, DNN-based acoustic models can easily match state-of-the-art performance on the Aurora 4 task (Parihar and Picone, 2002), which is a standard noise-robustness large-vocabulary speech recognition task, without any explicit noise compensation. The CD-DNN-HMM is expected to make further progress on noise-robust ASR due to the DNN's ability to handle heterogeneous data (Li et al., 2012; Seltzer et al., 2013). Although the CD-DNN-HMM is a modeling technology, its layer-by-layer setup provides a feature extraction strategy that automatically derives powerful noise-resistant features from primitive raw data for senone classification.

From the architecture point of view, a DNN can be considered as a conventional multi-layer perceptron (MLP) with many hidden layers (thus deep) as illustrated in Figure 2.1, in which the input and output of the DNN are denoted as $\mathbf{x}$ and $\mathbf{o}$, respectively. Let us denote the input vector at layer l as $\mathbf{v}^l$ (with $\mathbf{v}^0 = \mathbf{x}$), the weight matrix as $\mathbf{A}^l$, and bias vector as $\mathbf{b}^l$. Then, for a DNN with L hidden layers, the output of the lth hidden layer can be written as

$$\mathbf{v}^{l+1} = \sigma(z(\mathbf{v}^l)), \quad 0 \le l < L, \tag{2.29}$$

where

$$\mathbf{u}^l = z(\mathbf{v}^l) = \mathbf{A}^l \mathbf{v}^l + \mathbf{b}^l \tag{2.30}$$

and

$$\sigma(\mathbf{x}) = 1/(1 + e^{\mathbf{x}}) \tag{2.31}$$

is the sigmoid function applied element-wise. The posterior probability is

$$P(o = s|\mathbf{x}) = \text{softmax}(z(\mathbf{v}^L)), \tag{2.32}$$

where s belongs to the set of senones (also known as the tied triphone states) . We compute the HMM's state emission probability density function $p(\mathbf{x}|o = s)$ by converting the state posterior probability $P(o = s|\mathbf{x})$ to

$$p(\mathbf{x}|o = s) = \frac{P(o = s|\mathbf{x})}{P(o = s)} p(\mathbf{x}), \tag{2.33}$$

where $P(o = s)$ is the prior probability of state s, and $p(\mathbf{x})$ is independent of state and can be dropped during evaluation.

Although recent studies (Senior et al., 2014; Zhang and Woodland, 2014) started the DNN training from scratch without using GMM-HMM systems, in most implementations the CD-DNN-HMM inherits the model structure, especially in the output layer including the phone set, the HMM topology, and senones, directly from the GMM-HMM system. The senone labels used to train the DNNs are extracted from the forced alignment generated by the GMM-HMM. The training criterion to be minimized is the cross entropy between the posterior distribution represented by the reference labels and the predicted distribution:

$$F_{\text{CE}} = -\sum_{t}\sum_{s=1}^{N} P_{\text{target}}(o = s|\mathbf{x}_t) log P(o = s|\mathbf{x}_t), \tag{2.34}$$

where N is the number of senones, $P_{\text{target}}(o = s|\mathbf{x}_t)$ is the target probability of senone s at time t, and $P(o = s|\mathbf{x}_t)$ is the DNN output probability calculated from Equation 2.32.

In the standard CE training of DNN, the target probabilities of all senones at time t are formed as a one-hot vector, with only the dimension corresponding to the reference senone assigned a value of 1 and the rest as 0. As a result, Equation 2.34 is reduced to minimize the negative log likelihood because every frame has only one target label s_t:

$$F_{CE} = -\sum_t \log P(o = s_t | \mathbf{x}_t). \tag{2.35}$$

This objective function is minimized by using error back propagation (Rumelhart et al., 1988) which is a gradient-descent based optimization method developed for neural networks. The weight matrix W and bias b of layer l are updated with:

$$\hat{\mathbf{A}}^l = \mathbf{A}^l + \alpha \mathbf{v}^l (\mathbf{e}^l)^T, \tag{2.36}$$

$$\hat{\mathbf{b}}^l = \mathbf{b}^l + \alpha \mathbf{e}^l, \tag{2.37}$$

where α is the learning rate. $\mathbf{v}^l$ and $\mathbf{e}^l$ are the input and error vector of layer l, respectively. $\mathbf{e}^l$ is calculated by back propagating the error signal from its upper layer with

$$e_i^l = \left[\sum_{k=1}^{N_{l+1}} A_{ik}^{l+1} e_k^{l+1} \right] \sigma'(u_i^l), \tag{2.38}$$

where A_{ik}^{l+1} is the element of weighting matrix $\mathbf{A}^{l+1}$ in the ith row and kth column for layer $l+1$, and e_k^{l+1} is the kth element of error vector $\mathbf{e}^{l+1}$ for layer $l+1$. N_{l+1} is the number of units in layer $l+1$. $\sigma'(u_i^l)$ is the derivative of sigmoid function. The error signal of the top layer (i.e., output layer) is defined as:

$$e_s^L = -\sum_t (\delta_{ss_t} - P(o = s | \mathbf{x}_t)), \tag{2.39}$$

where δ_{ss_t} is the Kronecker delta function. Then the parameters of the DNN can be efficiently updated with the back propagation algorithm.

Speech recognition is inherently a sequence classification problem. Therefore, the frame-based cross-entropy criterion is not optimal. The sequence training criterion has been explored to optimize DNN parameters for speech recognition. As the GMM parameter optimization with sequence training criterion, MMI, BMMI, and MBR criteria are typically used (Kingsbury, 2009; Mohamed et al., 2010; Su et al., 2013; Veselý et al., 2013). For example, the MMI objective function is

$$F_{MMI} = \sum_r \log P(\hat{\mathbf{S}}^r | \mathbf{X}^r), \tag{2.40}$$

where $\hat{\mathbf{S}}^r$ and $\mathbf{X}^r$ are the reference string and the observation sequence for rth utterances. Generally, $P(\mathbf{S}|\mathbf{X})$ is the posterior of path $\mathbf{S}$ given the current model:

$$P(\mathbf{S}|\mathbf{X}) = \frac{p^k(\mathbf{X}|\mathbf{S})P(\mathbf{S})}{\sum_{\mathbf{S}'} p^k(\mathbf{X}|\mathbf{S}')P(\mathbf{S}')} \tag{2.41}$$

$P(\mathbf{X}|\mathbf{S})$ is the acoustic score of the whole utterance, $P(\mathbf{S})$ is the language model score, and k is the acoustic weight. Then the error signal of MMI criterion for utterance r becomes

$$e_s^{(r,L)} = -k \sum_t \left(\delta_{s\hat{s}_t} - \sum_{\mathbf{S}^r} \delta_{ss_t} P(\mathbf{S}^r|\mathbf{X}^r) \right). \qquad (2.42)$$

There are different strategies to update the DNN parameters. The batch gradient descent updates the parameters with the gradient only once after each sweep through the whole training set and in this way parallelization can be easily conducted. However, the convergence of batch update is very slow and stochastic gradient descent (SGD) (Zhang, 2004) usually works better in practice where the true gradient is approximated by the gradient at a single frame and the parameters are updated right after seeing each frame. The compromise between the two, the mini-batch SGD (Dekel et al., 2012), is more widely used, as the reasonable size of mini-batches makes all the matrices fit into GPU memory, which leads to a more computationally efficient learning process. Recent advances in Hessian-free optimization (Martens, 2010) have also partially overcome this difficulty using approximated second-order information or stochastic curvature estimates. This second-order batch optimization method has also been explored to optimize the weight parameters in DNNs (Kingsbury et al., 2012; Wiesler et al., 2013).

Decoding of the CD-DNN-HMM is carried out by plugging the DNN into a conventional large vocabulary HMM decoder with the senone likelihood evaluated with Equation 2.33. This strategy was initially explored and established in Yu et al. (2010) and Dahl et al. (2011), and has soon become the standard industry practice because it allows the speech recognition industry to re-use much of the decoder software infrastructure built for the GMM-HMM system over many years.

2.4.4 ALTERNATIVE DEEP LEARNING ARCHITECTURES

In addition to the standard architecture of the DNN, there are plenty of studies of applying alternative nonlinear units and structures to speech recognition. Although sigmoid and tanh functions are the most commonly used nonlinearity types in DNNs, their limitations are well known. For example, it is slow to learn the whole network due to weak gradients when the units are close to saturation in both directions. Therefore, rectified linear units (ReLU) (Dahl et al., 2013; Jaitly and Hinton, 2011; Zeiler et al., 2013) and maxout units (Cai et al., 2013; Miao et al., 2013; Swietojanski et al., 2014) are applied to speech recognition to overcome the weakness of the sigmoidal units. ReLU refers to the units in a neural network that use the activation function of $f(x) = \max(0, x)$. Maxout refers to the units that use the activation function of getting the maximum output value from a group of input values.

Deep convolutional neural networks

The CNN, originally developed for image processing, can also be robust to distortion due to its invariance property (Abdel-Hamid et al., 2013, 2012, 2014; Sainath et al., 2013a). Figure 2.2 (after (Abdel-Hamid et al., 2014)) shows the structure of one CNN with full weight sharing. The first layer is called a convolution layer which consists of a number of feature maps. Each neuron in the convolution layer receives inputs from a local receptive field representing features of a limited frequency range. Neurons that belong to the same feature map share the same weights (also called filters or kernels) but receive different inputs shifted in frequency. As a result, the convolution layer carries out the convolution operation on the kernels with its lower layer activations. A pooling layer is added on top of the convolution layer to compute a lower resolution representation of the convolution layer activations through sub-sampling. The pooling function, which computes some statistics of the activations, is typically applied to the neurons along a window of frequency bands and generated from the same feature map in the convolution layer. The most popular pooling function is the maxout function. Then a fully connected DNN is built on top of the CNN to do the work of senone classification.

It is important to point out that the invariant property of the CNN to frequency shift applies when filter-bank features are used and it does not apply with the cepstral

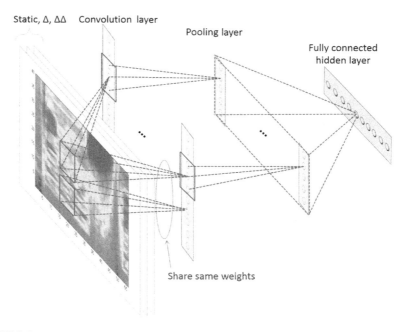

FIGURE 2.2

Illustration of the CNN in which the convolution is applied along frequency bands.

feature; see a detailed analysis in Deng et al. (2013a). Indeed using filter bank features as the input open a door for the CNN to exploit the structure in the features. It was shown that by using a CNN along the frequency axis they can normalize speaker differences and further reduce the phone error rate from 20.7% to 20.0% on the TIMIT phone recognition task (Abdel-Hamid et al., 2012). These results were later extended to large vocabulary speech recognition in 2013 with improved CNN architectures, pretraining techniques, and pooling strategies (Abdel-Hamid et al., 2013, 2014; Deng et al., 2013a; Sainath et al., 2013a,c). Further studies showed that the CNN helps mostly for the tasks in which the training set size or data variability is relatively small (Huang et al., 2015). For most other tasks the relative word error rate reduction is often in the range of 2-3%.

Deep recurrent neural networks

A more popular and effective deep learning architecture than the CNN in the recent speech recognition literature is a version of the recurrent neural network (RNN) which stacks one on another and which often contains a special cell structure called long short-term memory (LSTM). RNNs and LSTMs have been reported to work well specifically for robust speech recognition due to its powerful context modeling (Maas et al., 2012; Weng et al., 2014; Wöllmer et al., 2013a,b).

Here we briefly discuss the basics of the RNN as a class of neural network models where many connections among its neurons form a directed cycle. This gives rise to the structure of internal states or memory in the RNN, endowing it with the dynamic temporal behavior not exhibited by the basic DNN discussed earlier in this chapter.

An RNN is fundamentally different from the feed-forward DNN in that the RNN operates not only based on inputs, as for the DNN, but also on internal states. The internal states encode the past information in the temporal sequence that has already been processed by the RNN. In this sense, the RNN is a dynamic system, more general than the DNN that performs memoryless input-output transformations. The use of the state space in the RNN enables its representation to learn sequentially extended dependencies over a long time span, at least in principle.

Let us now formulate the simple one-hidden-layer RNN in terms of the (noise-free) nonlinear state space model commonly used in signal processing. At each time point t, let $\mathbf{x}_t$ be the $K \times 1$ vector of inputs, $\mathbf{h}_t$ be the $N \times 1$ vector of hidden state values, and $\mathbf{y}_t$ be the $L \times 1$ vector of outputs, the simple one-hidden-layer RNN can be described as

$$\mathbf{h}_t = f(\mathbf{W}_{xh}\mathbf{x}_t + \mathbf{W}_{hh}\mathbf{h}_{t-1}), \tag{2.43}$$

$$\mathbf{y}_t = g(\mathbf{W}_{hy}\mathbf{h}_t), \tag{2.44}$$

where $\mathbf{W}_{hy}$ is the $L \times N$ matrix of weights connecting the N hidden units to the L outputs, $\mathbf{W}_{xh}$ is the $N \times K$ matrix of weights connecting the K inputs to the N hidden units, and $\mathbf{W}_{hh}$ is the $N \times N$ matrix of weights connecting the N hidden units from time $t-1$ to time t, $\mathbf{u}_t = \mathbf{W}_{xh}\mathbf{x}_t + \mathbf{W}_{hh}\mathbf{h}_{t-1}$ is the $N \times 1$ vector of hidden layer potentials, $\mathbf{v}_t = \mathbf{W}_{hy}\mathbf{h}_t$ is the $L \times 1$ vector of output layer potentials, $f(\mathbf{u}_t)$ is the hidden layer activation function, and $g(\mathbf{v}_t)$ is the output layer activation function. Typical hidden layer activation functions are Sigmoid, tanh, and rectified linear units while the typical output layer activation functions are linear and softmax functions. Equations 2.43 and 2.44 are often called the observation and state equations, respectively. Note that, outputs from previous time frames can also be used to update the state vector, in which case the state equation becomes

$$\mathbf{h}_t = f(\mathbf{W}_{xh}\mathbf{x}_t + \mathbf{W}_{hh}\mathbf{h}_{t-1} + \mathbf{W}_{yh}\mathbf{y}_{t\text{-}1}), \qquad (2.45)$$

where $\mathbf{W}_{yh}$ denotes the weight matrix connecting output layer to the hidden layer. For simplicity purposes, most RNNs in speech recognition do not include output feedback.

It is important to note that before the recent rise of deep learning, especially deep recurrent neural networks, for speech modeling and recognition, a number of earlier attempts had been made to develop computational architectures that are "deeper" than the conventional GMM-HMM architecture. One prominent class of such models are hidden dynamic models where the internal representation of dynamic speech features is generated probabilistically from the higher levels in the overall deep speech model hierarchy (Bridle et al., 1998; Deng et al., 1997, 2006b; Ma and Deng, 2000; Togneri and Deng, 2003). Despite separate developments of the RNNs and of the hidden dynamic or trajectory models, they share a very similar motivation—representing aspects of dynamic structure in human speech. Nevertheless, a number of different ways in which these two types of deep dynamic models are constructed endow them with distinct pros and cons. Careful analysis of the contrast between these two model types and of the similarity to each other will help provide insights into the strategies for developing new types of deep dynamic models with the hidden representations of speech features superior to both existing RNNs and hidden dynamic models. This type of multi-faceted analysis has been provided in the recent book (Yu and Deng, 2014), which we refer the readers to for further studies on this topic.

While the RNN as well as the related nonlinear neural predictive models saw its early success in small ASR tasks (Robinson, 1994; Deng et al., 1994b), it was not easy to duplicate due to the intricacy in training, let alone to scale them up for larger speech recognition tasks. Learning algorithms for the RNN have been dramatically improved since these early days, however, and much stronger and practical results have been obtained recently using the RNN, especially when the bidirectional LSTM architecture is exploited (Graves et al., 2013a,b) or when the high-level DNN features are used as inputs to the RNN (Chen and Deng, 2014; Deng and Chen, 2014; Deng and Platt, 2014; Hannun et al., 2014). The LSTM was reported to give the lowest PER on the benchmark TIMIT phone recognition task in 2013 by Grave et al. (Graves et al., 2013a,b). In 2014, researchers published the results using the LSTM on large-scale tasks with applications to Google Now, voice search, and mobile dictation with excellent accuracy results (Sak et al., 2014a,b). To reduce the model size, the otherwise very large output vectors of LSTM units are linearly projected to smaller-dimensional vectors. Asynchronous stochastic gradient descent (ASGD) algorithm with truncated backpropagation through time (BPTT) is performed across hundreds of machines in CPU clusters. The best accuracy is obtained by optimizing the frame-level cross-entropy objective function followed by sequence discriminative training. With one LSTM stacking on top of another, this deep and recurrent LSTM model produced 9.7% WER on a large voice search task trained with 3 million utterances. This result is better than 10.7% WER achieved with frame-level cross entropy training criterion alone. It is also significantly better than the 10.4% WER obtained with the best DNN-HMM system using rectified linear units. Furthermore, this better accuracy is achieved while the total number of parameters is drastically reduced from 85 millions in the DNN system to 13 millions in the LSTM system. Some recent publications also showed that deep LSTMs are effective in speech recognition in reverberant multisource acoustic environments, as indicated by the strong results achieved by LSTMs in a recent ChiME Challenge task involving speech recognition in such difficult environments (Weninger et al., 2014).

2.5 SUMMARY

In this chapter, two major classes of acoustic models used for speech recognition are reviewed. In the first, generative-model class, we have the HMM where GMMs are used as the statistical distribution of speech features that is associated with each HMM state. This class also includes the hidden dynamic model that generalizes the GMM-HMM by incorporating aspects of deep structure in the human speech production process used as the internal representation of speech features.

Much of the robust speech recognition studies in the past, to be surveyed and analyzed at length in the following chapters of this book, have been carried out based on the generative models of speech, especially the GMM-HMM model, as reviewed in this chapter. One important advantage of generative models of speech in robust speech recognition is the straightforward way of thinking about the noise-robustness problem: noisy speech as the observations for robust speech recognizers can be viewed as the outcome of a further generative process combining clean speech and noise signals (Deng et al., 2000; Deng and Li, 2013; Frey et al., 2001b; Gales, 1995; Li et al., 2007a) using well-established distortion models. Some commonly used distortion models will be covered in later part of this book including the major publications (Deng, 2011; Li et al., 2014, 2008). Practical embodiment of such straightforward thinking to achieve noise-robust speech recognition systems based on generative acoustic models of distorted speech will also be presented (Huang et al., 2001a).

In the second, discriminative-model class of acoustic modeling for speech, we have the more recent DNN model as well as its convolutional and recurrent variants. These deep discriminative models have been shown to significantly outperform all previous versions of generative models of speech, shallow or deep, in speech recognition accuracy. The main sub-classes of these deep discriminative models are reviewed in some detail in this chapter. How to handle noise robustness within the framework of discriminative deep learning models of speech acoustics, which is less straightforward and more recent than the generative models of speech, will form the bulk part of later chapters.

REFERENCES

Abdel-Hamid, O., Deng, L., Yu, D., 2013. Exploring convolutional neural network structures and optimization techniques for speech recognition. In: Proc. Interspeech, pp. 3366-3370.

Abdel-Hamid, O., Mohamed, A., Jiang, H., Penn, G., 2012. Applying convolutional neural networks concepts to hybrid NN-HMM model for speech recognition. In: Proc. International Conference on Acoustics, Speech and Signal Processing (ICASSP).

Abdel-Hamid, O., Mohamed, A.r., Jiang, H., Deng, L., Penn, G., Yu, D., 2014. Convolutional neural networks for speech recognition. IEEE Trans. Audio Speech Lang. Process. 22 (10), 1533-1545.

Bahl, L.R., Brown, P.F., Souza, P.V.D., Mercer, R.L., 1997. Maximum mutual information estimation of hidden Markov model parameters for speech recognition. In: Proc. International Conference on Acoustics, Speech and Signal Processing (ICASSP), vol. 11, pp. 49-52.

Baker, J., 1976. Stochastic modeling for automatic speech recognition. In: Reddy, D. (Ed.), Speech Recognition. Academic, New York.

Baker, J., Deng, L., Glass, J., Khudanpur, S., Lee, C.H., Morgan, N., et al., 2009a. Research developments and directions in speech recognition and understanding, Part I. IEEE Signal Process. Mag. 26 (3), 75-80.

Baker, J., Deng, L., Glass, J., Khudanpur, S., Lee, C.H., Morgan, N., et al., 2009b. Updated MINDS report on speech recognition and understanding. IEEE Signal Process. Mag. 26 (4), 78-85.

Baum, L., Petrie, T., 1966. Statistical inference for probabilistic functions of finite state Markov chains. Ann. Math. Statist. 37 (6), 1554-1563.

Bengio, Y., 1991. Artificial Neural Networks and their Application to Sequence Recognition. McGill University, Montreal, Canada.

Bilmes, J., 2003. Buried markov models: A graphical modeling approach to automatic speech recognition. Comput. Speech Lang. 17, 213-231.

Bilmes, J., 2006. What HMMs can do. IEICE Trans. Informat. Syst. E89-D (3), 869-891.

Bilmes, J., 2010. Dynamic graphical models. IEEE Signal Process. Mag. 33, 29-42.

Bilmes, J., Bartels, C., 2005. Graphical model architectures for speech recognition. IEEE Signal Process. Mag. 22, 89-100.

Boulard, H., Morgan, N., 1993. Continuous speech recognition by connectionist statistical methods. IEEE Trans. Neural Networks 4 (6), 893-909.

Bridle, J., Deng, L., Picone, J., Richards, H., Ma, J., Kamm, T., et al., 1998. An investigation of segmental hidden dynamic models of speech coarticulation for automatic speech recognition. Final Report for 1998 Workshop on Language Engineering, CLSP, Johns Hopkins.

Cai, M., Shi, Y., Liu, J., 2013. Deep maxout neural networks for speech recognition. In: Proc. IEEE Workshop on Automatic Speech Recognition and Understanding (ASRU), pp. 291-296.

Chen, J., Deng, L., 2014. A primal-dual method for training recurrent neural networks constrained by the echo-state property. In: Proc. ICLR.

Chengalvarayan, R., Deng, L., 1998. Speech trajectory discrimination using the minimum classification error learning. IEEE Trans. Speech Audio Process. (6), 505-515.

Dahl, G., Sainath, T., Hinton, G., 2013. Improving deep neural networks for LVCSR using rectified linear units and dropout. In: Proc. International Conference on Acoustics, Speech and Signal Processing (ICASSP), pp. 8609-8613.

Dahl, G., Yu, D., Deng, L., Acero, A., 2011. Large vocabulary continuous speech recognition with context-dependent DBN-HMMs. In: Proc. International Conference on Acoustics, Speech and Signal Processing (ICASSP).

Dahl, G.E., Yu, D., Deng, L., Acero, A., 2012. Context-dependent pre-trained deep neural networks for large-vocabulary speech recognition. IEEE Trans. Audio Speech Lang. Process. 20 (1), 30-42.

Dekel, O., Gilad-Bachrach, R., Shamir, O., Xiao, L., 2012. Optimal distributed online prediction using mini-batches. J. Mach. Learn. Res. 13 (1), 165-202.

Dempster, A.P., Laird, N.M., Rubin, D.B., 1977. Maximum likelihood from incomplete data via the EM algorithm. J. R. Stat. Soc. 39 (1), 1-38.

Deng, L., 1992. A generalized hidden Markov model with state-conditioned trend functions of time for the speech signal. Signal Process. 27 (1), 65-78.

Deng, L., 1993. A stochastic model of speech incorporating hierarchical nonstationarity. IEEE Trans. Acoust. Speech Signal Process. 1 (4), 471-475.

Deng, L., 1998. A dynamic, feature-based approach to the interface between phonology and phonetics for speech modeling and recognition. Speech Commun. 24 (4), 299-323.

Deng, L., 1999. Computational models for speech production. In: Computational Models of Speech Pattern Processing. Springer-Verlag, New York, pp. 199-213.

Deng, L., 2003. Switching dynamic system models for speech articulation and acoustics. In: Mathematical Foundations of Speech and Language Processing. Springer-Verlag, New York, pp. 115-134.

Deng, L., 2006. Dynamic Speech Models—Theory, Algorithm, and Applications. Morgan and Claypool, San Rafael, CA.

Deng, L., 2011. Front-end, back-end, and hybrid techniques for noise-robust speech recognition. In: Robust Speech Recognition of Uncertain or Missing Data: Theory and Application. Springer, New York, pp. 67-99.

Deng, L., Abdel-Hamid, O., Yu, D., 2013.a. A deep convolutional neural network using heterogeneous pooling for trading acoustic invariance with phonetic confusion. In: Proc. International Conference on Acoustics, Speech and Signal Processing (ICASSP), Vancouver, Canada.

Deng, L., Acero, A., Plumpe, M., Huang, X., 2000. Large vocabulary speech recognition under adverse acoustic environment. In: Proc. International Conference on Spoken Language Processing (ICSLP), vol. 3, pp. 806-809.

Deng, L., Aksmanovic, M., Sun, D., Wu, J., 1994a. Speech recognition using hidden Markov models with polynomial regression functions as non-stationary states. IEEE Trans. Acoust. Speech Signal Process. 2 (4), 101-119.

Deng, L., Chen, J., 2014. Sequence classification using high-level features extracted from deep neural networks. In: Proc. International Conference on Acoustics, Speech and Signal Processing (ICASSP).

Deng, L., Droppo, J., A.Acero., 2003. Recursive estimation of nonstationary noise using iterative stochastic approximation for robust speech recognition. IEEE Trans. Speech Audio Process. 11, 568-580.

Deng, L., Droppo, J., Acero, A., 2002.a. A Bayesian approach to speech feature enhancement using the dynamic cepstral prior. In: Proc. International Conference on Acoustics, Speech and Signal Processing (ICASSP), vol. 1, pp. I-829-I-832.

Deng, L., Erler, K., 1992. Structural design of hidden Markov model speech recognizer using multivalued phonetic features: comparison with segmental speech units. J. Acoust. Soc. Amer. 92 (6), 3058-3067.

Deng, L., Hassanein, K., Elmasry, M., 1994b. Analysis of correlation structure for a neural predictive model with applications to speech recognition. Neural Networks 7, 331-339.

Deng, L., Hinton, G., Kingsbury, B., 2013.b. New types of deep neural network learning for speech recognition and related applications: An overview. In: Proc. International Conference on Acoustics, Speech and Signal Processing (ICASSP), Vancouver, Canada.

Deng, L., Hinton, G., Yu, D., 2009. Deep learning for speech recognition and related applications. In: NIPS Workshop, Whistler, Canada.

Deng, L., Kenny, P., Lennig, M., Gupta, V., Seitz, F., Mermelsten, P., 1991a. Phonemic hidden Markov models with continuous mixture output densities for large vocabulary word recognition. IEEE Trans. Acoust. Speech Signal Process. 39 (7), 1677-1681.

Deng, L., Lennig, M., Seitz, F., Mermelstein, P., 1991b. Large vocabulary word recognition using context-dependent allophonic hidden Markov models. Comput. Speech Lang. 4, 345-357.

Deng, L., Li, J., Huang, J.T., Yao, K., Yu, D., Seide, F., et al., 2013c. Recent advances in deep learning for speech research at Microsoft. In: Proc. International Conference on Acoustics, Speech and Signal Processing (ICASSP), Vancouver, Canada.

Deng, L., Li, X., 2013. Machine learning paradigms in speech recognition: An overview. IEEE Trans. Audio Speech Lang. Process. 21 (5), 1060-1089.

Deng, L., O'Shaughnessy, D., 2003. Speech Processing—A Dynamic and Optimization-Oriented Approach. Marcel Dekker Inc., New York.

Deng, L., Platt, J., 2014. Ensemble deep learning for speech recognition. In: Proc. Annual Conference of International Speech Communication Association (INTERSPEECH).

Deng, L., Ramsay, G., Sun, D., 1997. Production models as a structural basis for automatic speech recognition. Speech Commun. 33 (2-3), 93-111.

Deng, L., Sameti, H., 1996. Transitional speech units and their representation by regressive Markov states: Applications to speech recognition. IEEE Trans. Speech Audio Process. 4 (4), 301-306.

Deng, L., Seltzer, M., Yu, D., Acero, A., Mohamed, A., Hinton, G., 2010. Binary coding of speech spectrograms using a deep auto-encoder. In: Proc. Annual Conference of International Speech Communication Association (INTERSPEECH).

Deng, L., Wang, K., A. Acero, H.H., Huang, X., 2002b. Distributed speech processing in MiPad's multimodal user interface. IEEE Trans. Audio Speech Lang. Process. 10 (8), 605-619.

Deng, L., Yu, D., 2007. Use of differential cepstra as acoustic features in hidden trajectory modelling for phonetic recognition. In: Proc. International Conference on Acoustics, Speech and Signal Processing (ICASSP), pp. 445-448.

Deng, L., Yu, D., 2014. Deep Learning: Methods and Applications. NOW Publishers, Hanover, MA.

Deng, L., Yu, D., Acero, A., 2006a. A bidirectional target filtering model of speech coarticulation: two-stage implementation for phonetic recognition. IEEE Trans. Speech Audio Process. 14, 256-265.

Deng, L., Yu, D., Acero, A., 2006b. Structured speech modeling. IEEE Trans. Speech Audio Process. 14, 1492-1504.

Divenyi, P., Greenberg, S., Meyer, G., 2006. Dynamics of Speech Production and Perception. IOS Press, Santa Venetia, CA.

Frey, B., Deng, L., Acero, A., Kristjansson, T., 2001a. ALGONQUIN: iterating Laplace's method to remove multiple types of acoustic distortion for robust speech recognition. In: Proc. Interspeech, pp. 901-904.

Frey, B., Kristjansson, T., Deng, L., Acero, A., 2001b. ALGONQUIN—learning dynamic noise models from noisy speech for robust speech recognition. In: NIPS, pp. 1165-1171.

Gales, M.J.F., 1995. Model-based techniques for noise robust speech recognition. Ph.D. thesis, University of Cambridge.

Ghoshal, A., Swietojanski, P., Renals, S., 2013. Multilingual training of deep-neural networks. In: Proc. International Conference on Acoustics, Speech and Signal Processing (ICASSP).

Gibson, M., Hain, T., 2006. Hypothesis spaces for minimum Bayes risk training in large vocabulary speech recognition. In: Proc. Interspeech.

Gong, Y., Illina, I., Haton, J.P., 1996. Modeling long term variability information in mixture stochastic trajectory framework. In: Proc. International Conference on Spoken Language Processing (ICSLP).

Graves, A., Jaitly, N., Mahamed, A., 2013a. Hybrid speech recognition with deep bidirectional LSTM. In: Proc. International Conference on Acoustics, Speech and Signal Processing (ICASSP), Vancouver, Canada.

Graves, A., Mahamed, A., Hinton, G., 2013b. Speech recognition with deep recurrent neural networks. In: Proc. International Conference on Acoustics, Speech and Signal Processing (ICASSP), Vancouver, Canada.

Hannun, A.Y., Case, C., Casper, J., Catanzaro, B.C., Diamos, G., Elsen, E., et al., 2014. Deep speech: Scaling up end-to-end speech recognition. CoRR abs/1412.5567. URL http://arxiv.org/abs/1412.5567.

He, X., Deng, L., 2008. DISCRIMINATIVE LEARNING FOR SPEECH RECOGNITION: Theory and Practice. Morgan and Claypool, San Rafael, CA.

He, X., Deng, L., Chou, W., 2008. Discriminative learning in sequential pattern recognition—A unifying review for optimization-oriented speech recognition. IEEE Signal Process. Mag. 25 (5), 14-36.

Heigold, G., Vanhoucke, V., Senior, A., Nguyen, P., Ranzato, M., Devin, M., et al., 2013. Multilingual acoustic models using distributed deep neural networks. In: Proc. International Conference on Acoustics, Speech and Signal Processing (ICASSP).

Hinton, G., Deng, L., Yu, D., et al., G.D., 2012. Deep neural networks for acoustic modeling in speech recognition. IEEE Sig. Proc. Mag. 29 (6), 82-97.

Hinton, G., Osindero, S., Teh, Y., 2006. A fast learning algorithm for deep belief nets. Neural Comput. 18, 1527-1554.

Hinton, G., Salakhutdinov, R., 2006. Reducing the dimensionality of data with neural networks. Science 313 (5786), 504-507.

Holmes, W., Russell, M., 1999. Probabilistic-trajectory segmental HMMs. Comput. Speech Lang. 13, 3-37.

Huang, J.T., Li, J., Gong, Y., 2015. An analysis of convolutional neural networks for speech recognition. In: Proc. International Conference on Acoustics, Speech and Signal Processing (ICASSP).

Huang, J.T., Li, J., Yu, D., Deng, L., Gong, Y., 2013. Cross-language knowledge transfer using multilingual deep neural network with shared hidden layers. In: Proc. International Conference on Acoustics, Speech and Signal Processing (ICASSP).

Huang, X., Acero, A., Chelba, C., Deng, L., Droppo, J., Duchene, D., et al., 2001a. Mipad: a multimodal interaction prototype. In: Proc. International Conference on Acoustics, Speech and Signal Processing (ICASSP).

Huang, X., Acero, A., Hon, H.W., 2001b. Spoken Language Processing, vol. 18. Prentice Hall, Englewood Cliffs, NJ.

Huang, X., Jack, M., 1989. Semi-continuous hidden Markov models for speech signals. Comput. Speech Lang. 3 (3), 239-251.

Jaitly, N., Hinton, G., 2011. Learning a better representation of speech soundwaves using restricted Boltzmann machines. In: Proc. International Conference on Acoustics, Speech and Signal Processing (ICASSP), pp. 5884-5887.

Jelinek, F., 1976. Continuous speech recognition by statistical methods. Proc. IEEE 64 (4), 532-557.

Jiang, H., Li, X., Liu, C., 2006. Large margin hidden Markov models for speech recognition. IEEE Trans. Audio Speech Lang. Process. 14 (5), 1584-1595.

Juang, B.H., 1985. Maximum-likelihood estimation for mixture multivariate stochastic observations of Markov chains. AT&T Tech. J. 64 (6), 1235-1249.

Juang, B.H., Hou, W., Lee, C.H., 1997. Minimum classification error rate methods for speech recognition. IEEE Trans. Speech Audio Process. 5 (3), 257-265.

Juang, B.H., Levinson, S.E., Sondhi, M.M., 1986. Maximum likelihood estimation for mixture multivariate stochastic observations of Markov chains. IEEE Trans. Informat. Theory 32 (2), 307-309.

Kingsbury, B., 2009. Lattice-based optimization of sequence classification criteria for neural-network acoustic modeling. In: Proc. International Conference on Acoustics, Speech and Signal Processing (ICASSP), pp. 3761-3764.

Kingsbury, B., Sainath, T., Soltau, H., 2012. Scalable minimum Bayes risk training of deep neural network acoustic models using distributed Hessian-free optimization. In: Proc. Interspeech.

Lee, L.J., Fieguth, P., Deng, L., 2001. A functional articulatory dynamic model for speech production. In: Proc. International Conference on Acoustics, Speech and Signal Processing (ICASSP), vol. 2 , Salt Lake City, pp. 797-800.

Levinson, S., Rabiner, L., Sondhi, M., 1983. An introduction to the application of the theory of probabilistic functions of a Markov process to automatic speech recognition. Bell Syst. Tech. J. 62 (4), 1035-1074.

Li, J., Deng, L., Gong, Y., Haeb-Umbach, R., 2014. An overview of noise-robust automatic speech recognition. IEEE/ACM Trans. Audio Speech Lang. Process. 22 (4), 745-777.

Li, J., Deng, L., Yu, D., Gong, Y., Acero, A., 2007a. High-performance HMM adaptation with joint compensation of additive and convolutive distortions via vector Taylor series. In: Proc. IEEE Workshop on Automatic Speech Recognition and Understanding (ASRU), pp. 65-70.

Li, J., Deng, L., Yu, D., Gong, Y., Acero, A., 2008. HMM adaptation using a phase-sensitive acoustic distortion model for environment-robust speech recognition. In: Proc. International Conference on Acoustics, Speech and Signal Processing (ICASSP), pp. 4069-4072.

Li, J., Yu, D., Huang, J.T., Gong, Y., 2012. Improving wideband speech recognition using mixed-bandwidth training data in CD-DNN-HMM. In: Proc. IEEE Spoken Language Technology Workshop, pp. 131-136.

Li, J., Yuan, M., Lee, C.H., 2006. Soft margin estimation of hidden Markov model parameters. In: Proc. Interspeech, pp. 2422-2425.

Li, J., Yuan, M., Lee, C.H., 2007b. Approximate test risk bound minimization through soft margin estimation. IEEE Trans. Audio Speech Lang. Process. 15 (8), 2393-2404.

Li, X., Jiang, H., 2007. Solving large-margin hidden Markov model estimation via semidefinite programming. IEEE Trans. Audio Speech Lang. Process. 15 (8), 2383-2392.

Lin, H., Deng, L., Yu, D., Gong, Y., Acero, A., Lee, C.H., 2009. A study on multilingual acoustic modeling for large vocabulary ASR. In: Proc. International Conference on Acoustics, Speech and Signal Processing (ICASSP), pp. 4333-4336.

Liporace, L., 1982. Maximum likelihood estimation for multivariate observations of Markov sources. IEEE Trans. Informat. Theory 28 (5), 729-734.

Ma, J., Deng, L., 2000. A path-stack algorithm for optimizing dynamic regimes in a statistical hidden dynamic model of speech. Comput. Speech Lang. 14, 101-104.

Ma, J., Deng, L., 2003. Efficient decoding strategies for conversational speech recognition using a constrained nonlinear state-space model. IEEE Trans. Audio Speech Process. 11 (6), 590-602.

Ma, J., Deng, L., 2004. Target-directed mixture dynamic models for spontaneous speech recognition. IEEE Trans. Audio Speech Process. 12 (1), 47-58.

Maas, A.L., Le, Q.V., O'Neil, T.M., Vinyals, O., Nguyen, P., Ng, A.Y., 2012. Recurrent neural networks for noise reduction in robust ASR. In: Proc. Interspeech, pp. 22-25.

Martens, J., 2010. Deep learning via Hessian-free optimization. In: Proceedings of the 27th International Conference on Machine Learning, pp. 735-742.

Miao, Y., Metze, F., Rawat, S., 2013. Deep maxout networks for low-resource speech recognition. In: Proc. IEEE Workshop on Automatic Speech Recognition and Understanding (ASRU), pp. 398-403.

Mohamed, A., Dahl, G., Hinton, G., 2009. Deep belief networks for phone recognition. In: NIPS Workshop on Deep Learning for Speech Recognition and Related Applications.

Mohamed, A., Dahl, G.E., Hinton, G., 2012. Acoustic modeling using deep belief networks. IEEE Trans. Audio Speech Lang. Process. 20 (1), 14-22.

Mohamed, A., Yu, D., Deng, L., 2010. Investigation of full-sequence training of deep belief networks for speech recognition. In: Proc. Annual Conference of International Speech Communication Association (INTERSPEECH).

Morgan, N., Bourlard, H., 1990. Continuous speech recognition using multilayer perceptrons with hidden Markov models. In: Proc. International Conference on Acoustics, Speech and Signal Processing (ICASSP), pp. 413-416.

Neto, J., Almeida, L., Hochberg, M., Martins, C., Nunes, L., Renals, S., et al., 1995. Speaker-adaptation for hybrid HMM-ANN continuous speech recognition system. In: Proc. European Conference on Speech Communication and Technology (EUROSPEECH), pp. 2171-2174.

Ostendorf, M., Digalakis, V., Kimball, O., 1996. From HMM's to segment models: A unified view of stochastic modeling for speech recognition. IEEE Trans. Speech Audio Process. 4 (5), 360-378.

Ostendorf, M., Kannan, A., Kimball, O., Rohlicek, J., 1992. Continuous word recognition based on the stochastic segment model. Proc. DARPA Workshop CSR.

Parihar, N., Picone, J., Institute for Signal and Infomation Processing, Mississippi State Univ., 2002. Aurora working group: DSR front end LVCSR evaluation AU/384/02.

Picone, J., Pike, S., Regan, R., Kamm, T., Bridle, J., Deng, L., et al., 1999. Initial evaluation of hidden dynamic models on conversational speech. In: Proc. International Conference on Acoustics, Speech and Signal Processing (ICASSP).

Plahl, C., Schluter, R., Ney, H., 2011. Cross-lingual portability of Chinese and English neural network features for French and German LVCSR. In: Proc. IEEE Workshop on Automatic Speech Recognition and Understanding (ASRU), pp. 371-376.

Povey, D., Kanevsky, D., Kingsbury, B., Ramabhadran, B., Saon, G., Visweswariah, K., 2008. Boosted MMI for model and feature-space discriminative training. In: Proc. International Conference on Acoustics, Speech and Signal Processing (ICASSP), pp. 4057-4060.

Povey, D., Woodland, P.C., 2002. Minimum phone error and I-smoothing for improved discriminative training. In: Proc. International Conference on Acoustics, Speech and Signal Processing (ICASSP), pp. 105-108.

Rabiner, L., 1989. A tutorial on hidden Markov models and selected applications in speech recognition. Proc. IEEE 77 (2), 257-286.

Rabiner, L., Juang, B.H., 1993. Fundamentals of Speech Recognition. Prentice-Hall, Upper Saddle River, NJ.

Renals, S., Morgan, N., Boulard, H., Cohen, M., Franco, H., 1994. Connectionist probability estimators in HMM speech recognition. IEEE Trans. Speech Audio Process. 2 (1), 161-174.

Robinson, A., 1994. An application to recurrent nets to phone probability estimation. IEEE Trans. Neural Networks 5 (2), 298-305.

Rumelhart, D.E., Hinton, G.E., Williams, R.J., 1988. Learning representations by back-propagating errors. In: Polk, T.A., Seiferet, C.M. (Eds.), Cognitive Modeling. Chapter 8, pp. 213-220.

Russell, M., Jackson, P., 2005. A multiple-level linear/linear segmental HMM with a formant-based intermediate layer. Comput. Speech Lang. 19, 205-225.

Sainath, T., Kingsbury, B., Mohamed, A., Dahl, G., Saon, G., Soltau, H., et al., 2013a. Improvements to deep convolutional neural networks for LVCSR. In: Proc. IEEE Workshop on Automatic Speech Recognition and Understanding (ASRU), pp. 315-320.

Sainath, T., Kingsbury, B., Soltau, H., Ramabhadran, B., 2013b. Optimization techniques to improve training speed of deep neural networks for large speech tasks. IEEE Trans. Audio Speech Lang. Process. 21 (11), 2267-2276.

Sainath, T., Mohamed, A., Kingsbury, B., Ramabhadran, B., 2013c. Deep convolutional neural networks for LVCSR. In: Proc. International Conference on Acoustics, Speech and Signal Processing (ICASSP), pp. 8614-8618.

Sainath, T.N., Kingsbury, B., Ramabhadran, B., Fousek, P., Novák, P., Mohamed, A., 2011. Making deep belief networks effective for large vocabulary continuous speech recognition. In: Proc. IEEE Workshop on Automatic Speech Recognition and Understanding (ASRU), pp. 30-35.

Sak, H., Senior, A., Beaufays, F., 2014a. Long short-term memory recurrent neural network architectures for large scale acoustic modeling. In: Proc. Interspeech.

Sak, H., Vinyals, O., Heigold, G., Senior, A., McDermott, E., Monga, R., et al., 2014b. Sequence discriminative distributed training of long short-term memory recurrent neural networks. In: Proc. Annual Conference of International Speech Communication Association (INTERSPEECH).

Schultz, T., Waibel, A., 1998. Multilingual and crosslingual speech recognition. In: Proc. DARPA Workshop on Broadcast News Transcription and Understanding, pp. 259-262.

Seide, F., Fu, H., Droppo, J., Li, G., Yu, D., 2014. On parallelizability of stochastic gradient descent for speech DNNs. In: Proc. International Conference on Acoustics, Speech and Signal Processing (ICASSP).

Seide, F., Li, G., Yu, D., 2011. Conversational speech transcription using context-dependent deep neural networks. In: Proc. Annual Conference of International Speech Communication Association (INTERSPEECH), pp. 437-440.

Seltzer, M.L., Yu, D., Wang, Y., 2013. An investigation of deep neural networks for noise robust speech recognition. In: Proc. International Conference on Acoustics, Speech and Signal Processing (ICASSP), pp. 7398-7402.

Senior, A., Heigold, G., Bacchiani, M., Liao, H., 2014. GMM-free DNN training. In: Proc. International Conference on Acoustics, Speech and Signal Processing (ICASSP).

Sha, F., 2007. Large margin training of acoustic models for speech recognition. Ph.D. thesis, University of Pennsylvania.

Sha, F., Saul, L., 2006. Large margin Gaussian mixture modeling for phonetic classification and recognition. In: Proc. International Conference on Acoustics, Speech and Signal Processing (ICASSP).

Stevens, K., 2000. Acoustic Phonetics. MIT Press, Cambridge, MA.

Su, H., Li, G., Yu, D., Seide, F., 2013. Error back propagation for sequence training of context-dependent deep networks for conversational speech transcription. In: Proc. International Conference on Acoustics, Speech and Signal Processing (ICASSP), pp. 6664-6668.

Swietojanski, P., Li, J., Huang, J.T., 2014. Investigation of maxout networks for speech recognition. In: Proc. International Conference on Acoustics, Speech and Signal Processing (ICASSP).

Togneri, R., Deng, L., 2003. Joint state and parameter estimation for a target-directed nonlinear dynamic system model. IEEE Trans. Signal Process. 51 (12), 3061-3070.

Tuske, Z., Golik, P., Schluter, R., Ney, H., 2014. Acoustic modeling with deep neural networks using raw time signal for LVCSR. In: Proc. Annual Conference of International Speech Communication Association (INTERSPEECH).

Vanhoucke, V., Devin, M., Heigold, G., 2013. Multiframe deep neural networks for acoustic modeling. In: Proc. International Conference on Acoustics, Speech and Signal Processing (ICASSP).

Vanhoucke, V., Senior, A., Mao, M., 2011. Improving the speed of neural networks on CPUs. In: Proc. NIPS Workshop on Deep Learning and Unsupervised Feature Learning.

Veselỳ., K., Ghoshal, A., Burget, L., Povey, D., 2013. Sequence-discriminative training of deep neural networks. In: Proc. Interspeech, pp. 2345-2349.

Waibel, A., Hanazawa, T., Hinton, G., Shikano, K., Lang, K., 1989. Phoneme recognition using time-delay neural networks. IEEE Trans. Speech Audio Process. 37 (3), 328-339.

Weng, C., Yu, D., Watanabe, S., Juang, B., 2014. Recurrent deep neural networks for robust speech recognition. In: Proc. International Conference on Acoustics, Speech and Signal Processing (ICASSP).

Weninger, F., Geiger, J., Wöllmer, M., Schuller, B., Rigoll, G., 2014. Feature enhancement by deep LSTM networks for ASR in reverberant multisource environments. Comput. Speech Lang. 888-902.

Wiesler, S., Li, J., Xue, J., 2013. Investigations on hessian-free optimization for cross-entropy training of deep neural networks. In: Proc. Interspeech, pp. 3317-3321.

Wöllmer, M., Weninger, F., Geiger, J., Schuller, B., Rigoll, G., 2013a. Noise robust ASR in reverberated multisource environments applying convolutive NMF and long short-term memory. Comput. Speech Lang. 27 (3), 780-797.

Wöllmer, M., Zhang, Z., Weninger, F., Schuller, B., Rigoll, G., 2013b. Feature enhancement by bidirectional LSTM networks for conversational speech recognition in highly non-stationary noise. In: Proc. International Conference on Acoustics, Speech and Signal Processing (ICASSP), pp. 6822-6826.

Xiao, X., Li, J., Cheng, E.S., Li, H., Lee, C.H., 2010. A study on the generalization capability of acoustic models for robust speech recognition. IEEE Trans. Audio Speech Lang. Process. 18 (6), 1158-1169.

Yu, D., Deng, L., 2007. Speaker-adaptive learning of resonance targets in a hidden trajectory model of speech coarticulation. Comput. Speech Lang. 27, 72-87.

Yu, D., Deng, L., 2011. Deep learning and its applications to signal and information processing. In: IEEE Signal Processing Magazine., vol. 28, pp. 145-154.

Yu, D., Deng, L., 2014. Automatic Speech Recognition—A Deep Learning Approach. Springer, New York.

Yu, D., Deng, L., Acero, A., 2006. A lattice search technique for a long-contextual-span hidden trajectory model of speech. Speech Commun. 48, 1214-1226.

Yu, D., Deng, L., Dahl, G., 2010. Roles of pretraining and fine-tuning in context-dependent DBN-HMMs for real-world speech recognition. In: Proc. NIPS Workshop on Deep Learning and Unsupervised Feature Learning.

Yu, D., Deng, L., Liu, P., Wu, J., Gong, Y., Acero, A., 2009. Cross-lingual speech recognition under runtime resource constraints. In: Proc. International Conference on Acoustics, Speech and Signal Processing (ICASSP), pp. 4193-4196.

Zeiler, M., Ranzato, M., Monga, R., Mao, M., Yang, K., Le, Q., et al., 2013. On rectified linear units for speech processing. In: Proc. International Conference on Acoustics, Speech and Signal Processing (ICASSP), pp. 3517-3521.

Zen, H., Tokuda, K., Kitamura, T., 2004. An introduction of trajectory model into HMM-based speech synthesis. In: Proc. of ISCA SSW5, pp. 191-196.

Zhang, C., Woodland, P., 2014. Standalone training of context-dependent deep neural network acoustic models. In: Proc. International Conference on Acoustics, Speech and Signal Processing (ICASSP).

Zhang, L., Renals, S., 2008. Acoustic-articulatory modelling with the trajectory HMM. IEEE Signal Process. Lett. 15, 245-248.

Zhang, T., 2004. Solving large scale linear prediction problems using stochastic gradient descent algorithms. In: Proceedings of the twenty-first international conference on Machine learning.

Zhou, J.L., Seide, F., Deng, L., 2003. Coarticulation modeling by embedding a target-directed hidden trajectory model into HMM—model and training. In: Proc. International Conference on Acoustics, Speech and Signal Processing (ICASSP), vol. 1, Hongkong, pp. 744-747.

Background of robust speech recognition

3

CHAPTER OUTLINE

In this chapter, we provide the background of robust ASR and establish some fundamental concepts, which are most relevant to and form the basis for the discussions in the remainder of this book. In particular, we elaborate on a set of prominent speech distortion models as part of the overall generative model of speech (Deng, 2006; Deng and O'Shaughnessy, 2003) commonly used in machine learning paradigms for solving pattern recognition problems (Deng and Li, 2013).

Noise robustness in ASR is a vast topic, spanning research literature over 30 years. From this chapter to Chapter 8, we only consider single-channel input with additive noise and convolutive channel distortion, and assume that the impulse response of the channel is much shorter than the frame size of speech signals. This is the most popular case in robust ASR. The topics of reverberant speech recognition will be discussed in Chapter 9, and multi-channel speech recognition in Chapter 10.

3.1 STANDARD EVALUATION DATABASES

In the early years of developing noise-robust ASR technologies, it was very hard to decide which technology was better since different groups used different databases for evaluation. The introduction of standard evaluation databases and training recipes finally allowed noise-robustness methods developed by different groups to be compared fairly using the same task, thereby fast-tracking development of these methods. Among the standard evaluation databases, the most famous tasks are the Aurora series developed by the European Telecommunications Standards Institute (ETSI), although there are some other tasks such as SPINE (speech in noisy environments) (Schmidt-Nielsen et al., 2000), CHIL (computers in the human interaction loop) (Mostefa et al., 2007), AMI (Renals et al., 2008), COSINE (conversational speech in noisy environments) (Stupakov et al., 2012), and the 1st and 2nd CHiME (computational hearing in multisource environments) speech separation and recognition challenges (Christensen et al., 2010; Vincent et al., 2013). Among all of these tasks, Aurora 2 and Aurora 4 are the most popular tasks used to report the benefits of robust ASR technologies.

The first Aurora database is Aurora 2 (Hirsch and Pearce, 2000), a task of recognizing digit strings in noise and channel distorted environments. The data is artificially corrupted. Aurora 2 has both clean and multi-condition training sets, each of which consists of 8440 clean or multi-condition utterances. The test material consists of three sets of distorted utterances. The data in set A and set B contain eight different types of additive noise, while set C contains two different types of noise plus additional channel distortion. Each type of noise is added into a subset of clean speech utterances, with seven different levels of signal-to-noise ratios (SNRs). This generates seven subgroups of test sets for a specified noise type, with clean, that is, infinite SNR, and SNRs of 20, 15, 10, 5, 0, and -5 dB. Since the introduction of Aurora 2, noise-robustness methods can be evaluated and compared using a standard task. This significantly contributed to the development of noise-robustness technologies. The Aurora 3 task consists of noisy speech data recorded inside cars as part of the SpeechDatCar project (Moreno et al., 2000). Although still a digit recognition task, the utterances in Aurora 3 are collected in real noisy environments. The Aurora 4 task (Parihar and Picone, 2002) is a standard large vocabulary continuous speech recognition (LVCSR) task which is constructed by artificially corrupting the clean data from the Wall Street Journal (WSJ) corpus (Paul and Baker, 1992). Aurora 4 has two training sets: clean and multi-condition. Each of them consists of 7138 utterances (about 14 h of speech data). For the multi-condition training set, half of the data was recorded with a Sennheiser microphone and the other was with a secondary microphone. In addition, six types of noises (car, babble, restaurant, street, airport, and train) were added with SNRs from 10 to 20 dB. The subset recorded with the Sennheiser microphone was called as channel wv1 data and the other part as channel wv2 data. The test set contains 14 subsets. Two of them are clean and the other 12 are noisy. The noisy test sets were recorded with the same types of microphone as in multi-condition training set. Also, the same 6 types of noise as in multi-condition training set were added with SNRs between 5 and 15 dB. Aurora 5 (Hirsch, 2007) was mainly developed to investigate the influence of hands-free speech input on the performance of digit recognition in noisy room environments and over a cellular telephone network. The evaluation data is artificially simulated.

The progression of the Aurora tasks shows a clear trend: from digit recognition in artificial environments (Aurora 2), to digit recognition in real noisy environments (Aurora 3), to a LVCSR task (Aurora 4), and to working in the popular cellular scenario (Aurora 5). This is consistent with the need to develop noise-robust ASR technologies for real-world deployment.

3.2 MODELING DISTORTIONS OF SPEECH IN ACOUSTIC ENVIRONMENTS

Mel-frequency cepstral coefficients (MFCCs) (Davis and Mermelstein, 1980) are the most widely used acoustic features when the underlying acoustic model is GMM. The short-time discrete Fourier transform (STDFT) is applied to the speech signal, and the power or magnitude spectrum is generated. A set of Mel scale filters is applied to obtain the Mel-filter-bank output. Then the log operator is used to get the log-Mel-filter-bank output, which is the popular acoustical feature when the underlying acoustic model is a DNN. Finally, the discrete cosine transform (DCT) is used to generate MFCCs. In the following, we use MFCCs as the acoustic feature to elaborate on the relation between clean and distorted speech.

Figure 3.1 shows a time-domain model for speech degraded by both additive noise and convolutive channel distortions (Acero, 1993). The observed distorted speech signal $y[m]$, where m denotes the discrete time index, is generated from the clean speech signal $x[m]$ with the additive noise $n[m]$ and the convolutive channel distortions $h[m]$ according to

$$y[m] = x[m] * h[m] + n[m], \tag{3.1}$$

where "$*$" denotes the convolution operator.

After applying the STDFT, the following equivalent relation can be established in the spectral domain:

$$\dot{y}[k] = \dot{x}[k]\dot{h}[k] + \dot{n}[k]. \tag{3.2}$$

Here, k is the frequency bin index. Note that we left out the frame index for ease of notation. To arrive at Equation 3.2 we assume that the impulse response $h[m]$ is

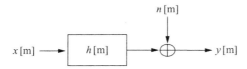

FIGURE 3.1

A model of acoustic environment distortion in the discrete-time domain relating the clean speech sample $x[m]$ to the distorted speech sample $y[m]$.

much shorter than the STDFT analysis window size. Then we can make use of the multiplicative transfer function approximation by which a convolution in the time domain corresponds to a multiplication in the STDFT domain (Cohen and Gannot, 2008). This approximation does not hold in the presence of reverberated speech, because the acoustic impulse response characterizing the reverberation is typically much longer than the STDFT window size. Thus, Equation 3.2 is not adequate to describe reverberated speech in the STDFT domain.

The power spectrum of the distorted speech can then be obtained as:

$$|\dot{y}[k]|^2 = |\dot{x}[k]|^2|\dot{h}[k]|^2 + |\dot{n}[k]|^2 + 2|\dot{x}[k]||\dot{h}[k]||\dot{n}[k]|\cos\beta_k, \quad (3.3)$$

where β_k denotes the (random) angle between two complex variables $\dot{n}[k]$ and $\dot{x}[k]\dot{h}[k]$. If $\cos\beta_k$ is set as 0, Equation 3.3 will become:

$$|\dot{y}[k]|^2 = |\dot{x}[k]|^2|\dot{h}[k]|^2 + |\dot{n}[k]|^2. \quad (3.4)$$

Neglecting this "phase" term is a common practice in the formulation of speech distortion in the power spectral domain; for example, in the spectral subtraction technique. So is approximating the phase term in the log-spectral domain (Deng et al., 2002, 2004b; Frey et al., 2001a,b). While achieving simplicity in developing speech enhancement algorithms, neglecting this term is a partial cause of the degradation of enhancement performance at low SNRs (around 0 dB) (Deng et al., 2004a; Droppo et al., 2003).

By applying a set of Mel-scale filters (L in total) to the power spectrum in Equation 3.3, we have the lth Mel-filter-bank energies for distorted speech, clean speech, noise, and channel:

$$|\breve{y}[l]|^2 = \sum_k w_k[l]|\dot{y}[k]|^2, \quad (3.5)$$

$$|\breve{x}[l]|^2 = \sum_k w_k[l]|\dot{x}[k]|^2, \quad (3.6)$$

$$|\breve{n}[l]|^2 = \sum_k w_k[l]|\dot{n}[k]|^2, \quad (3.7)$$

$$|\breve{h}[l]|^2 = \frac{\sum_k w_k[l]|\dot{x}[k]|^2|\dot{h}[k]|^2}{|\breve{x}[l]|^2}, \quad (3.8)$$

where the lth filter is characterized by the transfer function $w_k[l] \geq 0$ with $\sum_k w_k[l] = 1$.

The phase factor $\alpha[l]$ of the lth Mel-filter-bank is (Deng et al., 2004a)

$$\alpha[l] = \frac{\sum_k w_k[l]|\dot{x}[k]||\dot{h}[k]||\dot{n}[k]|\cos\beta_k}{|\breve{x}[l]||\breve{h}[l]||\breve{n}[l]|}. \quad (3.9)$$

Then, the following relation is obtained in the Mel-filter-bank domain for the lth Mel-filter-bank output

$$|\breve{y}[l]|^2 = |\breve{x}[l]|^2|\breve{h}[l]|^2 + |\breve{n}[l]|^2 + 2\alpha[l]|\breve{x}[l]||\breve{h}[l]||\breve{n}[l]|, \qquad (3.10)$$

By taking the log operation in both sides of Equation 3.10, we have the following in the log-Mel-filter-bank domain, using vector notation

$$\tilde{\mathbf{y}} = \tilde{\mathbf{x}} + \tilde{\mathbf{h}} + \log\left(1 + \exp(\tilde{\mathbf{n}} - \tilde{\mathbf{x}} - \tilde{\mathbf{h}}) + 2\boldsymbol{\alpha}.*\exp\left(\frac{\tilde{\mathbf{n}} - \tilde{\mathbf{x}} - \tilde{\mathbf{h}}}{2}\right)\right). \qquad (3.11)$$

Here, the bold face variables refer to vectors, for example, $\tilde{\mathbf{y}} = \left(y[1] \quad \dots \quad y[L]\right)^T$. The .* operation for two vectors denotes element-wise product, and taking the logarithm and exponentiation of a vector above are also element-wise operations.

By applying the DCT transform to Equation 3.11, we can get the distortion formulation in the cepstral domain as

$$\mathbf{y} = \mathbf{x} + \mathbf{h} + \mathbf{C}\log\left(1 + \exp(\mathbf{C}^{-1}(\mathbf{n} - \mathbf{x} - \mathbf{h})) + 2\boldsymbol{\alpha}.*\exp\left(\mathbf{C}^{-1}\frac{\mathbf{n} - \mathbf{x} - \mathbf{h}}{2}\right)\right), \qquad (3.12)$$

where $\mathbf{C}$ denotes the DCT matrix.

In Deng et al. (2004a), it was shown that the phase factor $\alpha[l]$ for each Mel-filter l can be approximated by a weighted sum of a number of independent zero-mean random variables distributed over $[-1, 1]$, where the total number of terms equals the number of STDFT bins. When the number of terms becomes large, the central limit theorem postulates that $\alpha[l]$ will be approximately Gaussian. A more precise statistical description has been developed in Leutnant and Haeb-Umbach (2009), where it is shown that all moments of $\alpha[l]$ of odd order are zero.

If we ignore the phase factor, Equations 3.11 and 3.12 can be simplified to

$$\tilde{\mathbf{y}} = \tilde{\mathbf{x}} + \tilde{\mathbf{h}} + \log(1 + \exp(\tilde{\mathbf{n}} - \tilde{\mathbf{x}} - \tilde{\mathbf{h}})), \qquad (3.13)$$

$$\mathbf{y} = \mathbf{x} + \mathbf{h} + \mathbf{C}\log(1 + \exp(\mathbf{C}^{-1}(\mathbf{n} - \mathbf{x} - \mathbf{h}))), \qquad (3.14)$$

which are the log-Mel-filter-bank and cepstral representations, respectively, corresponding to Equation 3.4 in the power spectral domain. Equations 3.13 and 3.14 are widely used in noise-robust ASR technologies as the basic formulation that characterizes the relationship between clean and distorted speech in the logarithmic domain. The effect of the phase factor is small if noise estimates are poor. However, with an increase in the quality of the noise estimates, the effect of the phase factor is shown experimentally to be stronger (Deng et al., 2004a).

The above nonlinear relationship can be simplified when only noise or channel is presented without considering the phase term. For example, when channel is not presented, the impact of noise is additive in the spectrum domain and the Mel-filter-bank domain. Equations 3.4 and 3.10 can be simplified as

$$|\dot{y}[k]|^2 = |\dot{x}[k]|^2 + |\dot{n}[k]|^2, \qquad (3.15)$$

$$|\breve{y}[l]|^2 = |\breve{x}[l]|^2 + |\breve{n}[l]|^2. \qquad (3.16)$$

In the absence of noise, the convolutive channel distortion in the time domain has an additive effect in the log-Mel-filter-bank domain and the cepstral domain. Equation 3.11 and 3.12 can be simplified as

$$\tilde{\mathbf{y}} = \tilde{\mathbf{x}} + \tilde{\mathbf{h}}, \qquad (3.17)$$

$$\mathbf{y} = \mathbf{x} + \mathbf{h}. \qquad (3.18)$$

These simplified linear formulations are widely used to provide an easy way to handling distortion from noise and channel.

We use the utterances in Aurora 2 to illustrate the impact of noise on clean speech. We select the segments corresponding to the word *oh* in Aurora 2 test set A with noise type 1 (subway noise). The starting and ending times of those segments are the same for all the SNR conditions. This enables us to have an apple-to-apple comparison of speech samples in different SNR conditions. In Figure 3.2(a), we plot the distribution of the 1st cepstral coefficient (C1) of word *oh* in Aurora 2 test set A with noise type 1. It is clear that the clean speech signal has the largest variance. With decreasing SNR the variance of the noisy signal is reduced. Another observation is that the mean value is also shifted.

Figure 3.2(b) shows the distribution of the 0th cepstral (C0) of word *oh* in Aurora 2 test set A with noise type 1. We can also observe that the mean value is shifted and the variance is reduced with the reduction of SNR levels.

The impact of channel distortion on the speech signal in the cepstral domain is relatively simple. In Figure 3.2(c), we plot the distribution of the 1st cepstral coefficient (C1) of the word *oh* in the Aurora 2 test set C with noise type 1. The difference between Aurora 2 test set A and C is the additional channel distortion. It is clear that in all SNR cases the only change is the mean value shift. The distribution shape doesn't change.

3.3 IMPACT OF ACOUSTIC DISTORTION ON GAUSSIAN MODELING

In ASR, Gaussian mixture models (GMMs) are widely used to characterize the distribution of speech in the log-Mel-filter-bank or cepstral domain. It is important to understand the impact of noise, which is additive in the spectral domain, on the distribution of noisy speech in the log-Mel-filter-bank and cepstral domains when the underlying model is Gaussian. Using Equation 3.13 to simulate the noisy speech in the log-Mel-filter-bank domain, while setting for simplicity $\tilde{\mathbf{h}} = \mathbf{0}$, we obtain

$$\tilde{\mathbf{y}} = \tilde{\mathbf{x}} + \log(1 + \exp(\tilde{\mathbf{n}} - \tilde{\mathbf{x}})). \qquad (3.19)$$

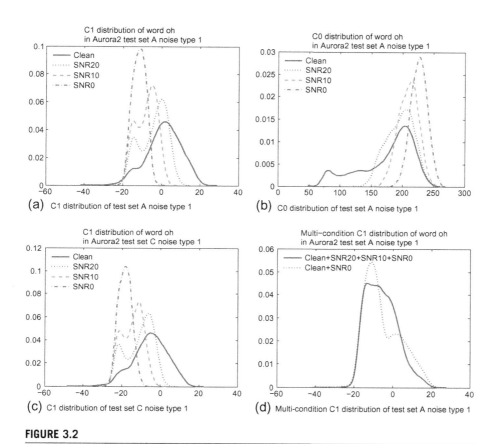

FIGURE 3.2

Cepstral distribution of word *oh* in Aurora 2.

Figure 3.3 shows the impact of noise on the clean speech signal in the log-Mel-filter-bank domain with increasing noise mean values, that is, decreasing SNRs. The clean speech shown with solid lines is Gaussian distributed, with a mean value of 25 and a standard deviation of 10. The noise $\tilde{n}$ is also Gaussian distributed, with different mean values and a standard deviation of 2. The noisy speech shown with dashed lines deviates from the Gaussian distribution to a varying degree. We can use a Gaussian distribution, shown with dotted lines, to make an approximation. The approximation error is large in the low SNR cases. When the noise mean is raised to 20 and 25, as in Figure 3.3(c) and (d), the distribution of noisy speech is skewed far away from a Gaussian distribution.

In Figure 3.4, we examine the impact of noise on the clean speech signal in the log-Mel-filter-bank domain as a function of the noise standard deviation values. The clean speech in solid line is the same as the one in Figure 3.3, with a mean value of 25 and a standard deviation of 10. The noise is also Gaussian distributed, with a mean

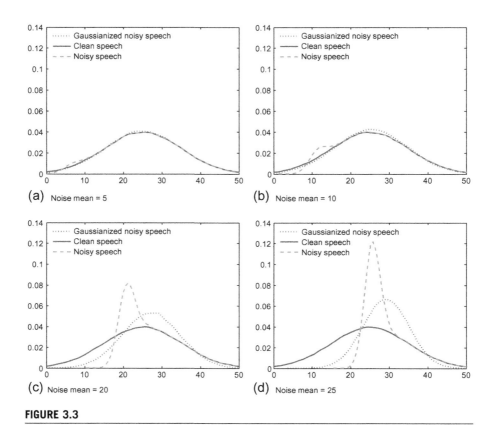

FIGURE 3.3

The impact of noise, with varying mean values from 5 in (a) to 25 in (d), in the log-Mel-filter-bank domain. The clean speech has a mean value of 25 and a standard deviation of 10. The noise has a standard deviation of 2.

value of 10 and different standard deviation values. When the standard deviation of the noise is small, as in Figure 3.4(a) and (b), the peak caused by the noise around 10 is sharp. As the deviation increases, the peak caused by the noise is gradually smoothed. In Figure 3.4(d) when the noise standard deviation reaches 10, the same as the clean speech, the noisy speech is almost Gaussian-distributed.

A natural way to deal with noise in acoustic environments is to use multi-style training (Lippmann et al., 1987), which trains the acoustic model with all available noisy speech data. The hope is that one of the noise types in the training set will appear in the deployment scenario. Generally speaking, this is a very effective technology to boost the ASR performance from a clean-trained model without requiring any advanced technology. For example, on the Aurora 2 task, the multi-style training model can have 86% average word accuracy while the clean training model only has 60% average word accuracy (Hirsch and Pearce, 2000). Recently,

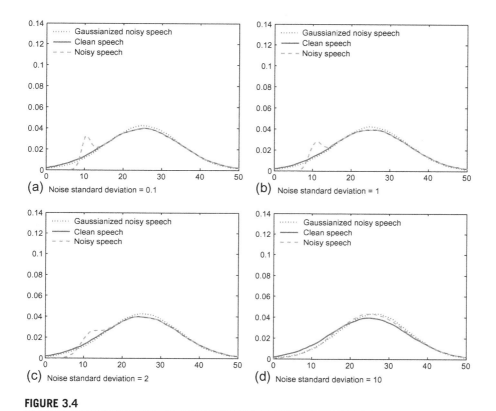

FIGURE 3.4

Impact of noise with different standard deviation values in the log-Mel-filter-bank domain. The clean speech has a mean value of 25 and a standard deviation of 10. The noise has a mean of 10.

Baidu has synthesized thousands of hours of noisy data to build a large training set for an end-to-end deep learning system which obtained a good noise-robustness property (Hannun et al., 2014). However, there are two major problems with multi-style training. The first is that during training it is hard to enumerate all noise types and SNRs encountered in test environments. The second is that the model trained with multi-style training has a very broad distribution because it needs to model all the environments. This can be verified by comparing Figure 3.2(d) with (a). In Figure 3.2(a), the clean signal and the SNR0 signal have only one mode while the SNR20 and SNR10 have two modes. If the clean signal and SNR0 signal are mixed together, a single Gaussian cannot model the multi-style distribution, as can be shown by the dotted line in Figure 3.2(d). This phenomenon becomes even worse when the signals from all conditions are mixed together. Then a GMM with more Gaussians is required to model the noisy speech appropriately. Given the unsatisfactory behavior of multi-style training, it is necessary to work on technologies that directly deal with the noise and channel impacts. In Section 3.5, we will lay down a general mathematical framework for robust speech recognition.

3.4 IMPACT OF ACOUSTIC DISTORTION ON DNN MODELING

With the excellent modeling power of DNNs (Dahl et al., 2012; Hinton et al., 2012), it was shown that the DNN-based acoustic models trained with multi-style data and without any explicit noise compensation can easily match the state-of-the-art performance of GMM systems with advanced noise-robust technologies (Seltzer et al., 2013). Why are DNNs more robust to noise? This is because many layers of simple nonlinear processing in DNNs can generate a complicated nonlinear transform. The output of layer l is

$$\mathbf{v}^{l+1} = \sigma(z(\mathbf{v}^l)), \tag{3.20}$$

with

$$z(\mathbf{v}^l) = \mathbf{A}^l \mathbf{v}^l + \mathbf{b}^l. \tag{3.21}$$

To show that the nonlinear transform is robust to small variations in the input features, let's assume the output of layer $l-1$, or equivalently the input to the layer l is changed from $\mathbf{v}^l$ to $\mathbf{v}^l + \delta^l$, where δ^l is a small change. This change will cause the output of layer l, or equivalently the input to the layer $l+1$ to change by

$$\delta^{l+1} = \sigma(z(\mathbf{v}^l + \delta^l)) - \sigma(z(\mathbf{v}^l)) \approx \text{diag}\left(\sigma'(z(\mathbf{v}^l))\right)(\mathbf{A}^l)^T \delta^l, \tag{3.22}$$

where $\sigma'(\cdot)$ is the derivative of the sigmoid function, Equation 2.31, and where $\text{diag}\left(\sigma'(z(\mathbf{v}^l))\right)$ expands the vector $\sigma'(z(\mathbf{v}^l))$ to a diagonal matrix. The norm of the change δ^{l+1} is

$$\begin{aligned}
\|\delta^{l+1}\| &\approx \|\text{diag}(\sigma'(z(\mathbf{v}^l)))((\mathbf{A}^l)^T \delta^l)\| \\
&\leq \|\text{diag}(\sigma'(z(\mathbf{v}^l)))(\mathbf{A}^l)^T\|\|\delta^l\| \\
&= \|\text{diag}(\mathbf{v}^{l+1}.*(1-\mathbf{v}^{l+1}))(\mathbf{A}^l)^T\|\|\delta^l\|,
\end{aligned} \tag{3.23}$$

where $.*$ refers to an element-wise product.

Note that the magnitude of the majority of the weights is typically very small if the size of the hidden layer is large (Yu et al., 2013). While each element in $\mathbf{v}^{l+1}.*(1-\mathbf{v}^{l+1})$ is less than or equal to 0.25, the actual value is typically much smaller. This means that a large percentage of hidden neurons will not be active, as shown in Figure 3.5 (after (Yu et al., 2013)). As a result, the average norm $\|\text{diag}(\mathbf{v}^{l+1}.*(1-\mathbf{v}^{l+1}))(\mathbf{A}^l)^T\|$ in Equation 3.23 is smaller than one in all layers, as indicated in Figure 3.6 (after (Yu et al., 2013)). Since all hidden layer values are bounded in the same range between 0 and 1, this indicates that when there is

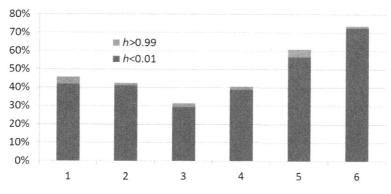

FIGURE 3.5

Percentage of saturated activations at each layer on a 6×2k DNN.

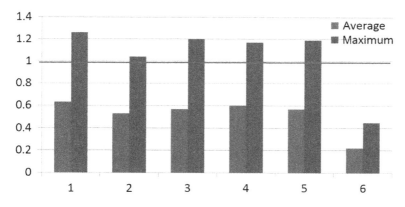

FIGURE 3.6

Average and maximum of $\|\mathrm{diag}(v^{l+1}. * (1 - v^{l+1}))(A^l)^T\|_2$ across layers on a 6×2k DNN.

a small perturbation on the input, the perturbation shrinks at each higher hidden layer. In other words, features generated by higher hidden layers are more invariant to variations than those represented by lower layers. Note that the maximum norm over the same development set is larger than one, as seen in Figure 3.6. This is necessary since the differences need to be enlarged around the class boundaries to have discrimination ability.

We use t-SNE (der Maaten and Hinton, 2008) to visualize the invariance property of DNN. t-SNE is a tool to visualize high-dimensional data in a low-dimensional space. It preserves neighbor relation so that the nearby points in the plot are also very close in the high-dimensional space. A DNN is first trained with multi-style training data of Aurora 4. The input feature to DNN is the mean-normalized log-Mel-filter-bank. A parallel pair of utterances from the training set of Aurora 4 are used for the illustration. The first one is a clean utterance, and the second one is synthesized from the clean one by adding restaurant noise with SNR 10 dB.

Figure 3.7 shows the t-SNE plots of the mean-normalized log-Mel-filter-bank feature and the 1st, 3rd, and 5th layer activation vectors of the corresponding multi-style DNN for this pair of stereo utterances. The green upper triangles (light gray in print versions) correspond to the frames of pure clean silence, and the black lower triangles are the frames of pure noise. The red stars (dark gray in print versions) refer to clean speech frames, and the blue circles (dark gray in print versions) are noisy speech frames. From Figure 3.7(a), we can clearly see that most clean speech and noisy speech samples are scattered away from each other. Also, the pure noise samples are mixed with some noisy speech samples. The DNN training provides a layer-by-layer feature extraction strategy that automatically derives powerful noise-resistant features from primitive raw data for senone classification. This invariance property can be observed in Figure 3.7(b)-(d), in which the noisy and clean speech samples become better aligned together when going through the DNN from the lower

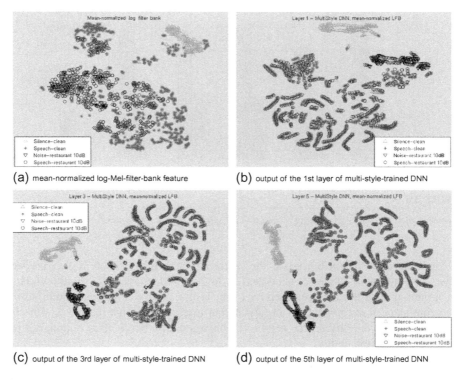

(a) mean-normalized log-Mel-filter-bank feature

(b) output of the 1st layer of multi-style-trained DNN

(c) output of the 3rd layer of multi-style-trained DNN

(d) output of the 5th layer of multi-style-trained DNN

FIGURE 3.7

t-SNE plot of a clean utterance and the corresponding noisy one with 10 dB SNR of restaurant noise from the training set of Aurora 4.

layer to the higher layer. If the noisy and clean samples are perfectly aligned in the t-SNE plot, that means these samples are very close to each other in the original high-dimension space. Therefore, when being presented to a classifier, similar results will be generated from these well-aligned samples. Note that the pure silence samples and the pure noise samples are not well aligned because the difference between the log-Mel-filter-bank values of pure silence and noise samples is much larger than that of the clean and noisy speech samples. As discussed before, the DNN's invariance property can only handle the input samples with moderate difference. Otherwise, even after going through multiple layers of a DNN, the invariance cannot be obtained.

By comparing Figure 3.7(a) and (d), it is clear that the activation vector in the 5th layer of the DNN is almost perfectly aligned for clean and noisy speech samples, much better than in the original log-Mel-filter-bank input feature space. As discussed in Li et al. (2014), a DNN can be decomposed into two parts. The first is from the input layer until the last hidden layer, working as a feature extraction component. The second is the top layer with a softmax classifier which has a linear classification boundary for each senone. Figure 3.7(d) clearly shows that the feature extracted at the last hidden layer (the 5th hidden layer in this example) is distortion invariant. In other words, the top-hidden-layer generated feature is very robust to distortion. Therefore, a simple classifier such as softmax classifier can work very well for senone classification. In contrast, the clean and noisy samples in Figure 3.7(a) are highly scattered, and a much more complicated classifier such as GMM classifier needs to be used.

We just showed that small perturbations in the input will be gradually shrunk as we move to the internal representation in the higher layers. However, we need to point out that the above result is only applicable to small perturbations around the training samples. When the test samples deviate significantly from the training samples, a DNN cannot accurately classify them. In other words, a DNN must see examples of representative variations in the data during training in order to generalize to similar variations in the test data. The utterances in Figure 3.7 come from the training set, which is used to train the DNN. Therefore, the DNN has been taught to work well on these utterances. In contrast, if we have utterances never seen during DNN training, the perfect invariance cannot be obtained. This is shown in Figure 3.8 in which a pair of stereo utterances from Aurora 4 test set are used for illustration. The first utterance is a clean utterance, and the second one is synthesized by adding restaurant noise with SNR 11 dB. Figures 3.8(a), (c), and (e) plot the 1st, 3rd, and 5th layer activation vectors from a DNN trained with only clean data, and Figures 3.8(b), (d), and (f) plot the 1st, 3rd, and 5th layer activation vectors from a DNN trained with multi-style data. The clean-trained DNN doesn't observe both the test utterances and the restaurant noise, and hence we can observe lots of clean and noisy speech samples are not aligned even with the 5th layer activation vector in Figure 3.8(e). In contrast, the multi-style-trained DNN has already observed the restaurant noise, although it doesn't see the test utterance. As a result, it has better invariance than the clean-trained DNN. However there are still some not-well-aligned samples, which

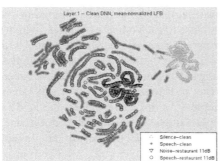
(a) output of the 1st layer of clean-trained DNN

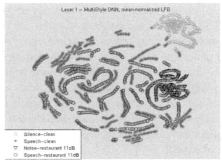
(b) output of the 1st layer of multi-style-trained DNN

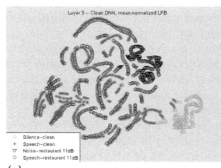
(c) output of the 3rd layer of clean-trained DNN

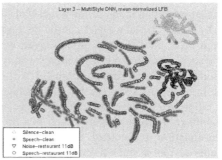
(d) output of the 3rd layer of multi-style-trained DNN

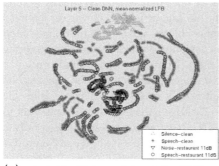
(e) output of the 5th layer of clean-trained DNN

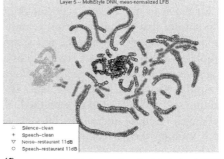
(f) output of the 5th layer of multi-style-trained DNN

FIGURE 3.8

t-SNE plot of a clean utterance and the corresponding noisy one with 11 dB SNR of restaurant noise from the test set of Aurora 4.

the DNN cannot perform good invariance on. Therefore, although DNN models can have better modeling invariance than GMM models, it is still necessary to apply robustness technologies to handle unseen samples.

3.5 A GENERAL FRAMEWORK FOR ROBUST SPEECH RECOGNITION

Following and elaborating on the analysis and formulation in earlier literature (Deng, 2011; Deng et al., 2005; Haeb-Umbach, 2011), here we outline a general Bayesian framework as the basis for discussing the problem of distortion-robust ASR. In Equation 2.1 from the preceding chapter, the recognized word sequence is obtained when the input acoustic signal is the clean speech sequence $\mathbf{X}$. When the distorted speech sequence $\mathbf{Y}$ is presented, the decision rule becomes

$$\hat{\mathbf{W}} = \underset{\mathbf{W}}{\operatorname{argmax}}\, P_{\Lambda,\Gamma}(\mathbf{W}|\mathbf{Y}) \tag{3.24}$$

to obtain the optimal word sequence $\hat{\mathbf{W}}$. Λ and Γ are the AM and LM parameters.

By introducing the clean speech sequence $\mathbf{X}$ as a hidden variable, we have

$$P_{\Lambda,\Gamma}(\mathbf{W}|\mathbf{Y}) = \int P_{\Lambda,\Gamma}(\mathbf{W}|\mathbf{X},\mathbf{Y})p(\mathbf{X}|\mathbf{Y})\,d\mathbf{X}$$
$$= \int P_{\Lambda,\Gamma}(\mathbf{W}|\mathbf{X})p(\mathbf{X}|\mathbf{Y})\,d\mathbf{X}. \tag{3.25}$$

In Equation 3.25 we exploited the fact that the distorted speech signal $\mathbf{Y}$ does not deliver additional information if clean speech signal $\mathbf{X}$ is given. With Equation 2.2, $P_{\Lambda,\Gamma}(\mathbf{W}|\mathbf{Y})$ can be re-written as

$$P_{\Lambda,\Gamma}(\mathbf{W}|\mathbf{Y}) = P_{\Gamma}(\mathbf{W})\int \frac{p_{\Lambda}(\mathbf{X}|\mathbf{W})}{p(\mathbf{X})}p(\mathbf{X}|\mathbf{Y})\,d\mathbf{X}. \tag{3.26}$$

Note that

$$p_{\Lambda}(\mathbf{X}|\mathbf{W}) = \sum_{\theta} p_{\Lambda}(\mathbf{X}|\theta,\mathbf{W})P_{\Lambda}(\theta|\mathbf{W}) \tag{3.27}$$

where θ belongs to the set of all possible state sequences for the transcription $\mathbf{W}$ and then Equation 3.26 becomes

$$P_{\Lambda,\Gamma}(\mathbf{W}|\mathbf{Y}) = P_{\Gamma}(\mathbf{W})\int \frac{\sum_{\theta} p_{\Lambda}(\mathbf{X}|\theta,\mathbf{W})P_{\Lambda}(\theta|\mathbf{W})}{p(\mathbf{X})}p(\mathbf{X}|\mathbf{Y})\,d\mathbf{X}. \tag{3.28}$$

Under some mild assumptions this can be simplified to (Haeb-Umbach (2011) and Ion and Haeb-Umbach (2008))

$$P_{\Lambda,\Gamma}(\mathbf{W}|\mathbf{Y}) = P_\Gamma(\mathbf{W}) \sum_\theta \prod_{t=1}^{T} \int \frac{p_\Lambda(\mathbf{x}_t|\theta_t)p(\mathbf{x}_t|\mathbf{Y})}{p(\mathbf{x}_t)} \, d\mathbf{x}_t P_\Lambda(\theta_t|\theta_{t-1}). \qquad (3.29)$$

One key component in Equation 3.29 is $p(\mathbf{x}_t|\mathbf{Y})$, the clean speech's posterior given distorted speech sequence $\mathbf{Y}$. In principle, it can be computed via

$$p(\mathbf{x}_t|\mathbf{Y}) \propto p(\mathbf{Y}|\mathbf{x}_t)p(\mathbf{x}_t), \qquad (3.30)$$

that is, employing an *a priori* model $p(\mathbf{x}_t)$ of clean speech and an observation model $p(\mathbf{Y}|\mathbf{x}_t)$, which relates the clean to the distorted speech features. Robustness techniques may be categorized by the kind of observation models used, also according to whether an explicit or an implicit distortion model is used, and according to whether or not prior knowledge about distortion is employed to learn the relationship between $\mathbf{x}_t$ and $\mathbf{y}_t$, as we will further develop in later chapters of the book.

For simplicity, many robust ASR techniques use a point estimate. That is, the back-end recognizer considers the cleaned or denoised signal $\hat{\mathbf{x}}_t(\mathbf{Y})$ as an estimate without uncertainty:

$$p(\mathbf{x}_t|\mathbf{Y}) = \delta(\mathbf{x}_t - \hat{\mathbf{x}}_t(\mathbf{Y})), \qquad (3.31)$$

where $\delta(\cdot)$ is a Kronecker delta function. Then, Equation 3.29 is reduced to

$$P_{\Lambda,\Gamma}(\mathbf{W}|\mathbf{Y}) = P_\Gamma(\mathbf{W}) \sum_\theta \prod_{t=1}^{T} \frac{p_\Lambda(\hat{\mathbf{x}}_t(\mathbf{Y})|\theta_t)}{p(\hat{\mathbf{x}}_t(\mathbf{Y}))} P_\Lambda(\theta_t|\theta_{t-1}). \qquad (3.32)$$

Because the denominator $p(\hat{\mathbf{x}}_t(\mathbf{Y}))$ is independent of the underlying word sequence, the decision is further reduced to

$$\hat{\mathbf{W}} = \underset{\mathbf{W}}{\operatorname{argmax}} \, P_\Gamma(\mathbf{W}) \sum_\theta \prod_{t=1}^{T} p_\Lambda(\hat{\mathbf{x}}_t(\mathbf{Y})|\theta_t) P_\Lambda(\theta_t|\theta_{t-1}). \qquad (3.33)$$

Equation 3.33 is the formulation most commonly used. Compared to Equation 2.4, the only difference is that $\hat{\mathbf{x}}_t(\mathbf{Y})$ is used to replace $\mathbf{x}_t$. In feature processing methods, only the distorted feature $\mathbf{y}_t$ is enhanced with $\hat{\mathbf{x}}_t(\mathbf{Y})$, without changing the acoustic model parameter, Λ. The general form of feature processing method is

$$\hat{\mathbf{X}} = \mathcal{G}(\mathbf{Y}), \qquad (3.34)$$

which generates the estimated clean feature sequence $\hat{\mathbf{X}}$ by cleaning the noisy feature sequence $\mathbf{Y}$ with function $\mathcal{G}(\cdot)$.

In contrast, there is another major category of model-domain processing methods, which adapts model parameters to fit the distorted speech signal

$$\hat{\Lambda} = \mathcal{F}(\Lambda, \mathbf{Y}) \tag{3.35}$$

and in this case the posterior used in the MAP decision rule is computed using

$$P_{\hat{\Lambda},\Gamma}(\mathbf{W}|\mathbf{Y}) = P_{\Gamma}(\mathbf{W}) \sum_{\theta} \prod_{t=1}^{T} p_{\hat{\Lambda}}(\mathbf{y}_t|\boldsymbol{\theta}_t) P_{\hat{\Lambda}}(\boldsymbol{\theta}_t|\boldsymbol{\theta}_{t-1}). \tag{3.36}$$

3.6 CATEGORIZING ROBUST ASR TECHNIQUES: AN OVERVIEW

The main theme of this book is to provide insights from multiple perspectives in organizing a multitude of robust ASR techniques. Based on the general framework in this chapter, we provide a comprehensive view, in a mathematically rigorous and unified manner, of robust ASR using five different ways of categorizing, analyzing, and characterizing major existing techniques for single-microphone non-reverberant speech. The categorization is based on the following key attributes of the algorithms: (1) the processing domain; (2) the availability of prior knowledge about distortion exploited; (3) the nature of the underlying speech distortion model in use; (4) the nature of uncertainty exploited in robust processing; and (5) the style of model training. We provide a brief overview on the above five ways of categorizing and analyzing the techniques of noise-robust ASR. Chapters 4–8 will be devoted to detailed exposition of each of the five ways of analysis.

3.6.1 COMPENSATION IN FEATURE DOMAIN VS. MODEL DOMAIN

As shown in Figure 3.9 (after (Lee, 1998)), the acoustic mismatch between training and test conditions can be viewed from either the feature domain or the model

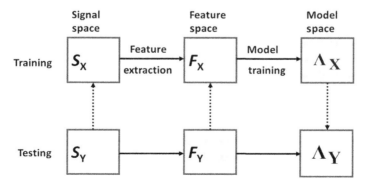

FIGURE 3.9

Noise-robust methods in feature and model domain.

domain, and noise or distortion can be compensated for in either space although there is a recent study normalizing the final acoustic likelihood score for robustness (Vincent et al., 2014). Some methods are formulated in both the feature and model domains, and can thus be categorized as "Hybrid." Feature-space approaches usually do not change the parameters of acoustic models. Most feature-space methods use Equation 3.33 to compute the posterior $P_{\Lambda,\Gamma}(\mathbf{W}|\mathbf{Y})$ after "plugging in" the enhanced signal $\hat{\mathbf{x}}_t(\mathbf{Y})$. On the other hand, model-domain methods modify the acoustic model parameters with Equation 3.35 to incorporate the effects of the distortion, as in Equation 3.36. In contrast with feature-space methods, the model-domain methods are closely linked with the objective function of acoustic modeling. While typically achieving higher accuracy than feature-domain methods, they usually incur significantly larger computational costs. We will discuss both the feature- and model-domain methods in detail in Sections 4.1 and 4.2. Specifically, noise-resistant features, feature moment normalization, and feature compensation methods are presented in Sections 4.1.1, 4.1.2, and 4.1.3, respectively. General adaptation technologies for GMMs and DNNs are presented in Sections 4.2.1 and 4.2.2, respectively, and robustness via better modeling is discussed in Section 4.2.3.

3.6.2 COMPENSATION USING PRIOR KNOWLEDGE ABOUT ACOUSTIC DISTORTION

This axis for categorizing and analyzing robust ASR techniques examines whether the method exploits prior knowledge about the distortion. Some of these methods, discussed in Section 5.1, learn the mapping between clean and distorted speech features when they are available as a pair of "stereo" data.

During decoding, with the pre-learned mapping, the clean speech feature $\hat{\mathbf{x}}_t(\mathbf{Y})$ can be estimated and plugged into Equation 3.33 to generate the word sequence. Another type of methods presented in Section 5.2 builds multiple models or dictionaries of speech and noise from multi-environment data. Some examples discussed in Section 5.2.1 collect and learn a set of models first, each corresponding to one specified environment in training. These pre-learned models are then combined online to form a new model $\hat{\Lambda}$ that fits the test environment best. The methods described in Section 5.2.2 are usually based on source separation—they build clean speech and noise exemplars from training data, and then reconstruct the speech signal $\hat{\mathbf{x}}_t(\mathbf{Y})$ only from the exemplars of clean speech. With variable-parameter methods, examined in Section 5.2.3, the acoustic model parameters or transforms are polynomial functions of an environment variable. In addition to training an HMM, all the methods analyzed in this category have the unique attribute of exploiting prior knowledge in the training stage. They then use such prior knowledge as a guide to either remove noise or adapt models in the test stage. In contrast, when only HMMs are obtained in the training stage, there will be no prior knowledge used in the test stage.

3.6.3 COMPENSATION WITH EXPLICIT VS. IMPLICIT DISTORTION MODELING

To adapt the model parameters in Equation 3.35, general technologies make use of a set of linear transformations to compensate for the mismatch between training and test conditions. This involves many parameters and thus typically requires a large amount of data for estimation. This difficulty can be overcome when exploiting an explicit distortion model which takes into account the way in which distorted speech features are produced. That is, the distorted speech features are represented by using a nonlinear function of clean speech features, additive noise, and convolutive distortion. This type of physical model enables structured transformations to be used, which are generally nonlinear and involve only a parsimonious set of free parameters to be estimated. In Section 3.2, we have already discussed the most popular type of such physical distortion model. There are many of its variations widely used in robust ASR. We refer to a robust method as an explicit distortion modeling one when a physical model for the generation of distorted speech features is employed. If no physical model is explicitly used, the method will be referred to as an implicit distortion modeling method. Since physical constraints are modeled, the explicit distortion modeling methods exhibit high performance and require a relatively small number of distortion parameters to be estimated. Explicit distortion models can also be applied to feature processing. With the guide of explicit modeling, the enhancement of speech often becomes more effective. Robust ASR techniques with explicit distortion modeling will be explored in Chapter 6.

3.6.4 COMPENSATION WITH DETERMINISTIC VS. UNCERTAINTY PROCESSING

Most robust ASR methods use a deterministic strategy; that is, the compensated feature is a point estimate from the corrupted speech feature with Equation 3.31, or the compensated model is a point estimate as adapted from the clean speech model with Equation 3.35. We refer to methods in this category as deterministic processing methods. However, strong noise and unreliably decoded transcriptions necessarily create inherent uncertainty in either the feature or the model space, which should be accounted for in MAP decoding. When a noise-robust method takes that uncertainty into consideration, we call it an uncertainty processing method. In the feature space, the presence of noise brings uncertainty to the enhanced speech signal, which is modeled as a distribution instead of a deterministic value. In the general case, $p(\mathbf{x}_t|\mathbf{Y})$ in Equation 3.29 is not a Kronecker delta function and there is uncertainty in the estimate of $\hat{\mathbf{x}}_t$ given $\mathbf{Y}$. Uncertainty can also be introduced in the model space when assuming the true model parameters are in a neighborhood of the trained model parameters Λ, or compensated model parameters $\tilde{\Lambda}$. We will study uncertainty methods in model and feature spaces in Sections 7.1 and 7.2, respectively. Then joint uncertainty decoding is described in Section 7.3 and missing feature approaches are discussed in Section 7.4.

3.6.5 COMPENSATION WITH DISJOINT VS. JOINT MODEL TRAINING

Finally, we can categorize most existing noise-robust techniques in the literature into two broad classes depending on whether or not the acoustic model, Λ, is trained jointly with the same process of feature enhancement or model adaptation used in the test stage. Among the joint model training methods, the most prominent set of techniques are based on a paradigm called noise adaptive training (NAT) which applies consistent processing during the training and test phases while eliminating any residual mismatch in an otherwise disjoint training paradigm. Further developments of NAT include joint training of a canonical acoustic model and a set of transforms under maximum likelihood estimation or a discriminative training criterion. In Chapter 8, these methods will be examined in detail.

Note that the chosen categories discussed above are by no means orthogonal. While it may be ambiguous under which category a particular robustness approach would fit the best, we have used our best judgment with a balanced view.

3.7 SUMMARY

In this chapter, we cover detailed derivations of a set of acoustic distortion models of speech, all of which are commonly used in robust ASR techniques when the speech models are of generative nature (e.g., the GMM-HMM), both in research and for practical uses in ASR systems (Huang et al., 2001). The effects of acoustic distortions, including additive noise and convolutive channel distortions, are simulated and analyzed. The analysis is carried on both speech models based on the GMM-HMM and on the DNN-HMM. The abstraction of such analyses leads to a general Bayeisan framework for handling distortion-robust ASR problems. Then, based partly on this general framework, we started the overview of five axes we have devised to categorize and characterize a vast set of distortion-robust ASR techniques in the literature, serving as a general introduction to the next five chapters of this book.

Before delving into the details of the next five chapters, a few caveats are in order. First, the selected five research axis for the categorization of the noise-robustness techniques in the literature are certainly not the only possible taxonomies. But we have tried our best to select the five most prominent attributes based on which we carried out our analyses. Second, noise robustness in ASR is a vast topic, spanning research literature over 30 years. In developing this overview, we necessarily have to limit its scope. In particular, we mainly consider single-channel inputs, thus leaving out the topics of acoustic beamforming, multi-channel speech enhancement and source separation in our analysis for the next five chapters. To limit the scope, we also assume that the noise can be considered more stationary than speech, thus disregarding most robust ASR techniques in the presence of music or other competing speakers in the next five chapters. Further, we assume that the channel impulse response is much shorter than the frame size in the next five chapters; that is, we do not consider the case of reverberation, but rather convolutional channel distortions caused by, for example, different microphone characteristics.

REFERENCES

Acero, A., 1993. Acoustical and Environmental Robustness in Automatic Speech Recognition. Cambridge University Press, Cambridge, UK.

Christensen, H., Barker, J., Ma, N., Green, P., 2010. The CHiME corpus: a resource and a challenge for computational hearing in multisource environments. In: Proc. Interspeech, pp. 1918-1921.

Cohen, I., Gannot, S., 2008. Spectral enhancement methods. In: Benesty, J., Sondhi, M.M., Huang, Y. (Eds.), Handbook of Speech Processing. Springer, New York.

Dahl, G., Yu, D., Deng, L., Acero, A., 2012. Context-dependent pre-trained deep neural networks for large-vocabulary speech recognition. IEEE Trans. Audio Speech Lang. Process. 20 (1), 30-42.

Davis, S.B., Mermelstein, P., 1980. Comparison of parametric representation for monosyllabic word recognition in continuously spoken sentences. IEEE Trans. Acoustics, Speech and Signal Processing 28 (4), 357-366.

Deng, L., 2006. Dynamic Speech Models—Theory, Algorithm, and Applications. Morgan and Claypool, San Rafael, CA.

Deng, L., 2011. Front-end, back-end, and hybrid techniques for noise-robust speech recognition. In: Robust Speech Recognition of Uncertain or Missing Data: Theory and Application. Springer, New York, pp. 67-99.

Deng, L., Droppo, J., Acero, A., 2002. A Bayesian approach to speech feature enhancement using the dynamic cepstral prior. In: Proc. International Conference on Acoustics, Speech and Signal Processing (ICASSP), vol. 1, pp. I-829-I-832.

Deng, L., Droppo, J., Acero, A., 2004a. Enhancement of Log Mel power spectra of speech using a phase-sensitive model of the acoustic environment and sequential estimation of the corrupting noise. IEEE Trans. Speech Audio Process. 12 (2), 133-143.

Deng, L., Droppo, J., Acero, A., 2004b. Estimating cepstrum of speech under the presence of noise using a joint prior of static and dynamic features. IEEE Trans. Audio Speech Lang. Process. 12 (3), 218-233.

Deng, L., Droppo, J., Acero, A., 2005. Dynamic compensation of HMM variances using the feature enhancement uncertainty computed from a parametric model of speech distortion. IEEE Trans. Speech Audio Process. 13 (3), 412-421.

Deng, L., Li, X., 2013. Machine learning paradigms in speech recognition: An overview. IEEE Trans. Audio Speech and Lang. Process. 21 (5), 1060-1089.

Deng, L., O'Shaughnessy, D., 2003. Speech Processing—A Dynamic and Optimization-Oriented Approach. Marcel Dekker Inc., New York.

der Maaten, L.V., Hinton, G., 2008. Visualizing data using t-SNE. J. Mach. Learning Res. 9 (11), 2579-2605.

Droppo, J., Deng, L., Acero, A., 2003. A comparison of three non-linear observation models for noisy speech features. In: Proc. European Conference on Speech Communication and Technology (EUROSPEECH), pp. 681-684.

Frey, B., Deng, L., Acero, A., Kristjansson, T., 2001a. ALGONQUIN: iterating Laplace's method to remove multiple types of acoustic distortion for robust speech recognition. In: Proc. Interspeech, pp. 901-904.

Frey, B., Kristjansson, T., Deng, L., Acero, A., 2001b. ALGONQUIN—learning dynamic noise models from noisy speech for robust speech recognition. In: NIPS, pp. 1165-1171.

Haeb-Umbach, R., 2011. Uncertainty decoding and conditional Bayesian estimation. In: Robust Speech Recognition of Uncertain or Missing Data: Theory and Application. Springer, New York, pp. 9-34.

Hannun, A., Case, C., Casper, J., Catanzaro, B., Diamos, G., Elsen, E., et al., 2014. Deepspeech: Scaling up end-to-end speech recognition. arXiv preprint arXiv:1412.5567

Hinton, G., Deng, L., Yu, D., Dahl, G.E., Mohamed, A., Jaitly, N., et al., 2012. Deep neural networks for acoustic modeling in speech recognition: The shared views of four research groups. IEEE Signal Process. Mag. 29 (6), 82-97.

Hirsch, H.G., 2007. Aurora-5 experimental framework for the performance evaluation of speech recognition in case of a hands-free speech input in noisy environments. Niederrhein Univ. of Applied Sciences.

Hirsch, H.G., Pearce, D., 2000. The Aurora experimental framework for the performance evaluation of speech recognition systems under noisy conditions. In: ISCA ITRW ASR.

Huang, X., Acero, A., Chelba, C., Deng, L., Droppo, J., Duchene, D., et al., 2001. MiPad: a multimodal interaction prototype. In: Proc. International Conference on Acoustics, Speech and Signal Processing (ICASSP).

Ion, V., Haeb-Umbach, R., 2008. A novel uncertainty decoding rule with applications to transmission error robust speech recognition. IEEE Trans. Audio Speech Lang. Process. 16 (5), 1047-1060. ISSN 1558-7916. doi:10.1109/TASL.2008.925879.

Lee, C.H., 1998. On stochastic feature and model compensation approaches to robust speech recognition. Speech Commun. 25, 29-47.

Leutnant, V., Haeb-Umbach, R., 2009. An analytic derivation of a phase-sensitive observation model for noise-robust speech recognition. In: Proc. Interspeech, pp. 2395-2398.

Li, J., Zhao, R., Gong, Y., 2014. Learning small-size DNN with output-distribution-based criteria. In: Proc. Interspeech.

Lippmann, R., Martin, E., Paul, D., 1987. Multi-style training for robust isolated-word speech recognition. In: Proc. International Conference on Acoustics, Speech and Signal Processing (ICASSP), pp. 705-708.

Moreno, A., Lindberg, B., Draxler, C., Richard, G., Choukri, K., Euler, S., et al., 2000. SPEECHDAT-CAR. a large speech database for automotive environments. In: LREC.

Mostefa, D., Moreau, N., Choukri, K., Potamianos, G., Chu, S., Tyagi, A., et al., 2007. The CHIL audiovisual corpus for lecture and meeting analysis inside smart rooms. Language Resources and Evaluation 41 (3-4), 389-407.

Parihar, N., Picone, J., Institute for Signal and Information Processing, Mississippi State Univ., 2002. Aurora working group: DSR front end LVCSR evaluation AU/384/02.

Paul, D.B., Baker, J.M., 1992. The design for the Wall Street Journal-based CSR corpus. In: Proceedings of the workshop on Speech and Natural Language, pp. 357-362.

Renals, S., Hain, T., Bourlard, H., 2008. Interpretation of multiparty meetings: The AMI and AMIDA projects. In: Proc. HSCMA.

Schmidt-Nielsen, A., Marsh, E., Tardelli, J., Gatewood, P., Kreamer, E., Tremain, T., et al., 2000. Speech in noisy environments (SPINE) evaluation audio. In: Linguistic Data Consortium.

Seltzer, M.L., Yu, D., Wang, Y., 2013. An investigation of deep neural networks for noise robust speech recognition. In: Proc. International Conference on Acoustics, Speech and Signal Processing (ICASSP), pp. 7398-7402.

Stupakov, A., Hanusa, E., Vijaywargi, D., Fox, D., Bilmes, J., 2012. The design and collection of COSINE., a multi-microphone in situ speech corpus recorded in noisy environments. Comput. Speech Lang. 26 (1), 52-66.

Vincent, E., Barker, J., Watanabe, S., Le Roux, J., Nesta, F., Matassoni, M., 2013. The second CHiME speech separation and recognition challenge: Datasets, tasks and baselines.

In: Proc. International Conference on Acoustics, Speech and Signal Processing (ICASSP), pp. 126-130.

Vincent, E., Gkiokas, A., Schnitzer, D., Flexer, A., 2014. An investigation of likelihood normalization for robust ASR. In: Proc. Interspeech.

Yu, D., Seltzer, M., Li, J., Huang, J.T., Seide, F., 2013. Feature learning in deep neural networks-studies on speech recognition tasks. In: Proc. International Conference on Learning Representations.

Processing in the feature and model domains

4

CHAPTER OUTLINE

In this chapter, we delve into the most popular way of categorizing and analyzing a full range of noise-robust ASR methods in the literature. Our categorizing is based on the most prominent attribute with which these methods are developed—whether the methods are applied to the feature domain or to the model domain. In the former case, no change of the parameters of the acoustic model is made. The processing operation typically takes the general form of

$$\hat{\mathbf{X}} = \mathcal{G}(\mathbf{Y}) \tag{4.1}$$

converting the distorted speech feature vector $\mathbf{Y}$ into its pseudo-clean version $\hat{\mathbf{X}}$ via a transform as denoted by the operation $\mathcal{G}$ above. Note that there is no

Robust Automatic Speech Recognition. http://dx.doi.org/10.1016/B978-0-12-802398-3.00004-0

model parameter in Equation 4.1; hence feature-domain processing. The techniques in this category rely either on auditory features **X** that are inherently robust to distortions such as noise or on modification of the features at the test time to match the training features. In the former case, $\mathcal{G}$ is an identity transform. For the latter case, it often takes either linear, piecewise linear, or nonlinear transformations. Because the transformations are not related to the back-end, the computational cost of these feature-domain processing methods is usually low.

On the other hand, model-domain methods modify the acoustic model parameters to incorporate the effects of distortion, taking the general form of Equation 4.2:

$$\hat{\Lambda} = \mathcal{F}(\Lambda, \mathbf{Y}), \tag{4.2}$$

where Λ denotes the set of parameters associated with the acoustic models in speech recognition systems. While typically achieving higher accuracy than feature-domain methods because model-domain methods are closely linked with the objective function of acoustic modeling, they usually incur a significantly larger computational cost. The model-domain approaches can be classified further into two sub-categories: general adaptation and noise-specific compensation. The general adaptation technology compensates for the mismatch between training and test conditions by using generic transformations to convert the acoustic model parameters. Some model adaptation methods are quite general, applicable not only to noise compensation but also to other types of acoustic variations. The dominating methods in this case use linear transforms for $\mathcal{F}$. Some methods, on the other hand, are noise-specific and usually modify model parameters by explicitly addressing the nature of distortions caused by the presence of noise. Because the distortion modeling in the cepstral or log-Mel-filter-bank domain is nonlinear as described in section 3.2, $\mathcal{F}$ usually has a nonlinear form in this case.

4.1 **FEATURE-SPACE APPROACHES**

Feature-space methods can be classified further into three sub-categories:

- noise-resistant features, where robust signal processing is employed to reduce the sensitivity of the speech features to environment conditions that don't match those used to train the acoustic model;
- feature normalization, where the features are normalized such that they have similar statistical moments, irrespective of whether the input signal is clean or noisy;
- feature compensation, where the effects of distortion embedded in the observed speech features are removed.

4.1.1 **NOISE-RESISTANT FEATURES**

Noise-resistant feature methods focus on the effect of noise rather than on the removal of noise. One of the advantages of these techniques is that they make only weak or no assumptions about the noise. In general, no explicit estimation of the noise statistics is required. On the other hand, this can be a shortcoming since it is impossible to make full use of the characteristics specific to a particular noise type.

Auditory-based features

There is a long history of auditory-based features that have been in development for ASR applications (Deng, 1999; Deng et al., 1988; Hermansky, 1990; Hermansky et al., 1985; Sheikhzadeh and Deng, 1998, 1999). Most notably, perceptually based linear prediction (PLP) (Hermansky, 1990; Hermansky et al., 1985) filters the speech signal with a Bark-scale filter-bank. The output is converted into an equal-loudness representation. The resulting auditory spectrum is then modeled by an all-pole model. A cepstral analysis can also be performed.

There are plenty of other auditory-based feature extraction methods, such as zero crossing peak amplitude (ZCPA) (Kim et al., 1999), average localized synchrony detection (ALSD) (Ali et al., 2002), perceptual minimum variance distortionless response (PMVDR) (Yapanel and Hansen, 2008), power-normalized cepstral coefficients (PNCC) (Kim and Stern, 2010), invariant-integration features (IIF) (Müller and Mertins, 2011), amplitude modulation spectrogram (Moritz et al., 2011), Gammatone frequency cepstral coefficients (Shao et al., 2010), sparse auditory reproducing kernel (SPARK) (Fazel and Chakrabartty, 2012), and Gabor filter bank features (Moritz et al., 2013), to name a few. Stern and Morgan (Stern and Morgan, 2012) provides a relatively complete review on auditory-based features. All these methods are designed by utilizing some auditory knowledge. However, there is no universally-accepted theory about which kind of auditory information is most important for robust speech recognition. Therefore, it is hard to argue which one in theory is better than another.

Here, we just briefly describe the recently-proposed PNCC feature, which has shown advantages (Kelly and Harte, 2010; Kim and Stern, 2012; Sárosi et al., 2010) in terms of accuracy over several popular robust features, such as MFCC, RASTA-PLP, PMVDR, and ZCPA. It has been widely used as a baseline method for comparison with other noise-robustness features. As shown in Figure 4.1 (after (Kim, 2010)), PNCC, MFCC, and RASTA-PLP share some similar components: frequency integration with different kinds of filters and nonlinearity processing, which can find its ground in auditory theory. In particular, PNCC features (Kim and Stern, 2012) have:

- a power-law nonlinearity (1/15), which is carefully chosen to approximate the nonlinear relation between signal intensity and auditory-nerve firing rate;
- the "medium-time" processing (a duration of 50-120 ms) to analyze the parameters characterizing environmental degradation;

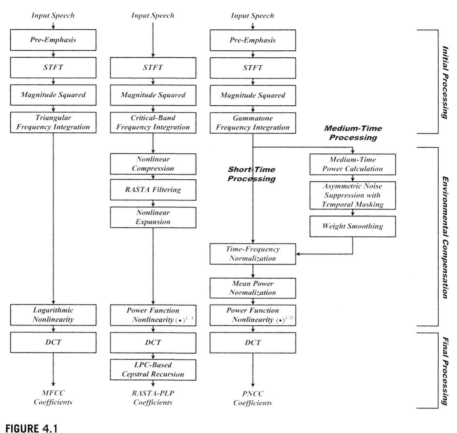

FIGURE 4.1

Comparison of the MFCC, RASTA-PLP, and PNCC feature extraction.

- a form of "asymmetric nonlinear filtering" to estimate the level of noise for each frame and frequency bin;
- a temporal masking.

Since there is no universally accepted auditory theory for robust speech recognition, it is sometimes very hard to set the right parameter values in auditory methods. Some parameters can be learned from data (Chiu et al., 2012), but this may not always be the case. Although the auditory-based features can usually achieve better performance than MFCC, they have a much more complicated generation process which sometimes prevents them from being widely used together with some noise-robustness technologies. For example, in Section 3.2, the relation between clean and noisy speech for MFCC features can be derived as Equation 3.12. However, it is very hard to derive such a relation for auditory-based features. As a result, MFCC is widely used as the acoustic feature for methods with explicit distortion modeling.

Temporal processing

In addition to the spectral characteristics of speech features, temporal characteristics are also important for speech recognition. When speech is distorted by noise, the temporal characteristics of feature trajectories are also distorted and need to be enhanced. Hence, temporal filters are proposed to improve the temporal characteristics of speech features by modifying the power spectral density (PSD) functions of the feature trajectories. The PSD functions are closely related to the modulation spectra of speech signals (Houtgast and Steeneken, 1973). Figure 4.2 (after (Xiao et al., 2008)) shows the computation of the modulation spectrum of a speech signal. The left panel is the speech spectrogram, and the modulation spectra on the right panel are the PSD functions of the spectrogram trajectories. It has been shown that the modulation spectra correlate with speech intelligibility, especially in some frequency bands. For example, the modulation frequency between 1 and 16 Hz is found to be most important to speech intelligibility and ASR (Houtgast and Steeneken, 1973, 1985). Temporal filters are usually designed to enhance the useful modulation frequency range for speech recognition and attenuate other modulation frequency ranges.

Many types of additive noise as well as most channel distortions vary slowly compared to the variations in speech signals. Filters that remove variations in the signal that are uncharacteristic of speech (including components with both slow and

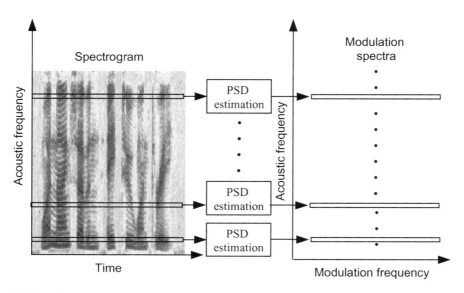

FIGURE 4.2

Computation of the modulation spectral of a speech signal.

fast modulation frequencies) improve the recognition accuracy significantly (Hanson and Applehaum, 1993). As one of the most popular temporal filters, relative spectral processing (RASTA) (Hermansky and Morgan, 1994; Hermansky et al., 1991) consists of suppressing constant additive offsets in each log spectral component of the short-term auditory-like spectrum. This analysis method can be applied to PLP parameters, resulting in RASTA-PLP (Hermansky and Morgan, 1994; Hermansky et al., 1991). Each frequency band is filtered by a noncausal infinite impulse response (IIR) filter that combines both high- and low-pass filtering. Assuming a frame rate of 100 Hz, the transfer function of RASTA

$$H(z) = 0.1z^4 \frac{2 + z^{-1} - z^{-3} - 2z^{-4}}{1 - 0.98z^{-1}} \tag{4.3}$$

is plotted in Figure 4.3. The filter has low cut-off frequency as 0.26 Hz. It also has sharp zeros at 28.9 and 50 Hz. The low-pass filtering helps to remove the fast frame-to-frame spectral changes which are not present in speech signal. The high-pass portion of the equivalent band-pass filter alleviates the effect of convolutional channel. While the design of the IIR-format RASTA filter in Equation 4.3 is based on auditory knowledge, the RASTA filter can also be designed as a linear finite impulse response (FIR) filter in a data-driven way using technology such as linear discriminant analysis (LDA) (Avendano et al., 1996).

As mentioned in Section 3.2, the additive noise is additive in the spectral domain while the convolutional distortion from the channel is additive in the log-spectral domain. To deal with both additive noise and convolutional distortion, J-RASTA has been further proposed (Morgan and Hermansky, 1992), which involves filtering the

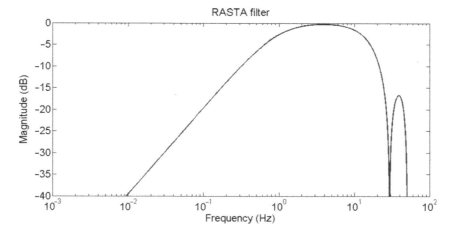

FIGURE 4.3

Frequency response of RASTA.

time series of some function of a spectrum and then performing the inverse of the function on the result. The function is roughly an identity in the case of significant additive noise and is roughly logarithmic in the case of convolutional distortion. The shape of the function is controlled by the value of J, which is dependent on the additive noise level.

Once designed, most temporal filters are fixed and do not change according to underlying environment conditions. However, as shown in Xiao et al. (2008), modulation spectra also get affected when the environment changes. As a solution, a temporal structure normalization (TSN) filter is designed in Xiao et al. (2008) to do environment-dependent temporal filtering. As shown in Figure 4.4 (after (Xiao et al., 2008)), the filter design process has two stages: in the first stage, the desired magnitude response of the temporal filter is estimated and then in the second stage the time-domain finite impulse response filter is designed as the implementation of the desired magnitude response. Specifically, the magnitude response is designed to modify the feature trajectory's PSD function towards the reference PSD function calculated in the first stage. In this way, the temporal characteristics of the features are normalized. This is extended in Xiao et al. (2012) where joint spectral and temporal normalization is performed.

Neural network approaches

Artificial neural network (ANN) based methods have a long history of providing effective features for ASR. For example, ANN-HMM hybrid systems (Bourlard and Morgan, 1994) replace the GMM acoustic model with an ANN when evaluating the likelihood score. The ANNs used before 2009 usually had the multi-layer perceptron (MLP) structure with one hidden layer. The input to the network is a context window of successive frames of the acoustic feature vector. The training target is obtained

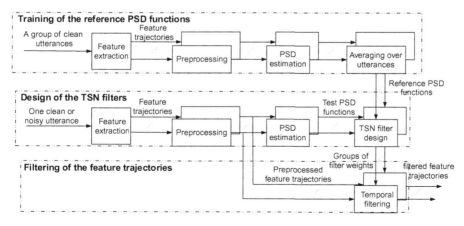

FIGURE 4.4

Illustration of the temporal structure normalization framework.

from the forced alignment of the training data with a previously-trained GMM model. The output of the ANN is a vector of posterior probabilities, with one element for each phone or subphone. Hybrid systems have been shown to have comparable performance to GMM-based systems.

The TANDEM system was later proposed in Hermansky et al. (2000) to combine ANN discriminative feature processing with a GMM, and it demonstrated strong performance on the Aurora 2 noisy continuous digit recognition task. Instead of using the posterior vector for decoding as in the hybrid system, the TANDEM system omits the final nonlinearity in the ANN output layer and applies a global decorrelation to generate a new set of features used to train a GMM system. One reason the TANDEM system has a very good performance on noise-robust tasks is that the ANN has the modeling power in small regions of feature space that lie on phone boundaries (Hermansky et al., 2000). Another reason is due to the nonlinear modeling power of the ANN, which can normalize data from different sources well.

Another way to obtain probabilistic features is temporal pattern (TRAP) processing (Hermansky and Sharma, 1998), which captures the appropriate temporal pattern with a long temporal vector of log-spectral energies from a single frequency band. The band-specific temporal pattern is classified by a band-specific MLP into a phoneme class. The outputs from all band-specific classifiers are then merged with a "merger" MLP to produce overall phoneme probabilities. One main reason for the noise-robustness of TRAP is the ability to handle band-specific processing (Jain et al., 2002). Even if one band of speech is polluted by noise, the phoneme classifier in another band can still work very well. TRAP processing works on temporal patterns, differing from the conventional spectral feature vector. Hence, TRAP can be combined very well with MFCC or PLP features to further boost performance (Chen et al., 2004). A multi-band system was proposed in Jain and Hermansky (2003) by combining three adjacent bands as the input to the band-specific classifier. However, as discussed in Grézl and Cernocký (2007), the multi-band system cannot outperform the original single-band system universally under all SNR conditions because the concatenation of adjacent bands spreads the noise from one band to other bands, especially in the low SNR case.

A new structure, hidden activation TRAPs (HATS), is shown in Chen et al. (2004) to have better performance than the TRAP with TANDEM structure. HATS uses the hidden activations of critical band MLPs instead of their outputs as the inputs to the "merger" MLP. The structure is further developed in Chen et al. (2005) with Tonotopic MLP, in which the first hidden layer is tonotopically organized: for each band, there is a disjoint set of hidden units using the long-term critical band log spectral energies as input. The hidden layers of MLP are fully connected with their inputs. In this way, the temporal features are hidden and the training process to generate them is unified with the overall MLP training to maximize the classification accuracy.

Building upon the aforementioned methods, bottle-neck (BN) features (Grézl et al., 2007) were developed as a new method to use ANNs for feature extraction. A five-layer MLP with a narrow layer in the middle (bottle-neck) is used to extract BN features. The fundamental difference between TANDEM and BN features is that the latter are not derived from the posterior vector. Instead, they are obtained as linear outputs of the bottle-neck layer. Principal component analysis (PCA) or heteroscedastic linear discriminant analysis (HLDA) (Kumar and Andreou, 1998) is used to decorrelate the BN features which then become inputs to the GMM-HMM system. Although current research of BN features is not focused on noise robustness, it has been shown that BN features outperform TANDEM features on some LVCSR tasks (Grézl et al., 2007; Tüske et al., 2012). Therefore, it is also possible that BN features can perform well on noise-robust tasks.

As shown in Section 3.4, the recently proposed DNNs provide a layer-by-layer feature extraction strategy that automatically derives powerful noise-resistant features from primitive raw data for senone classification, resulting in good noise-invariant property. In Huang et al. (2015), it is shown that CNNs have an advantage over standard DNNs in the following areas: channel-mismatched training-test conditions, noise robustness, and distant speech recognition. There are two possible factors in CNNs that might contribute to the robustness to environment/channel mismatch. One is that the weight-sharing structure of the frequency localized filters reduces the number of free parameters for the convolutional layer. This effective way of controlling model capacity could achieve better generalization, as opposed to adopting a fully-connected layer in the input layer. The other important factor is the use of max-pooling on top of the convolutional layer which makes CNNs more robust to translational variance in the input space. To be more specific, the channel or noise distortion affects the spectrum in some local filter banks. Due to the locality processing using frequency localized filters and the invariance processing via max-pooling, CNNs can generate better high-level distortion-robust features and give a better chance to higher network layers to do the classification job. Distant-talking speech has more distortion than close-talking speech because the speech signals are increasingly degraded by additive noise and room reverberation as the distance between the speaker and the microphones increases. In Huang et al. (2015), the gain of CNNs over DNNs increases as the distance between the speaker and microphone arrays increases, which suggests that the more distorted the signals are, the more effective CNNs become.

Even with the nice noise-invariant property, a single neural network may not handle all the acoustic conditions very well. In Li et al. (2014c), a long, deep, and wide neural network system is proposed to address this issue with a divide and conquer way. The speech signal is first decomposed into multiple frequency sub-bands, within which a sub-band ensemble network is trained to estimate the phoneme posterior probabilities. The sub-band ensemble network consists of several neural networks, each is trained from the speech with either clean, low, medium, or high SNR. Therefore, those neural networks cover conditions with all the SNRs. Finally, another ensemble neural network is used to merge the outputs from all the sub-band ensemble networks to get the final estimate of the phoneme posterior probabilities. The system is *long* in the sense that the input to each sub-band ensemble network is derived from 300 ms long spans of speech signal using the frequency domain linear prediction feature (Athineos et al., 2004) and the outputs from the first classification stage are concatenated over 200 ms. The system is *deep* because it includes three stages of neural networks for SNR-dependent processing and ensemble for each sub-band, and the final fusion of all the outputs from sub-band ensemble networks. The system is *wide* because the full band speech is divided into multiple sub-bands for processing.

Note that most aforementioned neural networks take a long context window of data as the input so that more information can be incorporated into the robust processing. Rather than being used as a single input of the network, the context window is split into several sub-contexts in Li and Sim (2014) as the input to train different DNNs. The last hidden layer activations from these DNNs are then combined to jointly predict the state posteriors through a single softmax output layer. Better noise-robustness has been achieved with this setup than the standard DNN.

4.1.2 FEATURE MOMENT NORMALIZATION

Feature moment normalization methods normalize the statistical moments of speech features. Cepstral mean normalization (CMN) (Atal, 1974) and cepstral mean and variance normalization (CMVN) (Viikki et al., 1998) normalize the first- and second-order statistical moments, respectively, while histogram equalization (HEQ) (Molau et al., 2003) normalizes the higher order statistical moments through the feature histogram.

Cepstral mean normalization

Cepstral mean normalization (CMN) (Atal, 1974) is the simplest feature moment normalization technique. Given a sequence of cepstral vectors $[\mathbf{x}_0, \mathbf{x}_1, ..., \mathbf{x}_{T-1}]$ where T is the number of frames in this sequence, CMN subtracts the mean value $\boldsymbol{\mu}_x$ from each cepstral vector $\mathbf{x}_t$ to obtain the normalized cepstral vector $\hat{\mathbf{x}}_t$.

$$\boldsymbol{\mu}_x = \frac{1}{T} \sum_{t=0}^{T-1} \mathbf{x}_t, \tag{4.4}$$

$$\hat{\mathbf{x}}_t = \mathbf{x}_t - \boldsymbol{\mu}_x. \tag{4.5}$$

After normalization, the mean of the cepstral sequence is 0. As shown in Equations 3.17 and 3.18, in the absence of noise, the convolutive channel distortion in the time domain has an additive effect in the log-Mel-filter-bank domain and the cepstral domain. Therefore, CMN is good at removing the channel distortion. It is also shown in Droppo and Acero (2008) that CMN can help to improve recognition in noisy environments even if there is no channel distortion.

A problem of CMN is that it does not discriminate between silence and speech when computing the utterance mean, and therefore the mean is affected by the ratio of the number of speech frames and the number of silence frames included in the calculation. Instead of using a single mean for the whole utterance, CMN can be extended to use multi-class normalization. Better performance is obtained with augmented CMN (Acero and Huang, 1995), where speech and silence frames are normalized to their own reference means rather than a global mean.

For real-time applications, CMN is unacceptable because the mean value is calculated using the features in the whole utterance. Hence, it needs to be modified as in the following for the deployment in a real-time system. CMN can be considered as a high-pass filter with a cutoff frequency that depends on T, the number of frames (Huang et al., 2001). An example of frequency response of CMN when $T = 200$ is plotted in Figure 4.5. Following this interpretation, it is reasonable to use other types of high-pass filters to approximate CMN. A widely used one is a first-order recursive filter, in which the cepstral mean is a function of time according to

$$\mu_{x_t} = \alpha \mathbf{x}_t + (1 - \alpha)\mu_{x_{t-1}}, \tag{4.6}$$

$$\hat{\mathbf{x}}_t = \mathbf{x}_t - \mu_{x_t}, \tag{4.7}$$

where α is chosen in the way that the filter has a time constant of at least 5 s of speech (Huang et al., 2001). Other types of filters can also be used. For example, the band-pass IIR filter of RASTA shown in Equation 4.3 performs similarly to CMN (Anastasakos et al., 1994). Its high-pass portion of the filter is used to compensate for channel convolution effects as with CMN, while its low-pass portion helps to smooth some of the fast frame-to-frame spectral changes which should not exist in speech.

Cepstral mean and variance normalization
Cepstral mean and variance normalization (CMVN) (Viikki et al., 1998) normalizes the mean and covariance together:

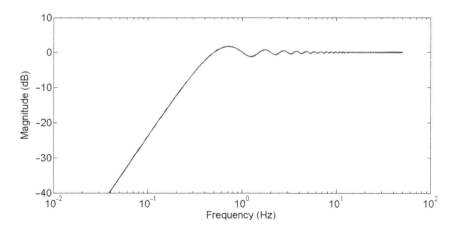

FIGURE 4.5

An example of frequency response of CMN when $T = 200$ at a frame rate of 10 Hz.

$$\sigma_x^2 = \frac{1}{T}\sum_t \mathbf{x}_t^2 - \mu_x^2, \tag{4.8}$$

$$\hat{\mathbf{x}}_t = \frac{\mathbf{x}_t - \mu_x}{\sigma_x}. \tag{4.9}$$

After normalization, the sample mean and variance of the cepstral sequence are 0 and 1, respectively. CMVN has been shown to outperform CMN in noisy test conditions. Droppo and Acero (Droppo and Acero, 2008) give a detailed comparison of CMN and CMVN, and discuss different strategies to apply them. Chen and Bilmes (2007) propose a method that combines mean subtraction, variance normalization, and autoregressive moving-average filtering (MVA) together. An analysis showing why MVA works well is also presented in Chen and Bilmes (2007).

Although mean normalization is directly related to removing channel distortion, variance normalization cannot be easily associated with removing any distortion explicitly. Instead, CMN and CMVN can be considered as ways to reduce the first- and second-order moment mismatches between the training and test conditions. In this way, the distortion brought by additive noise and a convolutive channel can be reduced to some extent. As an extension, third-order (Suk et al., 1999) or even higher-order (Hsu and Lee, 2004, 2009) moment normalization can be used to further improve noise robustness. Multi-class extensions can also be applied to CMVN. In Xiao et al. (2011b) it is shown that 8-class CMVN has 19% relative word error rate (WER) reduction from the standard CMVN on the Aurora 4 task. Standard CMVN is processed dimension-by-dimension, ignoring the correlation between feature dimensions. It can be considered as applying a diagonal transform to the speech feature. In Xiao et al. (2011b), a structured full transform is used for feature normalization. This method shows great advantages over the CMVN. With eight structured transforms, it is significantly better than the advanced feature extraction (AFE) front-end (ETSI, 2002) on the Aurora 4 task.

Histogram equalization

A natural extension to the moments normalization techniques is to normalize the distribution between training and test data. This normalizes all of the moments in the speech-feature distribution. This approach is called histogram equalization (HEQ) (de la Torre et al., 2005; Dharanipragada and Padmanabhan, 2000; Hilger and Ney, 2006; Molau et al., 2003). HEQ postulates that the transformed speech-feature distributions of the training and test data are the same. Each feature vector dimension is normalized independently. HEQ can be applied either in the log-Mel-filter-bank domain (Hilger and Ney, 2006; Molau et al., 2001) or the cepstral domain (de la Torre et al., 2005; Segura et al., 2004).

The following transformation function $f(\cdot)$ is applied to the test feature y:

$$f(y) = C_x^{-1}(C_y(y)), \tag{4.10}$$

where $C_y(\cdot)$ is the cumulative distribution function (CDF) of the test data and $C_x^{-1}(\cdot)$ is the inverse CDF of the training data. Afterwards, the transformed test feature will have the distribution of the training data. In this way, HEQ reduces the systematic statistical mismatch between test and training data.

While the underlying principle is rather straightforward, the problem is how to reliably estimate CDFs. When a large amount of data is available, the CDFs can be accurately approximated by the cumulative histogram. Such approximations become unreliable for short test utterances. Order-statistics based methods tend to be more accurate and reliable when there is an insufficient amount of data (de la Torre et al., 2005; Segura et al., 2004).

There are several implementation methods for HEQ. Table-based HEQ (THEQ) is a popular method (Dharanipragada and Padmanabhan, 2000) that uses a cumulative histogram to estimate the corresponding CDF value of the feature vector elements. This is done by dividing the training data into K non-overlapped equal-size histogram bins. For every histogram bin B_i, its probability can be approximated by dividing the number of training samples in that bin, n_i, by the total number of training samples, N:

$$P_x(B_i) = \frac{n_i}{N}. \tag{4.11}$$

The CDF $C_x(x)$ can then be estimated as

$$C_x(x) = \sum_{j=0}^{i} P_x(B_j), \tag{4.12}$$

where B_i is the bin to which x belongs.

Then, a look-up table is built with all the distinct reference pairs $\left(C_x(x), x_{B_i}\right)$, where x_{B_i} is the restored value calculated as the mean of the training samples in bin B_i. During testing, the CDF estimation of the test utterance can be obtained in the same way as the training CDF. The restored value of the test sample y is obtained by calculating its approximate CDF value $C_y(y)$ as the key for finding the corresponding value in the pre-built look-up table. In Molau et al. (2003), it was shown that normalizing both the training and test data can achieve better performance. This is in line with the feature-domain noise adaptive training (Deng et al., 2000), which will be discussed in Chapter 8.

In THEQ, a look-up table is used as the implementation of C_x^{-1} in Equation 4.10. This requires that all of the look-up tables in every feature dimension are kept in memory, causing a large deployment cost that applications with limited resources may find unaffordable. Also, the test CDF is not as reliable as the training CDF because limited data is available for the estimation. Therefore, several methods are proposed to work with only limited test data, such as quantile-based HEQ (QHEQ) (Hilger et al., 2002; Hilger and Ney, 2001) and polynomial-fit HEQ (PHEQ) (Lin et al., 2006, 2007).

Instead of fully matching the training and test CDF, QHEQ calibrates the test CDF to the training CDF in a quantile-corrective manner. To achieve this goal, it uses a transformation function which is estimated by minimizing the mismatch between the quantiles of the test and training data. To apply the transformation function, the feature in the test utterance is first scaled to the interval [0, 1]. Then the transformation function is applied and the transformed value is scaled back to the original value range. Because only a small number of quantiles are needed for reliable estimation, the data amount requirement is significantly reduced. However, an exhaustive online grid search is required to find the optimum transformation function for every feature dimension. Another alternative is cepstral shape normalization (CSN) proposed in Du and Wang (2008). The requirement on the amount of data is significantly reduced because only one shape parameter is needed to be estimated in CSN.

It is cost expensive for THEQ to store all the look-up tables in runtime as the implementation C_x^{-1}. In PHEQ, a polynomial function is used to fit C_x^{-1}. The polynomial coefficients are learned by minimizing the squared error between the input feature and the approximated feature for all the training data. Due to its robustness to limited data, rank statistics instead of a cumulative histogram is used to obtain the approximated CDF. During testing, the test feature is sorted to obtain its approximated CDF value, which can then be taken into the polynomial function to get the restored value. The polynomial function is simple and computationally efficient. Other functions can also be used as the basis function, for example, a sigmoid function is used in Xiao et al. (2011a).

Instead of using the training data to get a reference distribution, it is also possible to use a zero-mean unit-variance Gaussian as the reference distribution. In this way, the feature is normalized to a standard normal distribution. This Gaussianization HEQ (GHEQ) (de la Torre et al., 2005) is compared with THEQ and PHEQ in Lin et al. (2007).

One HEQ assumption is that the distributions of acoustic classes (e.g., phones) should be identical or similar for both training and test data. However, a test utterance is usually too short for the acoustic class distribution to be similar enough to the training distribution. To remedy this problem, two-class (García et al., 2006; Liu et al., 2004) or multi-class HEQ (García et al., 2012; Suh et al., 2007; Xiao et al., 2013) can be used. All of these methods equalize different acoustic classes separately according to their corresponding class-specific distribution.

Conventional HEQ always equalizes the test utterance after visiting the whole utterance. This is not a problem for offline processing. However, for commercial systems with real-time requirements this is not acceptable. Initially proposed to address the time-varying noise issue, progressive HEQ (Tsai and Lee, 2004) is a good candidate to meet the real-time processing requirement by equalizing with respect to a short interval around the current frame. The processing delay can be reduced from the length of the whole utterance to just half of the reference interval. Another weakness of HEQ is that it does not take into account the underlying acoustic model used for recognition, resulting in a possible mismatch between HEQ-processed features and the acoustic model. In Xiao et al. (2011a), HEQ online adaptation has been proposed to maximize the likelihood of the HEQ-processed feature for the given acoustic model, with a constraint on the HEQ parameters.

4.1.3 **FEATURE COMPENSATION**

Feature compensation aims to remove the effect of noise from the observed speech features. Most of the feature-domain methods developed in the past belong to this sub-category, with prominent examples including spectral subtraction (Boll, 1979), Wiener Filtering (Lim and Oppenheim, 1979), stereo piecewise linear compensation for environment (SPLICE) (Deng et al., 2000), and feature vector Taylor series (Moreno, 1996), to name just a few. In this chapter, we will introduce several methods in this class, but some will be discussed in later chapters.

Spectral subtraction

The spectral subtraction (Boll, 1979) method assumes that noise and clean speech are uncorrelated and additive in the time domain. Assuming the absence of channel distortions in Equation 3.4, the power spectrum of the noisy signal is the sum of the noise and the clean speech power spectrum:

$$|\dot{y}[k]|^2 = |\dot{x}[k]|^2 + |\dot{n}[k]|^2. \tag{4.13}$$

The method assumes that the noise characteristics change slowly relative to those of the speech signal. Therefore, the noise spectrum estimated during a non-speech period can be used for suppressing the noise contaminating the speech. The simplest way to get the estimated noise power spectrum, $|\hat{\dot{n}}[k]|^2$, is to average the noise power spectrum in N non-speech frames:

$$|\hat{\dot{n}}[k]|^2 = \frac{1}{N} \sum_{i=0}^{N-1} |\dot{y}_i[k]|^2, \tag{4.14}$$

where $|\dot{y}_i[k]|^2$ denotes the kth bin of the speech power spectrum in the ith frame.

Then the clean speech power spectrum can be estimated by subtracting $|\hat{\dot{n}}[k]|^2$ from the noisy speech power spectrum:

$$|\hat{\dot{x}}[k]|^2 = |\dot{y}[k]|^2 - |\hat{\dot{n}}[k]|^2, \tag{4.15}$$

$$= |\dot{y}[k]|^2 G^2[k], \tag{4.16}$$

where

$$
\begin{aligned}
G[k] &= \sqrt{\frac{|\dot{y}[k]|^2 - |\hat{\dot{n}}[k]|^2}{|\dot{y}[k]|^2}} \\
&= \sqrt{\frac{SNR(k)}{1 + SNR(k)}}
\end{aligned}
\tag{4.17}
$$

is a real-valued gain function, and

$$SNR(k) = \frac{|\dot{y}[k]|^2 - |\hat{\dot{n}}[k]|^2}{|\hat{\dot{n}}[k]|^2} \tag{4.18}$$

is the frequency-dependent signal-to-noise ratio estimate.

While the method is simple and efficient for stationary or slowly varying additive noise, it comes with several problems:

- The estimation of the noise spectrum from noisy speech is not an easy task. The simple scheme outlined in Equation 4.14 relies on a voice activity detector (VAD). However, voice activity detection in low SNR is known to be error-prone. Alternatives have therefore been developed to estimate the noise spectrum without the need of a VAD. A comprehensive performance comparison of state of the art noise trackers can be found in Taghia et al. (2011).
- The instantaneous noise power spectral density will fluctuate around its temporally and spectrally smoothed estimate, resulting in amplification of random time frequency bins, a phenomenon known under the name musical noise (Berouti et al., 1979), which is not only annoying to a human listener but also leads to word errors in a machine recognizer.
- The subtraction in Equation 4.15 may result in negative power spectrum values because $|\hat{n}[k]|^2$ is an estimated value and may be greater than $|\hat{y}[k]|^2$. If this happens, the numerator of Equation 4.18 should be replaced by a small positive constant.

Many highly sophisticated gain functions have been proposed that are derived from statistical optimization criteria.

Wiener filtering

Wiener filtering is similar to spectral subtraction in that a real-valued gain function is applied in order to suppress the noise. Without considering the impact of channel distortion, Equation 3.1 is simplified to only consider the impact of additive noise:

$$y[m] = x[m] + n[m]. \tag{4.19}$$

Wiener filtering aims at finding a linear filter $g[m]$ such that the sequence

$$\hat{x}[m] = y[m] * g[m] = \sum_{i=-\infty}^{\infty} g[i]y[m - i] \tag{4.20}$$

has the minimum expected squared error from $x[m]$:

$$E\left[x[m] - \sum_{i=-\infty}^{\infty} g[i]y[m - i] \right]^2 \tag{4.21}$$

results in the Wiener-Hopf equation (Haykin, 2001):

$$R_{xy}[l] = \sum_{i=-\infty}^{\infty} g[i]R_{yy}[l - i], \tag{4.22}$$

where R_{xy} is the cross covariance between x and y, and R_{yy} is the auto-covariance of y.

$$R_{xy}[l] = \sum_i x_i y_{i+l}, \tag{4.23}$$

$$R_{yy}[k] = \sum_i y_i y_{i+k}. \tag{4.24}$$

By taking Fourier transforms, this results in the frequency domain filter

$$G[k] = \frac{S_{xy}[k]}{S_{yy}[k]}. \tag{4.25}$$

Here, S_{xy} and S_{yy} are the cross power spectral density between clean and noisy speech and the power spectral density of noisy speech, respectively. The clean speech signal $x[m]$ and the noise signal $n[m]$ are usually considered to be uncorrelated. Then

$$S_{xy}[k] = S_{xx}[k], \tag{4.26}$$

$$S_{yy}[k] = S_{xx}[k] + S_{nn}[k], \tag{4.27}$$

with S_{xx} and S_{nn} being the power spectral density of clean speech and noise, respectively.

Using Equations 4.26 and 4.27, Equation 4.25 becomes

$$G[k] = \frac{S_{xx}[k]}{S_{xx}[k] + S_{nn}[k]}, \tag{4.28}$$

which is referred to as the Wiener filter (Haykin, 2001; Lim and Oppenheim, 1979; Quatieri, 2001), and can be realized only if $S_{xx}(f)$ and $S_{nn}(f)$ are known. Figure 4.6 shows an example of the Wiener filter G (solid line) as a function of the spectral density S_{xx} and S_{nn}, which are denoted by the dashed line and dot line, respectively. G reaches its maximum value of 1 when the noise spectral density is 0, which means no filtering needs to be done in this frequency region.

In practice the power spectra have to be estimated, for example, via the periodograms $|\hat{y}[k]|^2$ and $|\hat{n}[k]|^2$. Plugging them in Equation 4.28 we obtain

$$\begin{aligned} G[k] &= \frac{|\hat{y}[k]|^2 - |\hat{n}[k]|^2}{|\hat{y}[k]|^2} \\ &= \frac{SNR(k)}{1 + SNR(k)}, \end{aligned} \tag{4.29}$$

which shows that Wiener filtering and spectral subtraction are closely related.

From Equation 4.29, it is easy to see that the Wiener filter attenuates low SNR regions more than high SNR regions. If the speech signal is very clean with very large SNR approaching to ∞, $G[k]$ is close to 1, resulting in no attenuation. In contrast, if the speech is buried in the noise with very low SNR approaching 0, $G[k]$ is close to 0, resulting in total attenuation. Similar reasoning also applies to spectral subtraction.

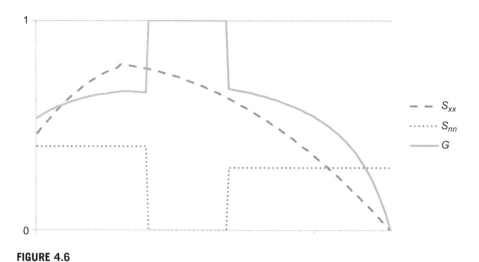

FIGURE 4.6

An example of the Wiener filtering gain G with respect to the spectral density S_{xx} and S_{nn}.

Advanced front-end

In 2002, the advanced front-end (AFE) for distributed speech recognition (DSR) was standardized by ETSI (ETSI, 2002). It obtained 53% relative word error rate reduction from the MFCC baseline on the Aurora 2 task (Macho et al., 2002). The AFE is one of the most popular methods for comparison in the noise robustness literature. It integrates several noise robustness methods to remove additive noise with two-stage Mel-warped Wiener filtering (Macho et al., 2002) and SNR-dependent waveform processing (Cheng and Macho, 2001), and to mitigate the channel effect with blind equalization (Mauuary, 1998). The input speech signal is first processed with two-stage Mel-warped Wiener filtering, followed by SNR-dependent waveform processing. Then the cepstrum is calculated and blind equalization is applied to reduce the channel effect.

The two-stage Mel-warped Wiener filtering algorithm is the main body of the noise reduction module of AFE and accounts for the major gain of noise reduction. It is a combination of the two-stage Wiener filter scheme from Agarwal and Cheng (1999) and the time domain noise reduction proposed in Noé et al. (2001). As shown in Figure 4.7, the algorithm has two stages of Mel-warped Wiener filtering. The denoised signal in the first stage is passed to the second stage, which is used to further reduce the residual noise.

The process of the first stage is described as follows:

- The spectrum estimation module calculates the STDFT spectrum.
- Then, the power spectral density (PSD) mean module averages two consecutive power spectra to reduce the variance of spectral estimation.

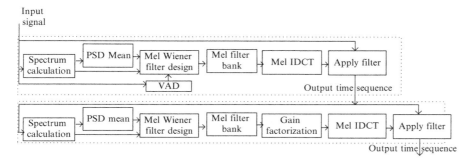

FIGURE 4.7

Two-stage Wiener filter in advanced front-end.

- An energy-based VAD is used to make the speech/non-speech decision for the noise spectrum estimation.
- The Wiener filter is then constructed with Equation 4.29 using the estimated $\hat{S}_{xx}(f)$ and $\hat{S}_{nn}(f)$.
- Then, the Mel-filter-bank module transforms the Wiener filter from the linear frequency domain to the Mel-filter-bank domain. This may also be considered as smoothing along the frequency axis.
- The following Mel Inverse DCT (IDCT) module obtains the impulse response of the Mel Wiener filter, truncates it to a length of 17 and windows it with a Hanning window. Truncation and windowing of the impulse response results in a smoothed filter, highly beneficial to ASR performance (Noé et al., 2001).
- Finally, the original signal is convolved with the impulse response obtained in the previous step to get the filtered signal in the time domain.

The second stage of Mel-warped Wiener filtering is similar to the first stage of Mel-warped Wiener filtering, with the following difference:

- There is no VAD module in the second stage because the noise power spectrum estimation method differs from that in the first stage.
- A gain factorization module is used in the second stage. It performs noise reduction in a SNR-dependent way: more aggressive noise reduction is applied to purely noise frames while moderate noise reduction is used in noisy speech frames. This processing is reasonable due to the fact that the filtering inevitably brings distortion to the speech signal. Therefore, it is better to only apply the aggressive noise reduction to pure noise frames. Another reason is that the noise-reduced signal has better SNR properties in the second stage.

The basic idea behind SNR-dependent waveform processing (Cheng and Macho, 2001) is that the speech waveform exhibits periodic maxima and minima in the voiced speech segments due to the glottal excitation while the additive noise energy is relatively constant. Therefore, the overall SNR of the voiced speech segments can be boosted if one can locate the high (or low) SNR period portions and increase (or decrease) their energy. This module contains the following three parts:

- Smoothed energy contour: A smoothed instant energy contour is first computed.
- Peak picking: Peaks of the smoothed energy contour are found by a peak-picking strategy.
- Waveform SNR weighting: The original waveform is modified by using a windowing function and a weight parameter which determines the degree of attenuation of low SNR portions with respect to high SNR portions. A function is used to ensure the total frame energy does not change after the processing.

Blind equalization (Mauuary, 1998) reduces convolutional distortion by minimizing the mean square error between the current and target cepstrum. The target cepstrum corresponds to the cepstrum of a flat spectrum. Blind equalization is an online method to remove convolutional distortion without the need to first visit the whole utterance as the standard CMN does. As shown in Mauuary (1998), it can obtain almost the same performance as the conventional offline cepstral subtraction approach. Therefore, it is preferred in real-time applications.

Although having outstanding performance, the two-stage Mel-warped Wiener filtering algorithm has large computational load due to the following two processes. First, the construction of Wiener filter in the linear frequency domain requires the power spectrum to be calculated at both stages of the algorithm, introducing the power spectrum re-calculations at both the second stage and the cepstrum calculation part. Second, the Wiener filter is applied in time domain by time-consuming convolution operations. For some low computational resource devices, such computational load may be unacceptable.

In order to reduce the runtime cost, a new structure for Wiener filtering algorithm, called two-stage Mel-warped filter-bank Wiener filtering, is proposed in Li et al. (2004). The block diagram of the proposed algorithm is shown in Figure 4.8.

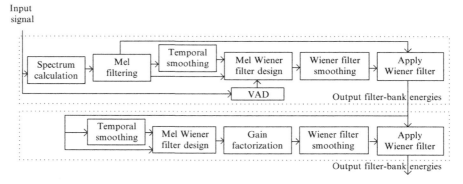

FIGURE 4.8

Complexity reduction for two stage Wiener filter.

The new algorithm is based on the Mel-warped filter-bank energies and reduces the computational load from three aspects. First, the Mel-warped Wiener filter coefficients are directly computed using the Mel-warped filter-bank energies. Since the number of the Mel-warped filter-banks is much less than the number of bins of the linear-frequency FFT power spectrum, the number of computations is effectively reduced. A similar idea is also used later in the cepstral minimum mean square error (CMMSE) processing (Yu et al., 2008), where the operation is on filter-bank instead of linear frequency. Second, Mel-warped Wiener filter coefficients are smoothed and applied back on Mel-warped filter-bank energies, because frequency-domain Wiener filter coefficients can also be viewed as the gains of the spectrum. This measure makes the time-domain convolution operations of applying the Wiener filter completely unnecessary. Third, since the de-noised Mel-warped filter-bank energies, instead of the de-noised time-domain signal, are fed into the next stage, the re-calculations of power spectrum are also avoided. The power spectrum is calculated only once in the whole algorithm. With all these changes, no accuracy degradation is observed, and more than two thirds of the computational load of the original algorithm is saved.

4.2 MODEL-SPACE APPROACHES

Most model-domain methods modify the acoustic model parameters to incorporate the effects of noise and typically achieve higher accuracy than feature-domain methods because the parameter optimization is directly linked with the ASR objective function. However, they usually incur significantly higher computational cost. The model-domain approaches can be further classified into two sub-categories: general adaptation and noise-specific compensation. General adaptation methods compensate for the mismatch between training and test conditions by using generic transforms to convert the acoustic model parameters. The popular methods in this sub-category are maximum likelihood linear regression (MLLR) (Leggetter and Woodland, 1995), maximum a posteriori (MAP) (Gauvain and Lee, 1994), constrained MLLR (CM-LLR) (Gales, 1998), etc. These methods are general, applicable not only to noise compensation but also to other types of acoustic variations.

4.2.1 GENERAL MODEL ADAPTATION FOR GMM

As in Equation 3.35, model-domain methods only adapt the model parameters to fit the distorted speech signal. The model adaptation can operate in either supervised or unsupervised mode. In supervised mode, the correct transcription of the adapting utterances is available. It is used to guide model adaptation to obtain the adapted model, $\hat{\Lambda}$, used to decode the incoming utterances. In unsupervised mode, the correct transcription is not available, and usually two-pass decoding is used. In the first pass, the initial model Λ is used to decode the utterance to generate a hypothesis. Usually one hypothesis is good enough. The gain from using a lattice or N-best list to represent multiple hypotheses is limited (Padmanabhan et al., 2000). Then, $\hat{\Lambda}$ is obtained with the model adaptation process and used to re-decode the utterance to generate the final result.

Popular speaker adaptation methods such as maximum a *posteriori* (MAP) (Gauvain and Lee, 1994) and its extensions such as structural MAP (SMAP) (Shinoda and Lee, 2001), MAP linear regression (MAPLR) (Siohan et al., 2001), and SMAP linear regression (SMAPLR) (Siohan et al., 2002), may not be a good fit for most noise-robust speech recognition scenarios where only a very limited amount of adaptation data is available, for example, when only the utterance itself is used for unsupervised adaptation. Most popular methods use the maximum likelihood estimation (MLE) criterion (Gales, 1998; Leggetter and Woodland, 1995). Discriminative adaptation is also investigated in some studies (He and Chou, 2003; Wu and Huo, 2002; Yu et al., 2009). Unlike MLE adaptation, discriminative adaptation is very sensitive to hypothesis errors (Wang and Woodland, 2004). As a result, most discriminative adaptation methods only work in supervised mode (He and Chou, 2003; Wu and Huo, 2002). Special processing needs to be used for unsupervised discriminative adaptation. In Yu et al. (2009), a speaker-independent discriminative mapping transformation (DMT) is estimated during training. During testing, a speaker-specific transform is estimated with unsupervised MLE, and the speaker-independent DMT is then applied. In this way, discriminative adaptation is implicitly applied without the strict dependency on a correct transcription.

In the following, popular MLE adaptation methods will be reviewed. All these general adaptation methods are widely used in robustness tasks. Maximum likelihood linear regression (MLLR) is proposed in Leggetter and Woodland (1995) to adapt model mean parameters with a class-dependent linear transform

$$\boldsymbol{\mu}_y(m) = \mathbf{A}(r_m)\boldsymbol{\mu}_x(m) + \mathbf{b}(r_m),\tag{4.30}$$

where $\boldsymbol{\mu}_y(m)$ and $\boldsymbol{\mu}_x(m)$ are the clean and distorted mean vectors for Gaussian component m, and r_m is the corresponding regression class. $\mathbf{A}(r_m)$ and $\mathbf{b}(r_m)$ are the regression-class-dependent transform and bias to be estimated, which can be put together as $\mathbf{W}(r_m) = [\mathbf{A}(r_m)\mathbf{b}(r_m)]$. One extreme case is that if every regression class only contains one Gaussian, then $r_m = m$. Another extreme case is that if there is only a single regression class, then all the Gaussians share the same transform as $\mathbf{A}(r_m) = \mathbf{A}$ and $\mathbf{b}(r_m) = \mathbf{b}$.

The expectation-maximization (EM) algorithm (Dempster et al., 1977) is used to get the maximum likelihood solution of $\mathbf{W}(r_m)$. First, an auxiliary Q function for an utterance is defined

$$Q(\hat{\Lambda}; \Lambda) = \sum_{t,m} \gamma_t(m) \log p_{\hat{\Lambda}}(\mathbf{y}_t|m),\tag{4.31}$$

where $\hat{\Lambda}$ denotes the adapted model, and $\gamma_t(m)$ is the posterior probability for Gaussian component m at time t. $\mathbf{W}(r_m)$ can be obtained by setting the derivative of Q w.r.t. $\mathbf{W}(r_m)$ to 0. A special case of MLLR is the signal bias removal algorithm

(Rahim and Juang, 1996), where the only single transform is simply a bias. The MLE criterion is used to estimate this bias, and it is shown that signal bias removal is better than CMN (Rahim and Juang, 1996).

The variance of the noisy speech signal also changes with the introduction of noise as shown in Section 3.2. Hence, in addition to transforming Gaussian mean parameters with Equation 4.30, it is better to also transform Gaussian covariance parameters (Gales, 1998; Gales and Woodland, 1996) as

$$\Sigma_y(m) = \mathbf{H}(r_m)\Sigma_x(m)\mathbf{H}^T(r_m). \tag{4.32}$$

A two-stage optimization is usually used. First, the mean transform $\mathbf{W}(r_m)$ is obtained, given the current variance. Then, the variance transform $\mathbf{H}(r_m)$ is computed, given the current mean. The whole process can be done iteratively. The EM method is used to obtain the solution, which is done in a row-by-row iterative format.

Constrained MLLR (CMLLR) (Gales, 1998) is a very popular model adaptation method in which the transforms of the mean and covariance, $\mathbf{A}(r_m)$ and $\mathbf{H}(r_m)$, are constrained to be the same:

$$\boldsymbol{\mu}_y(m) = \mathbf{H}(r_m)(\boldsymbol{\mu}_x(m) - \mathbf{g}(r_m)), \tag{4.33}$$

$$\Sigma_y(m) = \mathbf{H}(r_m)\Sigma_x(m)\mathbf{H}^T(r_m). \tag{4.34}$$

Rather than adapting all model parameters, CMLLR can be efficiently implemented in the feature space with the following relation

$$\mathbf{y} = \mathbf{H}(r_m)(\mathbf{x} - \mathbf{g}(r_m)), \tag{4.35}$$

or

$$\mathbf{x} = \mathbf{A}(r_m)\mathbf{y} + \mathbf{b}(r_m), \tag{4.36}$$

with $\mathbf{A}(r_m) = \mathbf{H}(r_m)^{-1}$ and $\mathbf{b}(r_m) = \mathbf{g}(r_m)$. The likelihood of the distorted speech $\mathbf{y}$ can now be expressed as

$$p(\mathbf{y}|m) = |\mathbf{A}(r_m)|\mathcal{N}(\mathbf{A}(r_m)\mathbf{y} + \mathbf{b}(r_m); \boldsymbol{\mu}_x(m), \Sigma_x(m)) \tag{4.37}$$

As a result, CMLLR is also referred to as feature space MLLR (fMLLR) in the literature. Note that signal bias removal is a special form of CMLLR with a unit scaling matrix.

In Saon et al. (2001a), fMLLR and its projection variant (fMLLR-P) (Saon et al., 2001b) are used to adapt the acoustic features in noisy environments. Adaptation needs to accumulate sufficient statistics for the test data of each speaker, which requires a relatively large number of adaptation utterances.

As reported in Cui and Alwan (2005) and Saon et al. (2001a), general adaptation methods such as MLLR and fMLLR in noisy environments yield moderate improvement, but with a large gap to the performance of noise-specific methods (Li et al., 2007a, 2009) on the same task. Noise-specific compensation methods usually modify model parameters by explicitly addressing the nature of the distortions caused by the presence of noise. Therefore, they can address the noise-robustness issue better. The representative methods in this sub-category are parallel model combination (PMC) (Gales, 1995) and model-domain vector Taylor series (Moreno, 1996). Some representative noise-specific compensation methods will be discussed in detail in Chapter 6.

4.2.2 GENERAL MODEL ADAPTATION FOR DNN

As discussed in earlier chapters, deep neural networks or DNNs have been making tremendous impact on ASR in recent years (Dahl et al., 2011, 2012; Deng and Yu, 2014; Hinton et al., 2012; Yu and Deng, 2014). One straightforward way to improve the robustness of a DNN is to adapt the current DNN with the data in new environments. There are several types of methods to adapt neural networks. The first type of method, linear input network (LIN) (Li and Sim, 2010; Neto et al., 1995), applies affine transforms to the input of a neural network to map the speaker-dependent input feature to the speaker-independent feature. Similarly, the linear output network (LON) adds a linear layer at the output layer of the neural network, right before the softmax functions are applied. However, the LON is reported to give worse results than the baseline neural network (Li and Sim, 2010). In the context of DNN, the feature discriminative linear regression (fDLR) (Seide et al., 2011), an example of the LIN, is proposed to adapt a DNN with decent gains. The second type of method, linear hidden network (LHN) (Gemello et al., 2007), adds a linear transform network before the output layer. The rationale behind the LHN is that the added linear layer generates discriminative features of the input pattern suitable for the classification performed at the output of the neural network. This LHN method is used in Yao et al. (2012) as the output-feature discriminative linear regression (oDLR) method to adapt a DNN by transforming the outputs of the final hidden layer of a DNN, but with worse performance than the fDLR. The additional linear layer can also be added on top of any hidden layer in a DNN. The basic idea of these methods is shown in Figure 4.9(a), where a new linear layer is inserted into the network.

Low-footprint DNN adaptation

Although the aforementioned adaptation approaches improve robustness of the DNN system against speaker variability, they need to update and store a large amount of parameters due to the high dimensionality of each layer. A large number of adaptation parameters usually requires a large number of adaptation utterances, which are not available in most scenarios. Therefore, several methods have been proposed as low-footprint adaptation techniques, which require a small number of speaker-dependent parameters.

The first method (Siniscalchi et al., 2012, 2013; Zhao et al., 2015) changes the shape of the hidden activation function instead of the network weights to better fit the speaker-specific features. As the number of parameters in the activation function is much smaller than that of the network weights, the goal of low-footprint adaptation can be reached. Although the Hermitian polynomial function in (Siniscalchi et al.,

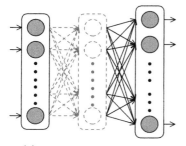

(a) Adaptation of one DNN layer

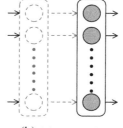

(b) Adaptation of
slopes and biases in
activation functions

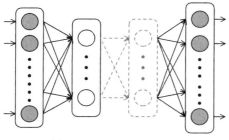

(c) SVD bottleneck adaptation

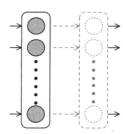

(d) Adaptation of am-
plitudes in activation
functions

FIGURE 4.9

Illustration of network structures of different adaptation methods. Shaded nodes denote
nonlinear units, unshaded nodes for linear units. Red dashed links (gray dashed links in
print versions) indicate the transformations that are introduced during adaptation.

2012, 2013) is not used to adapt a DNN, a similar idea can be applied to adapt a
DNN by adjusting the sigmoid function in every hidden layer with nice gain (Zhao
et al., 2015). Adapting the slopes and biases of a sigmoid activation function is equal
to adding a linear layer right before the activation functions with the one-to-one
correspondence, as shown in Figure 4.9(b).

The second method (Xue et al., 2014a) works on a low-dimensional space by
utilizing a low-rank structure of a DNN (Xue et al., 2013). With this method, the
original large full-rank DNN model is converted to a much smaller low-rank DNN
model without loss of accuracy. Given an $m \times n$ weight matrix **A** in a DNN, it is
approximated as the product of two low-rank matrices by applying singular value
decomposition (SVD)

$$\mathbf{A}_{m \times n} \approx \mathbf{U}_{m \times k} \mathbf{N}_{k \times n}. \tag{4.38}$$

If $\mathbf{A}_{m \times n}$ is a low-rank matrix, k will be much smaller than n and therefore the number of parameters of matrix $\mathbf{U}_{m \times k}$ and $\mathbf{N}_{k \times n}$ is much smaller than that of matrix $\mathbf{A}_{m \times n}$. Applying this decomposition to the DNN model, it acts as adding a linear bottleneck layer with fewer units between the original layers. If the number of parameters is reduced too aggressively, fine tuning can be used to recover accuracy.

Then, SVD bottleneck adaptation is done by applying the linear transform to the bottleneck layer with an additional layer of k units, as illustrated in Figure 4.9(c). Applying this change to Equation 4.38, we get

$$\mathbf{A}_{s,m \times n} = \mathbf{U}_{m \times k}\mathbf{S}_{s,k \times k}\mathbf{N}_{k \times n}, \tag{4.39}$$

where $\mathbf{S}_{s,k \times k}$ is the transform matrix for speaker s and is initialized to be an identity matrix $\mathbf{I}_{k \times k}$. The advantage of this approach is that only a few small matrices with the dimension of $k \times k$ need to be updated for each speaker. This dramatically reduces the deployment cost for speaker personalization.

In the third method, a DNN can be adapted by augmenting the unit activations with the amplitude parameter and is named learning hidden unit contribution (LHUC) (Swietojanski and Renals, 2014). Likewise, this is equivalent to adding a linear layer after the activation functions are employed, as shown in Figure 4.9(d). Note in Swietojanski and Renals (2014), the amplitudes are transformed with a sigmoid to constrain the amplitude range.

Last, in Abdel-Hamid and Jiang (2013), Abdel-Hamid and Jiang (2014), and Xue et al. (2014b) a method called speaker code is proposed to adapt a DNN by putting all the speakers together to train individual speaker codes and several adaptation layers which transform speaker-dependent features into speaker-independent ones before feeding them into the original DNN. Because speaker codes are of low dimension, the number of speaker-dependent parameters is small. Therefore, the speaker code method serves well for fast speaker adaptation with limited number of utterances. Instead of learning the speaker codes in the framework of DNN optimization, another type of methods directly use i-vectors (Dehak et al., 2011) to characterize different speakers and use it as an augmented input to the original DNN (Miao et al., 2014; Saon et al., 2013; Senior and Lopez-Moreno, 2014). I-vector is a popular method for speaker verification and speaker recognition because it encapsulates all the relevant information about a speaker's identity in a low-dimensional fixed-length representation. Therefore it is a good alternative to speaker code.

Adaptation criteria

A straightforward approach to adapt a DNN is to estimate the speaker-dependent parameters with the adaptation data using the regular cross entropy criterion in Equation 2.34. However, doing so may over-fit the model to the adaptation data, especially when the adaptation set is small and the supervision hypotheses are erroneous. A regularized adaptation method was proposed to address this issue (Yu et al., 2013). The idea is that the posterior senone distribution estimated from the adapted model should not deviate too far from the one estimated with the speaker independent model. By adding the Kullback-Leibler divergence (KLD) as a regularization term to Equation 2.34, we get a regularized optimization criterion, which has the same form as Equation 2.34 except that the target probability distribution $P_{\text{target}}(o = s|\mathbf{x}_t)$ is substituted by

$$\hat{P}(o = s|\mathbf{x}_t) = (1 - \rho)P_{\text{target}}(o = s|\mathbf{x}_t) + \rho P^{\text{SI}}(o = s|\mathbf{x}_t), \qquad (4.40)$$

where ρ is the regularization weight and $P^{\text{SI}}(o = s|\mathbf{x}_t)$ is the posterior probability estimated from the speaker independent model. It can be seen that $\hat{P}(o = s|\mathbf{x}_t)$ is a linear interpolation of the distribution estimated from the speaker independent model and the ground truth alignment of the adaptation data. This interpolation prevents the adapted model from deviating far away from the speaker independent model, when the adaptation data are limited. This KLD regularization is very effective to adapt even a whole DNN with only limited adaptation data. However, when the amount of to-be-adapted parameters is small as in the sigmoid function adaptation (Zhao et al., 2015) or LHUC adaptation (Swietojanski and Renals, 2014), KLD regularization is not necessary.

While originally designed for speaker recognition, all the aforementioned methods could be applied to noise-robust ASR tasks. However, these methods usually only adapt or add the network weights without differentiating the underlying factors that cause the mismatch between training and testing. In Li et al. (2014c), a DNN adaptation method was proposed by taking into account the underlying factors that contribute to the distorted speech signal. This will be discussed in Section 6.4.3.

4.2.3 ROBUSTNESS VIA BETTER MODELING

As discussed in Section 3.4, the layer-by-layer setup of a DNN provides a feature extraction strategy that automatically derives powerful distortion-resistant features from primitive raw data for senone classification. A GMM can also be noise-resistant when trained with a margin-based criterion—by securing a margin from the decision boundary to the nearest training sample, a correct decision can still be made if the mismatched test sample falls within a tolerance region around the original training samples defined by the margin. The most famous technology with the concept of margin is support vector machine (SVM) (Vapnik, 1995, 1998). Figure 4.10 shows the simplest case of SVM: a linear machine trained on separable data. In such a case, the margin is defined as the shortest distance from the separating hyperplane to the closest positive or negative samples, which are support vectors. A SVM is trained to choose an optimal separating hyperplane with the maximum margin value. For training with non-separable data, slack variables are introduced in the optimization constraints.

This nice property of margin is also shown effective for noise-robust ASR in Li and Lee (2008) and Xiao et al. (2009) by using soft margin estimation (SME) (Li et al., 2006, 2007b). SME optimizes a combination of empirical risk minimization and margin maximization by minimizing the objective function

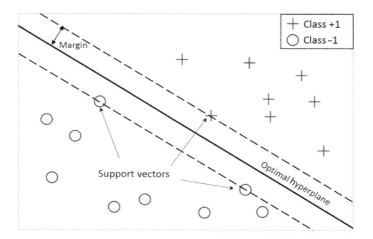

FIGURE 4.10

The illustration of support vector machines.

$$L^{SME}(\rho, \Lambda) = \frac{\lambda}{\rho} + \frac{1}{N} \sum_{i=1}^{N} l(\mathbf{Y}_i, \Lambda), \tag{4.41}$$

where ρ is the margin, N is the total number of utterances, and λ is a coefficient to balance the margin maximization and the empirical risk minimization. A larger λ favors a larger margin value, which corresponds to better generalization and improves the robustness in mismatched training and test conditions. Λ denotes the model parameter, and $l(\mathbf{Y}_i, \Lambda)$ is the empirical risk for utterance $\mathbf{Y}_i$.

The underlying models of SME are still GMMs, which are generative models. Although they are widely used in ASR, it is well known that the underlying model assumption is not correct. Therefore, there are increased interests in directly modeling sentence posteriors or decision boundaries given a set of features extracted from the observation sequence with discriminative models. Log linear models and SVMs are typical discriminative models. However, discriminative models generally are hard to adapt, compared to generative models which can be adapted to noisy environments with plenty of methods introduced in this book. To have advantages from both models, methods that combine them together are proposed in recent years (Gales and Flego, 2010; Ragni and Gales, 2011a,b; Zhang and Gales, 2013, 2011). These methods are called discriminative classifiers with adaptive kernels, and their frameworks are shown in Figure 4.11. Λ denotes the initial HMM model parameters, and can be adapted according to the data in test environments to address the acoustic mismatch. The adapted HMM parameters $\hat{\Lambda}$ can be passed into the ASR recognizer to generate the hypotheses. This is the general framework of recognition with generative classifier, which is the upper block in Figure 4.11. To combine with the discriminative classifier, test data Y, recognized hypotheses, and the adapted parameters $\hat{\Lambda}$ are

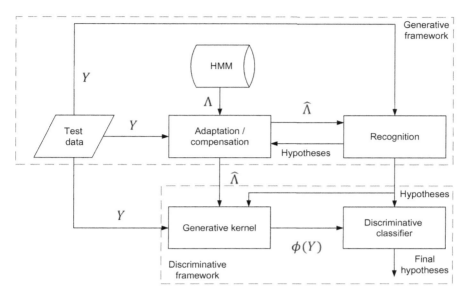

FIGURE 4.11

The framework to combine generative and discriminative classifiers.

passed to the generative kernel generation module to get the feature vector $\phi(\mathbf{Y})$ for the discriminative classifier which re-scores the hypotheses generated from the generative classifier and gets the best out as the final hypotheses. This process is the bottom block in Figure 4.11.

There are several choices of the feature vector $\phi(\mathbf{Y})$. In Ragni and Gales (2011a,b), the feature space derived from the generative models is referred to as a score space, and several score spaces can be used. The first is called appended-all score space which incorporates the log-likelihoods of all models, including the correct class w_i,

$$\phi_a(\mathbf{Y}|w_i) = \begin{bmatrix} \log(p(\mathbf{Y}|w_1)) \\ \vdots \\ \log(p(\mathbf{Y}|w_K)) \end{bmatrix}. \tag{4.42}$$

The second is the derivative score space which has the following form

$$\phi_b^n(\mathbf{Y}|w_i) = \begin{bmatrix} \log(p(\mathbf{Y}|w_i)) \\ \nabla_\Lambda \log(p(\mathbf{Y}|w_i)) \\ \vdots \\ \nabla_\Lambda^n \log(p(\mathbf{Y}|w_i)) \end{bmatrix}. \tag{4.43}$$

In addition to the correct class log-likelihood $\log(p(\mathbf{Y}|w_i))$, the feature vector in Equation 4.43 also incorporates up to the nth order derivatives of $\log(p(\mathbf{Y}|w_i))$ with respect to the generative model parameters Λ. When the generative model is a GMM, taking the first-order derivative with respect to the model parameter generates a term depending on the whole observation sequence. Therefore, the use of derivatives as in Equation 4.43 introduces additional dependencies into the feature vector for discriminative classifiers.

Examples of discriminative classifier used in Figure 4.11 are log-linear models (Ragni and Gales, 2011a,b), SVMs (Gales and Flego, 2010), and structured SVMs (Zhang and Gales, 2013, 2011). The training criteria for log-linear models can be conditional maximum likelihood training (Nadas, 1983) or minimum Bayes' risk (Byrne, 2006). SVMs and structured SVMs are optimized with the maximum margin criterion which helps to improve the generalization property of discriminative classifiers. Similar to SME, better generalization to unseen samples means better robustness to noise distortion which causes the mismatch between training and test samples.

4.3 SUMMARY

In this chapter, we first discussed the feature-domain methods in detail. We then introduced a few general model-domain adaptation methods. The coverage of the latter is not as comprehensive as the former since most model-domain methods are noise-specific and we will leave them for discussions in later chapters. We now summarize the representative methods originally proposed for GMM and DNN models in Tables 4.1 and 4.2, respectively, in a chronological order so that we can have a clear view of technology development. We can see that the common development route for all these methods is that more delicate modeling gives better performance. Separating the methods into two tables is to get a better picture for different underlying acoustic model. It doesn't mean that the methods originally proposed for GMM cannot be used for DNN, and vice versa. For example, CMN and CMVN are still widely used to process acoustic features before feeding them into DNNs.

- As a representative of auditory feature, PLP was proposed in 1985 and was combined with RASTA in 1991. Later, more and more auditory features were studied such as PNCC proposed in 2010 with more refined auditory modeling.
- CMN was proposed in 1974 to normalize the cepstral mean, and improved by CMVN in 1998 to normalize both cepstral mean and variance. It was natural to extend the normalization to feature distribution and resulted in HEQ in 2000. HEQ is further refined with methods such as QHEQ, PHEQ, and multi-class HEQ in 2000s.
- Proposed in 1970s, spectral subtraction and Wiener filtering are pretty simple. Their performance is significantly boosted by two-stage Mel-warped Wiener

Table 4.1 Feature- and Model-Domain Methods Originally Proposed for GMMs in Chapter 4, Arranged Chronologically

Method	Proposed Around	Characteristics
Cepstral mean normalization (CMN) (Atal, 1974)	1974	Handles convolutive distortion by removing cepstral mean
Spectral subtraction (Boll, 1979)	1979	Removes the estimated noise spectrum from noisy speech spectrum
Wiener filter (Lim and Oppenheim, 1979)	1979	Removes additive noise with a SNR-dependent filtering
Perceptually based linear prediction (PLP) (Hermansky et al., 1985)	1985	Auditory feature based on Bark-scale filter-bank
Relative spectral processing (RASTA) (Hermansky et al., 1991)	1991	A band-pass filter that removes the fast frame-to-frame spectral changes and alleviates the effect of convolutional channel
J-RASTA (Morgan and Hermansky, 1992)	1992	A RASTA filter that handles both additive noise and convolutive channel distortion
ANN-HMM hybrid systems (Bourlard and Morgan, 1994)	1994	Replaces the GMM with an ANN when evaluating the state likelihood score
Maximum likelihood linear regression (MLLR) (Leggetter and Woodland, 1995)	1995	Adapts GMM model mean vectors and covariance matrices using a set of linear transforms
Constrained MLLR (CMLLR) (Gales, 1998)	1998	Adapts GMM model mean and covariance which are constrained with the same linear transform, and can be effectively implemented in the feature space
Cepstral mean and variance normalization (CMVN) (Viikki et al., 1998)	1998	Normalizes cepstral mean and variance
TempoRAL Pattern (TRAP) (Hermansky and Sharma, 1998)	1998	Captures the appropriate temporal pattern with a long temporal vector of critical band log spectral energies from a single frequency band
Histogram equalization (HEQ) (Dharanipragada and Padmanabhan, 2000)	2000	Normalizes the feature distribution of training and test data
TANDEM (Hermansky et al., 2000)	2000	Omits the final nonlinearity in the ANN output layer and applies a global decorrelation transform to generate feature for GMMs

Continued

Table 4.1 Feature- and Model-Domain Methods Originally Proposed for GMMs in Chapter 4, Arranged Chronologically—*cont'd*

Method	Proposed Around	Characteristics
Quantile HEQ (QHEQ) (Hilger and Ney, 2001)	2001	Calibrates the test CDF to the training CDF in a quantile manner for HEQ
Two-stage Mel-warped Wiener filtering (Macho et al., 2002)	2002	Mel-filter-bank is used to clean noise in two stages
ETSI advanced front-end (AFE) (ETSI, 2002)	2002	Combines two-stage Mel-warped Wiener filtering, SNR-dependent waveform processing, and blind equalization to remove the distortion of both additive noise and convolutive channel
Two-stage Mel-warped filter-bank Wiener filtering (Li et al., 2004)	2004	Significantly reduces the cost of two-stage Mel-warped Wiener filtering by designing and applying Wiener filter only in the Mel-filter-bank domain
Polynomial HEQ (PHEQ) (Lin et al., 2006)	2006	Uses polynomial function to fit the inverse training CDF for HEQ
Multi-class HEQ (Suh et al., 2007)	2007	Performs HEQ to different acoustic classes separately according to their individual distributions
Bottle-neck (BN) feature (Grézl et al., 2007)	2007	Obtains the feature from linear outputs of the bottle-neck layer of a MLP
Temporal structure normalization (Xiao et al., 2008)	2008	Performs environment-dependent temporal filtering to normalize temporal characteristics of the features
Soft margin estimation (SME) (Li and Lee, 2008)	2008	Secures a large margin from the decision boundary so that a correct decision can still be made if the mismatch between the training and testing caused by distortion is within the margin
Power-normalized cepstral coefficients (PNCC) (Kim and Stern, 2010)	2010	The auditory feature with nonlinearity processing, temporal and spectral integration, and temporal masking
Discriminative classifiers with adaptive kernels (Gales and Flego, 2010)	2010	Combines the advantages of generative models and discriminative models by adapting generative models with environment-dependent data to get kernels for discriminative classifiers

Table 4.2 Feature- and Model-Domain Methods Originally Proposed for DNNs in Chapter 4, Arranged Chronologically

Method	Proposed Around	Characteristics
Context-dependent deep neural network hidden Markov model (CD-DNN-HMM) (Yu et al., 2010)	2010	A hybrid ANN/HMM method with deep neural network, senone-modeling output, and long context window; this model can extract robust features at the top hidden layer before the classification stage
Feature discriminative linear regression (fDLR) (Seide et al., 2011)	2011	Applies linear transforms to the input of a DNN
Output-feature discriminative linear regression (oDLR) (Yao et al., 2012)	2012	Adapts a DNN by transforming the outputs of the final hidden layer of a DNN
SVD bottleneck adaptation (Xue et al., 2014a)	2014	Applies linear transforms to the SVD layers of a DNN and it is low-footprint adaptation because the dimension of SVD layers is much smaller than that of standard layers
Long, deep, and wide neural network (Li et al., 2014c)	2014	A divide-and-conquer method which handles each specific environment with a sub neural network, and the whole system can be considered as an ensemble of all the sub neural networks
Learning hidden unit contribution (LHUC) (Swietojanski and Renals, 2014)	2014	Adapts a DNN by augmenting the unit activations with a set of amplitude parameters
Hidden activation adaptation (Zhao et al., 2015)	2015	Adapts the shape of activation function by adjusting its slope and bias parameters

filtering which is the major component of ETSI AFE standardized in 2002 and is later improved with less computational cost with two-stage Mel-warped filter-bank Wiener filtering in 2004.

• Temporal processing is used to enhance the useful modulation frequency range for speech recognition and attenuate other modulation frequency ranges. The representative methods are RASTA, J-RASTA, and LDA-based RASTA, which are fixed once designed. Temporal structure normalization extends the work by performing environment-dependent temporal filtering to normalize temporal characteristics of the features.

- TRAP and TANDEM, proposed in 1998 and 2000, are further developed into HATS and Tonotopic MLP, and finally bottle-neck feature in 2007.
- The margin concept helps model robustness because a correct decision can still be made if the mismatched test samples fall within a tolerance region around the original training samples defined by the margin. This is demonstrated by the work of SME and structured SVM.
- While most noise-robustness methods work on generative models which are easy to be adapted, combining generative models and discriminative models together can utilize advantages from both models.
- In 2010, CD-DNN-HMM becomes a break-through of modeling technology with deep hidden layers and senone modeling although it is still a hybrid ANN/HMM system proposed in 1994. As analyzed in Section 3.4, the layer-by-layer setup of DNN provides a feature extraction strategy that automatically derives powerful distortion-resistant features from primitive raw data for senone classification.
- For general GMM adaptation, CMLLR is a natural extension to MLLR and is extremely popular since it can be effectively implemented in the feature space.
- For general DNN adaptation, most methods can be considered as linear adaptation methods, having their counterparts in GMM adaptation methods. In the beginning, most methods such as fDLR and oDLR adapt added linear layers. Later, with the requirement of fast adaptation with a limited number of utterances, low-footprint DNN adaptation methods such as SVD bottleneck adaptation, LHUC, and hidden activation adaptation began to emerge.

REFERENCES

Abdel-Hamid, O., Jiang, H., 2013. Fast speaker adaptation of hybrid NN/HMM model for speech recognition based on discriminative learning of speaker code. In: Proc. International Conference on Acoustics, Speech and Signal Processing (ICASSP), pp. 7942-7946.
Abdel-Hamid, O., Jiang, H., 2014. Direct adaptation of hybrid DNN/HMM model for fast speaker adaptation in LVCSR based on speaker code. In: Proc. Interspeech.
Acero, A., Huang, X., 1995. Augmented cepstral normalization for robust speech recognition. In: Proc. of the IEEE Workshop on Automatic Speech Recognition.
Agarwal, A., Cheng, Y.M., 1999. Two-stage mel-warped wiener filter for robust speech recognition. In: Proc. IEEE Workshop on Automatic Speech Recognition and Understanding (ASRU), pp. 67-70.
Ali, A.M.A., der Spiegel, J.V., Mueller, P., 2002. Robust auditory-based speech processing using the average localized synchrony detection. IEEE Trans. Speech Audio Process. 10 (5), 279-292.
Anastasakos, A., Kubala, F., Makhoul, J., Schwartz, R., 1994. Adaptation to new microphones using tied-mixture normalization. In: Proc. International Conference on Acoustics, Speech and Signal Processing (ICASSP), vol. I, pp. 433-436.
Atal, B., 1974. Effectiveness of linear prediction characteristics of the speech wave for automatic speaker identification and verification. J. Acoust. Soc. Amer. 55, 1304-1312.

Athineos, M., Hermansky, H., Ellis, D.P., 2004. LP-TRAP: Linear predictive temporal patterns. In: Proc. Interspeech, pp. 949-952.

Avendano, C., van Vuuren, S., Hermansky, H., 1996. Data-based RASTA-like filter design for channel normalization in ASR. In: Proc. International Conference on Spoken Language Processing (ICSLP), pp. 2087-2090.

Berouti, M., Schwartz, R., Makhoul, J., 1979. Enhancement of speech corrupted by additive noise. In: Proc. International Conference on Acoustics, Speech and Signal Processing (ICASSP), pp. 208-211.

Boll, S.F., 1979. Suppression of acoustic noise in speech using spectral subtraction. IEEE Trans. Acoust. Speech Signal Process. 27 (2), 113-120.

Bourlard, H., Morgan, N., 1994. Connectionist speech recognition—A Hybrid approach. Kluwer Academic Press, Boston, MA.

Byrne, W., 2006. Minimum Bayes risk estimation and decoding in large vocabulary continuous speech recognition. IEICE Trans. Informat. Syst. 89 (3), 900-907.

Chen, B., Zhu, Q., Morgan, N., 2004. Learning long-term temporal features in LVCSR using neural networks. In: Proc. Interspeech.

Chen, B., Zhu, Q., Morgan, N., 2005. Tonotopic multi-layered perceptron: A neural network for learning long-term temporal features for speech recognition. In: Proc. International Conference on Acoustics, Speech and Signal Processing (ICASSP), vol. IV, pp. 757-760.

Chen, C.P., Bilmes, J., 2007. Mva processing of speech features. IEEE Trans. Audio Speech Lang. Process. 15 (1), 257-270.

Chiu, Y.H., Raj, B., Stern, R.M., 2012. Learning-based auditory encoding for robust speech recognition. IEEE Trans. Audio Speech Lang. Process. 20 (3), 900-914.

Cui, X., Alwan, A., 2005. Noise robust speech recognition using feature compensation based on polynomial regression of utterance SNR. IEEE Trans. Speech Audio Process. 13 (6), 1161-1172.

Cheng, Y.M., Macho, D., 2001. SNR-dependent waveform processing for robust speech recognition. In: Proc. International Conference on Acoustics, Speech and Signal Processing (ICASSP), vol. 1, pp. 305-308.

Dahl, G., Yu, D., Deng, L., Acero, A., 2011. Large vocabulary continuous speech recognition with context-dependent DBN-HMMs. In: Proc. International Conference on Acoustics, Speech and Signal Processing (ICASSP).

Dahl, G., Yu, D., Deng, L., Acero, A., 2012. Context-dependent pre-trained deep neural networks for large-vocabulary speech recognition. IEEE Trans. Audio Speech Lang. Process. 20 (1), 30-42.

de la Torre, A., Peinado, A.M., Segura, J.C., Perez-Cordoba, J.L., Benitez, M.C., Rubio, A.J., 2005. Histogram equalization of speech representation for robust speech recognition. IEEE Trans. Speech Audio Process. 13 (3), 355-366.

Dehak, N., Kenny, P., Dehak, R., Dumouchel, P., Ouellet, P., 2011. Front-end factor analysis for speaker verification. IEEE Trans. Audio Speech Lang. Process. 19 (4), 788-798.

Dempster, A.P., Laird, N.M., Rubin, D.B., 1977. Maximum likelihood from incomplete data via the EM algorithm. J. R. Stat. Soc. 39 (1), 1-38.

Deng, L., 1999. Computational models for auditory speech processing. In: Computational Models of Speech Pattern Processing. Springer-Verlag, New York, pp. 67-77.

Deng, L., 2011. Front-end, back-end, and hybrid techniques for noise-robust speech recognition. In: Robust Speech Recognition of Uncertain or Missing Data: Theory and Application. Springer, New York, pp. 67-99.

Deng, L., Acero, A., Plumpe, M., Huang, X., 2000. Large vocabulary speech recognition under adverse acoustic environment. In: Proc. International Conference on Spoken Language Processing (ICSLP), vol. 3, pp. 806-809.

Deng, L., Geisler, D., Greenberg, S., 1988. A composite model of the auditory periphery for the processing of speech. In: Journal of Phonetics, special theme issue on Representation of Speech in the Auditory Periphery, vol. 16, pp. 93-108.

Deng, L., Yu, D., 2014. Deep Learning: Methods and Applications. Now Publishers, Hanover, MA.

Dharanipragada, S., Padmanabhan, M., 2000. A nonlinear unsupervised adaptation technique for speech recognition. In: Proc. International Conference on Acoustics, Speech and Signal Processing (ICASSP), pp. 556-559.

Droppo, J., Acero, A., 2008. Environmental robustness. In: Benesty, J., Sondhi, M.M., Huang, Y. (Eds.), Handbook of Speech Processing. Springer, New York.

Du, J., Wang, R.H., 2008. Cepstral shape normalization CSN for robust speech recognition. In: Proc. International Conference on Acoustics, Speech and Signal Processing (ICASSP), pp. 4389-4392.

ETSI, 2002. Speech processing, transmission and quality aspects (STQ); distributed speech recognition; advanced front-end feature extraction algorithm; compression algorithms. In: ETSI.

Fazel, A., Chakrabartty, S., 2012. Sparse auditory reproducing kernel (SPARK) features for noise-robust speech recognition. IEEE Trans. Audio Speech Lang. Process. 20 (4), 1362-1371.

Gales, M.J.F., 1995. Model-based techniques for noise robust speech recognition. Ph.D. thesis, University of Cambridge.

Gales, M.J.F., 1998. Maximum likelihood linear transformations for HMM-based speech recognition. Comput. Speech Lang. 12, 75-98.

Gales, M.J.F., Flego, F., 2010. Discriminative classifiers with adaptive kernels for noise robust speech recognition. Comput. Speech Lang. 24 (4), 648-662.

Gales, M.J.F., Woodland, P.C., 1996. Mean and variance adaptation within the MLLR framework. Comput. Speech Lang. 10, 249-264.

García, L., Ortuzar, C.B., de la Torre, A., Segura, J.C., 2012. Class-based parametric approximation to histogram equalization for ASR. IEEE Signal Process. Lett. 19 (7), 415-418.

García, L., Segura, J.C., Ramrez, J., de la Torre, A., Bentez, C., 2006. Parametric nonlinear feature equalization for robust speech recognition. In: Proc. International Conference on Acoustics, Speech and Signal Processing (ICASSP), vol. I, pp. 529-532.

Gauvain, J.L., Lee, C.H., 1994. Maximum a posteriori estimation for multivariate Gaussian mixture observations of Markov chains. IEEE Trans. Speech Audio Process. 2 (2), 291-298.

Gemello, R., Mana, F., Scanzio, S., Laface, P., Mori, R.D., 2007. Linear hidden transformations for adaptation of hybrid ANN/HMM models. Speech Commun. 49 (10), 827-835.

Grézl, F., Cernocký, J., 2007. TRAP-based techniques for recognition of noisy speech. In: Proc. 10th International Conference on Text Speech and Dialogue (TSD), LNCS. Springer Verlag, New York, pp. 270-277.

Grézl, F., Karafiát, M., Kontár, S., Cernocký, J., 2007. Probabilistic and bottle-neck features for LVCSR of meetings. In: Proc. International Conference on Acoustics, Speech and Signal Processing (ICASSP), vol. IV, pp. 757-760.

Hanson, B.A., Applehaum, T.H., 1993. Subband or cepstral domain filtering for recognition of Lombard and channel-distorted speech. In: Proc. International Conference on Acoustics, Speech and Signal Processing (ICASSP), vol. II, pp. 79-82.

Haykin, S., 2001. Adaptive Filter Theory, fourth ed. Prentice Hall, Upper Saddle River, NJ.

He, X., Chou, W., 2003. Minimum classification error linear regression for acoustic model adaptation of continuous density HMMs. In: Proc. International Conference on Acoustics, Speech and Signal Processing (ICASSP), vol. I, pp. 556-559.

Hermansky, H., 1990. Perceptual linear predictive (PLP) analysis of speech. J. Acoust. Soc. Amer. 87 (4), 1738-1752.

Hermansky, H., Ellis, D.P.W., Sharma, S., 2000. Tandem connectionist feature extraction for conventional HMM systems. In: Proc. International Conference on Acoustics, Speech and Signal Processing (ICASSP), vol. 3, pp. 1635-1638.

Hermansky, H., Hanson, B.A., Wakita, H., 1985. Perceptually based linear predictive analysis of speech. In: Proc. International Conference on Acoustics, Speech and Signal Processing (ICASSP), vol. I, pp. 509-512.

Hermansky, H., Morgan, N., 1994. RASTA processing of speech. IEEE Trans. Speech Audio Process. 2 (4), 578-589.

Hermansky, H., Morgan, N., Bayya, A., Kohn, P., 1991. Compensation for the effect of communication channel in auditory-like analysis of speech (RASTA-PLP). In: Proceedings of European Conference on Speech Technology, pp. 1367-1370.

Hermansky, H., Sharma, S., 1998. TRAPs—classifiers of temporal patterns. In: Proc. International Conference on Spoken Language Processing (ICSLP).

Hilger, F., Molau, S., Ney, H., 2002. Quantile based histogram equalization for online applications. In: Proc. International Conference on Spoken Language Processing (ICSLP), pp. 237-240.

Hilger, F., Ney, H., 2001. Quantile based histogram equalization for noise robust speech recognition. In: Proc. European Conference on Speech Communication and Technology (EUROSPEECH), vol. 2, pp. 1135-1138.

Hilger, F., Ney, H., 2006. Quantile based histogram equalization for noise robust large vocabulary speech recognition. IEEE Trans. Audio Speech Lang. Process. 14 (3), 845-854.

Hinton, G., Deng, L., Yu, D., Dahl, G.E., Mohamed, A., Jaitly, N., et al., 2012. Deep neural networks for acoustic modeling in speech recognition: The shared views of four research groups. IEEE Signal Process. Mag. 29 (6), 82-97.

Houtgast, T., Steeneken, H., 1973. The modulation transfer function in room acoustics as a predictor of speech intelligibility. Acta Acustica united with Acustica 28 (1), 66-73.

Houtgast, T., Steeneken, H., 1985. A review of the MTF concept in room acoustics and its use for estimating speech intelligibility in auditoria. J. Acoust. Soc. Amer. 77 (3), 1069-1077.

Hsu, C.W., Lee, L.S., 2004. Higher order cepstral moment normalization (HOCMN) for robust speech recognition. In: Proc. International Conference on Acoustics, Speech and Signal Processing (ICASSP), pp. 197-200.

Hsu, C.W., Lee, L.S., 2009. Higher order cepstral moment normalization for improved robust speech recognition. IEEE Trans. Audio Speech Lang. Process. 17 (2), 205-220.

Huang, J.T., Li, J., Gong, Y., 2015. An analysis of convolutional neural networks for speech recognition. In: Proc. International Conference on Acoustics, Speech and Signal Processing (ICASSP).

Huang, X., Acero, A., Hon, H.W., 2001. Spoken Language Processing. Prentice-Hall, Upper Saddle River, NJ.

Jain, P., Hermansky, H., 2003. Beyond a single critical-band in TRAP based ASR. In: Proc. Interspeech.

Jain, P., Hermansky, H., Kingsbury, B., 2002. Distributed speech recognition using noise-robust MFCC and TRAPS-estimated manner features. In: Proc. Interspeech.

Kelly, F., Harte, N., 2010. Auditory features revisited for robust speech recognition. In: International Conference on Pattern Recognition, pp. 4456-4459.

Kim, C., 2010. Signal processing for robust speech recognition motivated by auditory processing. Ph.D. thesis, Carnegie Mellon University.

Kim, C., Stern, R.M., 2010. Feature extraction for robust speech recognition based on maximizing the sharpness of the power distribution and on power flooring. In: Proc. International Conference on Acoustics, Speech and Signal Processing (ICASSP), pp. 4574-4577.

Kim, C., Stern, R.M., 2012. Power-normalized cepstral coefficients (PNCC) for robust speech recognition. In: Proc. International Conference on Acoustics, Speech and Signal Processing (ICASSP), pp. 4101-4104.

Kim, D.S., Lee, S.Y., Kil, R.M., 1999. Auditory processing of speech signals for robust speech recognition in real-world noisy environments. IEEE Trans. Speech Audio Process. 7 (1), 55-69.

Kumar, N., Andreou, A.G., 1998. Heteroscedastic discriminant analysis and reduced rank HMMs for improved speech recognition. Speech Commun. 26 (4), 283-297.

Leggetter, C., Woodland, P., 1995. Maximum likelihood linear regression for speaker adaptation of continuous density hidden Markov models. Comput. Speech Lang. 9 (2), 171-185.

Li, B., Sim, K.C., 2010. Comparison of discriminative input and output transformations for speaker adaptation in the hybrid NN/HMM systems. In: Proc. Interspeech, pp. 526-529.

Li, B., Sim, K.C., 2014. Modeling long temporal contexts for robust DNN-based speech recognition. In: Proc. Interspeech.

Li, J., Deng, L., Gong, Y., Haeb-Umbach, R., 2014a. An overview of noise-robust automatic speech recognition. IEEE/ACM Trans. Audio Speech Lang. Process. 22 (4), 745-777.

Li, J., Huang, J.T., Gong, Y., 2014b. Factorized adaptation for deep neural network. In: Proc. International Conference on Acoustics, Speech and Signal Processing (ICASSP).

Li, F., Nidadavolu, P., Hermansky, H., 2014c. A long, deep and wide artificial neural net for robust speech recognition in unknown noise. In: Proc. Interspeech.

Li, J., Deng, L., Yu, D., Gong, Y., Acero, A., 2007a. High-performance HMM adaptation with joint compensation of additive and convolutive distortions via vector Taylor series. In: Proc. IEEE Workshop on Automatic Speech Recognition and Understanding (ASRU), pp. 65-70.

Li, J., Deng, L., Yu, D., Gong, Y., Acero, A., 2009. A unified framework of HMM adaptation with joint compensation of additive and convolutive distortions. Comput. Speech Lang. 23 (3), 389-405.

Li, J., Lee, C.H., 2008. On a generalization of margin-based discriminative training to robust speech recognition. In: Proc. Interspeech, pp. 1992-1995.

Li, J., Liu, B., Wang, R.H., Dai, L., 2004. A complexity reduction of ETSI advanced front-end for DSR. In: Proc. International Conference on Acoustics, Speech and Signal Processing (ICASSP), vol. 1, pp. 61-64.

Li, J., Yuan, M., Lee, C.H., 2006. Soft margin estimation of hidden Markov model parameters. In: Proc. Interspeech, pp. 2422-2425.

Li, J., Yuan, M., Lee, C.H., 2007b. Approximate test risk bound minimization through soft margin estimation. IEEE Trans. Audio Speech Lang. Process. 15 (8), 2393-2404.

Lim, J.S., Oppenheim, A.V., 1979. Enhancement and bandwidth compression of noisy speech. Proc. IEEE 67 (12), 1586-1604.

Lin, S.H., Yeh, Y.M., Chen, B., 2006. Exploiting polynomial-fit histogram equalization and temporal average for robust speech recognition. In: Proc. International Conference on Spoken Language Processing (ICSLP), pp. 2522-2525.

Lin, S.H., Yeh, Y.M., Chen, B., 2007. A comparative study of histogram equalization (HEQ) for robust speech recognition. Int. J. Comput. Linguist. Chinese Lang. Process. 12 (2), 217-238.

Liu, B., Dai, L., Li, J., Wang, R.H., 2004. Double Gaussian based feature normalization for robust speech. In: International Symposium on Chinese Spoken Language Processing, pp. 705-708.

Macho, D., Mauuary, L., Noé., B., Cheng, Y.M., Ealey, D., Jouvet, D., et al., 2002. Evaluation of a noise-robust DSR front-end on Aurora databases. In: Proc. International Conference on Spoken Language Processing (ICSLP), pp. 17-20.

Mauuary, L., 1998. Blind equalization in the cepstral domain for robust telephone based speech recognition. In: EUSPICO, vol. 1, pp. 359-363.

Miao, Y., Zhang, H., Metze, F., 2014. Towards speaker adaptive training of deep neural network acoustic models. In: Proc. Interspeech.

Molau, S., Hilger, F., Ney, H., 2003. Feature space normalization in adverse acoustic conditions. In: Proc. International Conference on Acoustics, Speech and Signal Processing (ICASSP), vol. 1, pp. 656-659.

Molau, S., Pitz, M., Ney, H., 2001. Histogram based normalization in the acoustic feature space. In: Proc. IEEE Workshop on Automatic Speech Recognition and Understanding (ASRU), pp. 21-24.

Moreno, P.J., 1996. Speech recognition in noisy environments. Ph.D. thesis, Carnegie Mellon University.

Morgan, N., Hermansky, H., 1992. RASTA extensions: Robustness to additive and convolutional noise. In: ESCA Workshop Proceedings of Speech Processing in Adverse Conditions, pp. 115-118.

Moritz, N., Anemuller, J., Kollmeier, B., 2011. Amplitude modulation spectrogram based features for robust speech recognition in noisy and reverberant environments. In: Proc. International Conference on Acoustics, Speech and Signal Processing (ICASSP), pp. 5492-5495.

Moritz, N., Schädler, M., Adiloglu, K., Meyer, B., Jürgens, T., Gerkmann, T., et al., 2013. noise robust distant automatic speech recognition utilizing NMF based source separation and auditory feature extraction. In: the 2nd CHiME workshop on machine listening and multisource environments.

Müller, F., Mertins, A., 2011. Contextual invariant-integration features for improved speaker-independent speech recognition. Speech Commun. 53 (6), 830-841.

Nadas, A., 1983. A decision theoretic formulation of a training problem in speech recognition and a comparison of training by unconditional versus conditional maximum likelihood. IEEE Trans. Acoust. Speech Signal Process. 31 (4), 814-817.

Neto, J., Almeida, L., Hochberg, M., Martins, C., Nunes, L., Renals, S., et al., 1995. Speaker-adaptation for hybrid HMM-ANN continuous speech recognition system. In: Proc. European Conference on Speech Communication and Technology (EUROSPEECH).

Noé., B., Sienel, J., Jouvet, D., Mauuary, L., Boves, L., Veth, J.D., et al., 2001. Noise reduction for noise robust feature extraction for distributed speech recognition. In: Proc. European Conference on Speech Communication and Technology (EUROSPEECH).

Padmanabhan, M., Saon, G., Zweig, G., 2000. Lattice-based unsupervised MLLR for speaker adaptation. In: Proc. ISCA ITRW ASR, pp. 128-131.

Quatieri, T.F., 2001. Discrete-Time Speech Signal Processing: Principles and Practice. Prentice Hall, Upper Saddle River, NJ.

Ragni, A., Gales, M.J.F., 2011a. Derivative kernels for noise robust ASR. In: Proc. IEEE Workshop on Automatic Speech Recognition and Understanding (ASRU), pp. 119-124.

Ragni, A., Gales, M.J.F., 2011b. Structured discriminative models for noise robust continuous speech recognition. In: Proc. International Conference on Acoustics, Speech and Signal Processing (ICASSP), pp. 4788-4791.

Rahim, M.G., Juang, B.H., 1996. Signal bias removal by maximum likelihood estimation for robust telephone speech recognition. IEEE Trans. Speech Audio Process. 4 (1), 19-30.

Saon, G., Huerta, J.M., Jan, E.E., 2001a. Robust digit recognition in noisy environments: the IBM Aurora 2 system. In: Proc. Interspeech, pp. 629-632.

Saon, G., Soltau, H., Nahamoo, D., Picheny, M., 2013. Speaker adaptation of neural network acoustic models using i-vectors. In: Proc. IEEE Workshop on Automatic Speech Recognition and Understanding (ASRU), pp. 55-59.

Saon, G., Zweig, G., Padmanabhan, M., 2001b. Linear feature space projections for speaker adaptation. In: Proc. International Conference on Acoustics, Speech and Signal Processing (ICASSP), vol. 1, pp. 325-328.

Sárosi, G., Mozsolics, T., Tarján, B., Balog, A., Mihajlik, P., Fegyó, T., 2010. Recognition of multiple language voice navigation queries in traffic situations. In: COST 2102 Conference, pp. 199-213.

Segura, J.C., Benitez, C., Torre, A., Rubio, A.J., Ramirez, J., 2004. Cepstral domain segmental nonlinear feature transformations for robust speech recognition. IEEE Signal Process. Lett. 11 (5), 517-520.

Seide, F., Li, G., Chen, X., Yu, D., 2011. Feature engineering in context-dependent deep neural networks for conversational speech transcription. In: Proc. IEEE Workshop on Automatic Speech Recognition and Understanding (ASRU), pp. 24-29.

Senior, A., Lopez-Moreno, I., 2014. Improving DNN speaker independence with i-vector inputs. In: Proc. International Conference on Acoustics, Speech and Signal Processing (ICASSP), pp. 225-229.

Shao, Y., Srinivasan, S., Jin, Z., Wang, D., 2010. A computational auditory scene analysis system for speech segregation and robust speech recognition. Comput. Speech Lang. 24 (1), 77-93.

Sheikhzadeh, H., Deng, L., 1998. Speech analysis and recognition using interval statistics generated from a composite auditory model. In: IEEE Trans. on Speech and Audio Processing, vol. 6, pp. 50-54.

Sheikhzadeh, H., Deng, L., 1999. A layered neural network interfaced with a cochlear model for the study of speech encoding in the auditory system. In: Comput. Speech Lang., 13, 39-64.

Shinoda, K., Lee, C.H., 2001. A structural Bayes approach to speaker adaptation. IEEE Trans. Speech Audio Process. 9 (3), 276-287.

Siniscalchi, S.M., Li, J., Lee, C.H., 2012. Hermitian-based hidden activation functions for adaptation of hybrid HMM/ANN models. In: Proc. Interspeech.

Siniscalchi, S.M., Li, J., Lee, C.H., 2013. Hermitian polynomial for speaker adaptation of connectionist speech recognition systems. IEEE Trans. Audio Speech Lang. Process. 21 (10), 2152-2161.

Siohan, O., Chesta, C., Lee, C.H., 2001. Joint maximum a posteriori adaptation of transformation and HMM parameters. IEEE Trans. Speech Audio Process. 9 (4), 417-428.

Siohan, O., Myrvoll, T.A., Lee, C.H., 2002. Structural maximum a posteriori linear regression for fast HMM adaptation. Comput. Speech Lang. 16 (1), 5-24.

Stern, R., Morgan, N., 2012. Features based on auditory physiology and perception. In: Techniques for Noise Robustness in Automatic Speech Recognition. John Wiley & Sons, West Sussex, UK.

Suh, Y., Ji, M., Kim, H., 2007. Probabilistic class histogram equalization for robust speech recognition. IEEE Signal Process. Lett. 14 (4), 287-290.

Suk, Y.H., Choi, S.H., Lee, H.S., 1999. Cepstrum third-order normalization method for noisy speech recognition. Electron. Lett. 35 (7), 527-528.

Swietojanski, P., Renals, S., 2014. Learning hidden unit contributions for unsupervised speaker adaptation of neural network acoustic models. In: Proc. IEEE Spoken Language Technology Workshop.

Taghia, J., Taghia, J., Mohammadiha, N., Sang, J., Bouse, V., Martin, R., 2011. An evaluation of noise power spectral density estimation algorithms in adverse acoustic environments. In: Proc. International Conference on Acoustics, Speech and Signal Processing (ICASSP), pp. 4640-4643.

Tsai, S.N., Lee, L.S., 2004. A new feature extraction front-end for robust speech recognition using progressive histogram equalization and mulit-eigenvector temporal filtering. In: Proc. International Conference on Spoken Language Processing (ICSLP), pp. 165-168.

Tüske, Z., Schlüter, R., Hermann, N., Sundermeyer, M., 2012. Context-dependent MLPs for LVCSR: Tandem, hybrid or both? In: Proc. Interspeech, pp. 18-21.

Vapnik, V., 1995. The nature of statistical learning theory. Springer, New York.

Vapnik, V., 1998. Statistical learning theory. Wiley, West Sussex, UK.

O. Viikki, D. Bye, K. Laurila, 1998. A recursive feature vector normalization approach for robust speech recognition in noise. In: Proc. IEEE Intl. Conf. on Acoustic, Speech and Signal Processing, pp. 733-736.

Wang, L., Woodland, P.C., 2004. MPE-based discriminative linear transform for speaker adaptation. In: Proc. International Conference on Acoustics, Speech and Signal Processing (ICASSP), pp. 321-324.

Wu, J., Huo, Q., 2002. Supervised adaptation of MCE-trained CDHMMs using minimum classification error linear regression. In: Proc. International Conference on Acoustics, Speech and Signal Processing (ICASSP), vol. I, pp. 605-608.

Xiao, X., Chng, E.S., Li, H., 2008. Normalization of the speech modulation spectra for robust speech recognition. IEEE Trans. Audio Speech Lang. Process. 16 (8), 1662-1674.

Xiao, X., Chng, E.S., Li, H., 2012. Joint spectral and temporal normalization of features for robust recognition of noisy and reverberated speech. In: Proc. International Conference on Acoustics, Speech and Signal Processing (ICASSP), pp. 4325-4328.

Xiao, X., Chng, E.S., Li, H., 2013. Attribute-based histogram equalization (HEQ) and its adaptation for robust speech recognition. In: Proc. Interspeech, pp. 876-880.

Xiao, X., Li, J., Chng, E.S., Li, H., 2011a. Maximum likelihood adaptation of histogram equalization with constraint for robust speech recognition. In: Proc. International Conference on Acoustics, Speech and Signal Processing (ICASSP), pp. 5480-5483.

Xiao, X., Li, J., Chng, E.S., Li, H., Lee, C.H., 2009. A study on hidden Markov model's generalization capability for speech recognition. In: Proc. IEEE Workshop on Automatic Speech Recognition and Understanding (ASRU), pp. 255-260.

Xiao, X., Li, J., Siong, C.E., Li, H., 2011b. Feature normalization using structured full transforms for robust speech recognition. In: Proc. Interspeech, pp. 693-696.

Xue, J., Li, J., Gong, Y., 2013. Restructuring of deep neural network acoustic models with singular value decomposition. In: Proc. Interspeech, pp. 2365-2369.

Xue, J., Li, J., Yu, D., Seltzer, M., Gong, Y., 2014a. Singular value decomposition based low-footprint speaker adaptation and personalization for deep neural network. In: Proc. International Conference on Acoustics, Speech and Signal Processing (ICASSP), pp. 6359-6363.

Xue, S., Abdel-Hamid, O., Jiang, H., Dai, L., Liu, Q., 2014b. Fast adaptation of deep neural network based on discriminant codes for speech recognition. Audio, Speech, and Language Processing, IEEE/ACM Transactions on 22 (12), 1713-1725.

Yao, K., Yu, D., Seide, F., Su, H., Deng, L., Gong, Y., 2012. Adaptation of context-dependent deep neural networks for automatic speech recognition. In: Proc. IEEE Spoken Language Technology Workshop, pp. 366-369.

Yapanel, U.H., Hansen, J.H.L., 2008. A new perceptually motivated MVDR-based acoustic front-end (PMVDR) for robust automatic speech recognition. Speech Commun. 50 (2), 142-152.

Yu, D., Deng, L., 2014. Automatic Speech Recognition—A Deep Learning Approach. Springer, New York.

Yu, D., Deng, L., Dahl, G., 2010. Roles of pretraining and fine-tuning in context-dependent DBN-HMMs for real-world speech recognition. In: Proc. NIPS Workshop on Deep Learning and Unsupervised Feature Learning.

Yu, D., Deng, L., Droppo, J., Wu, J., Gong, Y., Acero, A., 2008. Robust speech recognition using a cepstral minimum-mean-square-error-motivated noise suppressor. IEEE Trans. Audio Speech Lang. Process. 16 (5), 1061-1070.

Yu, D., Yao, K., Su, H., Li, G., Seide, F., 2013. KL-divergence regularized deep neural network adaptation for improved large vocabulary speech recognition. In: Proc. International Conference on Acoustics, Speech and Signal Processing (ICASSP), pp. 7893-7897.

Yu, K., Gales, M.J.F., Woodland, P.C., 2009. Unsupervised adaptation with discriminative mapping transforms. IEEE Trans. Audio Speech Lang. Process. 17 (4), 714-723.

Zhang, S.X., Gales, M., 2013. Structured SVMs for automatic speech recognition. IEEE Trans. Audio Speech Lang. Process. 21 (3), 544-555.

Zhang, S.X., Gales, M.J.F., 2011. Structured support vector machines for noise robust continuous speech recognition. In: Proc. Interspeech, pp. 989-992.

Zhao, Y., Li, J., Xue, J., Gong, Y., 2015. Investigating online low-footprint speaker adaptation using generalized linear regression and click-through data. In: Proc. International Conference on Acoustics, Speech and Signal Processing (ICASSP).

Compensation with prior knowledge

5

CHAPTER OUTLINE

In this chapter, we explore an alternative way of categorizing and analyzing existing robust ASR techniques, where we use the attribute of whether or not they make use of prior knowledge and information about the acoustic distortion before applying formal compensation procedures. This contrasts the previous chapter when the attribute was whether the operations were applied on the feature domain or on the model domain.

Major noise-robust methods that use the prior knowledge about acoustic distortions learn the generally nonlinear mapping functions between the clean and distorted speech features when they are available in the form of stereo data in the training phase. By modeling the differences between the features or models of the stereo data, a distortion model can be learned in training and then used in testing to perform feature enhancement or model compensation. The distortion model can be a deterministic mapping function. It can also be formulated probabilistically as in $p(\mathbf{y}|\mathbf{x})$. A collection of these methods are known as stereo-data mapping methods.

In addition to stereo-based methods, another collection of methods exploiting prior knowledge are based on first establishing or sampling a set of simple models for the acoustic environments, each corresponding to one specific environment during

Robust Automatic Speech Recognition. http://dx.doi.org/10.1016/B978-0-12-802398-3.00005-2

training. These models are then combined online to form the final acoustic model of distorted speech that best fits the test environment.

More recently, new methods based on clean speech and noise exemplar dictionaries learned from training data for source separation appeared in the literature. Using non-negative matrix factorization (NMF), these methods restore clean speech by constructing the noisy speech with pre-trained clean speech and noise exemplars and only keeping the clean speech exemplars. How to generalize to the unseen acoustic conditions is very important to robust ASR. Variable-parameter modeling presented in this chapter will provide a better solution by modeling the acoustic model parameters with a set of polynomial functions of the environment variable. The model parameters can be extrapolated from the learned polynomial functions if the test environments are not observed during training.

5.1 LEARNING FROM STEREO DATA

Many methods use stereo data to learn the mapping from the distorted speech to the clean speech. The stereo data consists of time-aligned speech samples that have been simultaneously recorded in training environments and in representative test environments. Stereo data can also be obtained by digitally introducing (e.g., adding noise) distortion to the clean speech. The success of these methods usually depends on how well the representative distorted samples during training really match test samples.

5.1.1 EMPIRICAL CEPSTRAL COMPENSATION

One group of methods is called empirical cepstral compensation (Stern et al., 1996), developed at Carnegie Mellon University (CMU). Let us recap Equation 3.14 in Equation 5.1 which is the cepstral representation of the relationship between the clean speech feature and the distorted speech feature as

$$\mathbf{y} = \mathbf{x} + \mathbf{h} + \mathbf{C}\log(1 + \exp(\mathbf{C}^{-1}(\mathbf{n} - \mathbf{x} - \mathbf{h}))). \tag{5.1}$$

Then, with

$$\mathbf{v} = \mathbf{h} + \mathbf{C}\log(1 + \exp(\mathbf{C}^{-1}(\mathbf{n} - \mathbf{x} - \mathbf{h}))). \tag{5.2}$$

the distorted speech cepstrum $\mathbf{y}$ is expressed as the clean speech cepstrum $\mathbf{x}$ plus a bias $\mathbf{v}$. In empirical cepstral compensation, this bias $\mathbf{v}$ can be formulated to depend on the SNR, the location of vector quantization (VQ) cluster k, the presumed phoneme identity p, and the specific test environment e. Hence, Equation 5.1 can be re-written as

$$\mathbf{y} = \mathbf{x} + \mathbf{v}(\text{SNR}, k, p, e). \tag{5.3}$$

$\mathbf{v}(\text{SNR}, k, p, e)$ can be learned from stereo training data. During testing, the clean speech cepstrum can be recovered from the distorted speech with

$$\hat{\mathbf{x}} = \mathbf{y} - \mathbf{v}(\text{SNR}, k, p, e). \tag{5.4}$$

Depending on how $\mathbf{v}(\text{SNR}, k, p, e)$ is defined, there are different cepstral compensation methods. If SNR is the only factor for $\mathbf{v}$, it is called SNR-dependent cepstral normalization (SDCN) (Acero and Stern, 1990). During training, frame pairs in the stereo data are allocated into different subsets according to SNR. Then, the compensation vector $\mathbf{v}(\text{SNR})$ corresponding to a range of SNRs is estimated by averaging the difference between the cepstral vectors of the clean and distorted speech features for all frames in that range. During testing, the SNR for each frame of the input speech is first estimated, and the corresponding compensation vector is then applied to the cepstral vector for that frame with Equation 5.4.

Fixed codeword-dependent cepstral normalization (FCDCN) (Acero, 1993) is a refined version of SDCN with the compensation vector as $\mathbf{v}(\text{SNR}, k)$, which depends on both SNR and VQ cluster location. For each SNR range, there is a VQ cluster trained from the utterances representative for the testing. During training, the frame pairs in the stereo data are allocated into different subsets according to the SNR and the VQ cluster location of the distorted feature. The compensation vector is calculated by averaging the difference between the cepstral vectors of the clean and distorted speech features for the SNR-specific VQ cluster location. During testing, both SNR and VQ cluster locations are estimated, and the corresponding compensation vector is then applied to the cepstral vector for that frame. Phone-dependent cepstral normalization (PDCN) (Liu et al., 1994) is another empirical cepstral compensation method in which the compensation vector depends on the presumed phoneme the current frame belongs to. During testing, the phoneme hypotheses can be obtained by a first pass HMM decoding. It can also be extended to include SNR as a factor, and is called SNR-dependent PDCN (SPDCN) (Liu et al., 1994). Environment is also a factor of the compensation vector. FCDCN and PDCN can be extended to multiple FCDCN (MFCDCN) and multiple PDCN (MPDCN) when multiple environments are used in training (Liu et al., 2004). The test utterance is first classified into one specific environment e, and then the compensation vector $\mathbf{v}(\text{SNR}, k, e)$ (in MFCDCN) or $\mathbf{v}(p, e)$ (in MPDCN) will be applied to the distorted speech cepstral vector. Another alternative is to interpolate the compensation vectors from those of multiple environments instead of making the hard decision of the specific environment. The corresponding methods are called interpolated FCDCN and interpolated PDCN (Liu et al., 1994).

5.1.2 SPLICE

Stereo-based piecewise linear compensation for environments (SPLICE), proposed originally in Deng et al. (2000) and elaborated in Deng et al. (2001, 2003); Droppo et al. (2001b, 2002), is a popular method to learn from stereo data and is more advanced than the aforementioned empirical cepstral compensation methods. In SPLICE, the noisy speech data, $\mathbf{y}$, is modeled by a mixture of Gaussians

$$p(\mathbf{y}, k) = P(k)p(\mathbf{y}|k) = P(k)\mathcal{N}(\mathbf{y}; \boldsymbol{\mu}(k), \boldsymbol{\Sigma}(k)), \tag{5.5}$$

and the *a posteriori* probability of the clean speech vector $\mathbf{x}$ given the noisy speech $\mathbf{y}$ and the mixture component k is modeled using an additive correction vector $\mathbf{b}(k)$:

$$p(\mathbf{x}|\mathbf{y}, k) = \mathcal{N}(\mathbf{x}; \mathbf{y} + \mathbf{b}(k), \boldsymbol{\Psi}(k)), \tag{5.6}$$

where $\boldsymbol{\Psi}(k)$ is the covariance matrix of the mixture component dependent posterior distribution, representing the prediction error. The dependence of the additive (linear) correction vector on the mixture component gives rise to a piecewise linear relationship between the noisy speech observation and the clean speech, hence the name of SPLICE. The feature compensation formulation can be described by

$$\hat{\mathbf{x}} = \sum_{k=1}^{K} P(k|\mathbf{y})(\mathbf{y} + \mathbf{b}(k)). \tag{5.7}$$

The prediction bias vector, $\mathbf{b}(k)$, is estimated by minimizing the mean square error (MMSE) as the weighted mean square error between the clean speech vector and the predicted clean speech vector in the mixture component k:

$$E = \sum_{t} P(k|\mathbf{y}_t)(\mathbf{x}_t - \mathbf{y}_t - \mathbf{b}(k))^2. \tag{5.8}$$

By setting $\frac{\partial E}{\partial \mathbf{b}(k)} = 0$, the estimation of the prediction bias vector, $\mathbf{b}(k)$, is obtained as

$$\mathbf{b}(k) = \frac{\sum_t P(k|\mathbf{y}_t)(\mathbf{x}_t - \mathbf{y}_t)}{\sum_t P(k|\mathbf{y}_t)}, \tag{5.9}$$

and $\boldsymbol{\Psi}(k)$ can be obtained as

$$\boldsymbol{\Psi}(k) = \frac{\sum_t P(k|\mathbf{y}_t)(\mathbf{x}_t - \mathbf{y}_t)(\mathbf{x}_t - \mathbf{y}_t)^T}{\sum_t P(k|\mathbf{y}_t)} - \mathbf{b}(k)\mathbf{b}^T(k). \tag{5.10}$$

To reduce the runtime cost, the following simplification can be used

$$\hat{k} = \operatorname*{argmax}_{k} p(\mathbf{y}, k),$$
$$\hat{\mathbf{x}} = \mathbf{y} + \mathbf{b}_{\hat{k}}. \tag{5.11}$$

Note that for implementation simplicity, a fundamental assumption is made in the above SPLICE algorithm that the expected clean speech vector $\mathbf{x}$ is a shifted version of the noisy speech vector $\mathbf{y}$. In reality, when $\mathbf{x}$ and $\mathbf{y}$ are Gaussians given component k, their joint distribution can be modeled as

$$\mathcal{N}\left(\left[\begin{array}{c} \mathbf{x} \\ \mathbf{y} \end{array} \right]; \left[\begin{array}{c} \boldsymbol{\mu}_x(k), \\ \boldsymbol{\mu}_y(k) \end{array} \right], \left[\begin{array}{cc} \boldsymbol{\Sigma}_x(k) & \boldsymbol{\Sigma}_{xy}(k) \\ \boldsymbol{\Sigma}_{yx}(k) & \boldsymbol{\Sigma}_y(k) \end{array} \right] \right). \tag{5.12}$$

and a rotation on $\mathbf{y}$ is needed for the conditional mean as

$$E(\mathbf{x}|\mathbf{y}, k) = \boldsymbol{\mu}_x(k) + \boldsymbol{\Sigma}_{xy}(k)\boldsymbol{\Sigma}_y^{-1}(k)(\mathbf{y} - \boldsymbol{\mu}_y(k)) \tag{5.13}$$

$$= \mathbf{A}(k)\mathbf{y} + \mathbf{b}(k), \tag{5.14}$$

where

$$\mathbf{A}(k) = \boldsymbol{\Sigma}_{xy}(k)\boldsymbol{\Sigma}_y^{-1}(k), \tag{5.15}$$

$$\mathbf{b}(k) = \boldsymbol{\mu}_x(k) - \boldsymbol{\Sigma}_{xy}(k)\boldsymbol{\Sigma}_y^{-1}(k)\boldsymbol{\mu}_y(k). \tag{5.16}$$

The feature compensation formulation in this case is

$$\hat{\mathbf{x}} = \sum_{k=1}^{K} P(k|\mathbf{y})(\mathbf{A}(k)\mathbf{y} + \mathbf{b}(k)). \tag{5.17}$$

It is interesting that feature space minimum phone error (fMPE) training (Povey et al., 2005), a very popular feature space discriminative training method, can be linked to SPLICE to some extent (Deng et al., 2005). Originally derived with the MMSE criterion, SPLICE can be improved with the maximum mutual information criterion (Bahl et al., 1997) by discriminative training $\mathbf{A}(k)$ and $\mathbf{b}(k)$ (Droppo and Acero, 2005). In Droppo et al. (2001b), dynamic SPLICE is proposed to not only minimize the static deviation from the clean to noisy cepstral vectors, but to also minimize the deviation between the delta parameters. This is implemented by using a simple zero-phase, non-causal IIR filter to smooth the cepstral bias vectors.

In addition to SPLICE, MMSE-based stereo mapping is studied in Cui et al. (2008a), and MAP-based stereo mapping is formulated in Afify et al. (2007, 2009). Most stereo mapping methods use a GMM to construct a joint space of the clean and noisy speech feature. This is extended in Cui et al. (2008b), where a HMM is used. The mapping methods can also be extended into a discriminatively trained feature space, such as the fMPE space (Cui et al., 2009a).

One concern for learning with stereo data is the requirement of stereo data, which may not be available in real-world application scenarios. In Droppo et al. (2002), it is shown that a small amount of real noise synthetically mixed into a large, clean corpus is enough to achieve significant benefits for the FCDCN method. In Du et al. (2010), the pseudo-clean features generated with a HMM-based synthesis method (Tokuda et al., 2000) are used to replace the clean features which are usually hard to get in real deployment. It is shown that this pseudo-clean feature is even more effective than the ideal clean feature (Du et al., 2010).

5.1.3 DNN FOR NOISE REMOVAL USING STEREO DATA

Both the empirical cepstral compensation and SPLICE are piecewise linear compensation methods, in which the noisy feature $\mathbf{y}$ and the estimated clean feature $\hat{\mathbf{x}}$ have an environment-dependent linear relationship. If putting them into the context of neural network with $\mathbf{y}$ as the input and $\hat{\mathbf{x}}$ as the output, all of these methods may be considered as a shallow neural network to learn the mapping of $\mathbf{y}$ and $\hat{\mathbf{x}}$ as $\hat{\mathbf{x}} = \mathcal{G}(\mathbf{y})$. From the success of DNNs, we learn that a deep neural network usually has more modeling power than a shallow neural network. Hence it is natural to use a DNN to better learn the mapping function $\mathcal{G}$, and this method has been recently very successful in both speech enhancement and speech recognition tasks (Du et al., 2014a,b; Feng et al., 2014; Gao et al., 2015; Lu et al., 2013; Maas et al., 2012; Narayanan and Wang, 2013, 2014a,b; Tu et al., 2015; Wang et al., 2014; Weninger et al., 2014a,b; Wöllmer et al., 2013).

As shown in Figure 5.1, a DNN can be trained to generate the clean feature from the noisy feature $\mathbf{y}$ by minimizing the mean squared error between the DNN output $\hat{\mathbf{x}} = \mathcal{G}(\mathbf{y})$ and the reference clean features $\mathbf{x}$ (Du et al., 2014a,b; Feng et al., 2014; Lu et al., 2013):

$$F_{\text{MSE}} = \sum_t \|\hat{\mathbf{x}}_t - \mathbf{x}_t\|^2 \tag{5.18}$$

Usually, the input noisy feature $\mathbf{y}$ is with a context window of consecutive frames, while the reference clean features $\mathbf{x}$ only corresponds to the current frame. The enhancement function $\mathcal{G}$ is realized with a DNN in Figure 5.1. This noise-removal

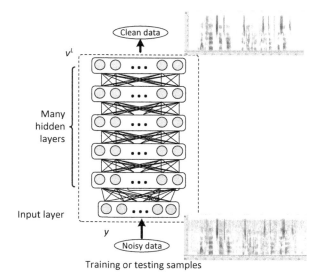

FIGURE 5.1

Generate clean feature from noisy feature with DNN.

strategy is very effective. Evaluated in Du et al. (2014b), when the underlying DNN model used for recognition is trained with clean data and the noisy test data is cleaned with the noise-removal DNN, huge WER reduction can be achieved. If the underlying DNN model used for recognition is trained with multi-condition data, remarkable WER reduction still can be achieved. Note that if the underlying model for recognition is a GMM, the improvement of using a noise-removal DNN is even larger. Also, the improvement is much larger than that obtained with AFE (ETSI, 2002) described in Section 4.1.3. This is due to the power of DNNs which learn the mapping between noisy and clean feature, while AFE is a traditional front-end without learning from the stereo data.

In addition to using a standard feed-forward DNN, a recurrent neural network (RNN) has also been proposed to predict the clean speech from noisy speech (Maas et al., 2012) by modeling temporal signal dependencies in an explicit way because a RNN directly uses its time-recurrent structure to model the long-range context of speech which cannot be approximated by the feature stacking with a context window in the standard feed-forward DNN. Standard RNNs have a known problem of weight decaying (or blowing up) during training. This issue can be solved by replacing the sigmoid units with long short-term memory (LSTM) units and bidirectional LSTM (BLSTM) units (Weninger et al., 2014a,b; Wöllmer et al., 2013) which allow for a more efficient exploitation of temporal context, leading to an improved feature mapping from noisy speech to clean speech. The LSTM units have an internal memory cell whose content is modified in every time step by input, output, and forget gates so that the network memory is modeled explicitly.

While most studies (Du et al., 2014a; Feng et al., 2014; Lu et al., 2013; Maas et al., 2012; Weninger et al., 2014a; Wöllmer et al., 2013) use clean speech features as the DNN training target, there are also some works (Narayanan and Wang, 2013, 2014a,b; Wang et al., 2014) using the time-frequency (T-F) masks such as ideal binary mask (IBM) or ideal ratio mask (IRM) as the training target. For each T-F unit, the corresponding IBM value is set to 1 if the local SNR is greater than a local criterion, otherwise it is set to 0. IRM is defined as the energy ratio of clean speech to noisy speech at each T-F unit with the assumption that noise is uncorrelated with clean speech, and can be written as a function of SNR:

$$\text{IRM}(t,k) = \left(\frac{\text{SNR}(t,k)}{1 + \text{SNR}(t,k)} \right)^{\beta}, \tag{5.19}$$

where β is a tunable parameter to scale the mask. This is closely related to the frequency-domain Wiener filter in Equation 4.29. The training of IBM or IRM estimation with a DNN is done by replacing the clean speech target in Figure 5.1 with either IBM or IRM as the target. During testing, the estimate of the T-F mask $\hat{\mathbf{m}}_t$ is obtained by forward propagating the learned DNN, and the estimated clean speech spectrum is obtained as

$$\hat{\mathbf{x}}_t = \hat{\mathbf{m}}_t .* \mathbf{y}_t, \tag{5.20}$$

where .∗ is the element-wise multiplication.

It is shown in Narayanan and Wang (2014a) that IRM is superior to IBM for the speech recognition task. However, it is still arguable whether using IRM is better than using clean speech feature as the target for noise removal. Suppose a clean utterance is corrupted by different types of noise with various SNRs, using clean speech feature as target directly maps the features from all the utterances with different distortion to the features from the same clean utterance. A DNN needs to learn this challenging many-to-one mapping. In contrast, by using IRM as the training target, the DNN learning is pretty simple—only the one-to-one mapping needs to be learned. Moreover, the target IRM value is between 0 and 1, which makes the learning avoid estimation of unbounded values. Wang et al. (Wang et al., 2014) also provides other arguments why IRM is better as the DNN training target for the task of speech separation. As a result, IRM as the training target is shown to outperform clean speech feature as the training target in speech separation tasks (Wang et al., 2014; Weninger et al., 2014b). On the other hand, using IRM in Equation 5.19 as the training target is supposed to remove only the noise distortion. If the distorted signal **y** is also impacted by the channel distortion, an additional feature mapping function has to be provided in Narayanan and Wang (2014a) to remove the channel distortion in the estimated clean speech feature from the noise-removal DNN. In contrast, using clean speech feature as the training target can directly map the noise and channel distorted feature to clean speech feature with its many-to-one mapping in one step.

In addition to noise removal with DNN based on the minimum square error criterion, similar methodology can separate multiple speakers by putting the mixed feature as the input and the target speaker feature as the output in Figure 5.1. This is done in Weng et al. (2014) where separate DNNs are trained to predict individual sources. Another solution is proposed in Huang et al. (2014) where a single DNN is trained to predict all the sources as in Figure 5.2. This is optimized by minimizing the objective function

$$F_{\text{MSE2}} = \sum_t \|\hat{\mathbf{x}}_{1t} - \mathbf{x}_{1t}\|^2 + \|\hat{\mathbf{x}}_{2t} - \mathbf{x}_{2t}\|^2. \quad (5.21)$$

One improvement proposed in Huang et al. (2014) is to refine the final speaker sources with the constraint that they can be combined to form the original mixed feature with

$$\tilde{\mathbf{x}}_{1t} = \frac{\|\hat{\mathbf{x}}_{1t}\|}{\|\hat{\mathbf{x}}_{1t}\| + \|\hat{\mathbf{x}}_{2t}\|} . * \mathbf{y}_t, \quad (5.22)$$

$$\tilde{\mathbf{x}}_{2t} = \frac{\|\hat{\mathbf{x}}_{2t}\|}{\|\hat{\mathbf{x}}_{1t}\| + \|\hat{\mathbf{x}}_{2t}\|} . * \mathbf{y}_t. \quad (5.23)$$

In this way, the reconstructed sources are more meaningful. The DNN optimization is still done with Equation 5.21 by replacing $\hat{\mathbf{x}}_{1t}$ and $\hat{\mathbf{x}}_{2t}$ with $\tilde{\mathbf{x}}_{1t}$ and $\tilde{\mathbf{x}}_{2t}$. This method

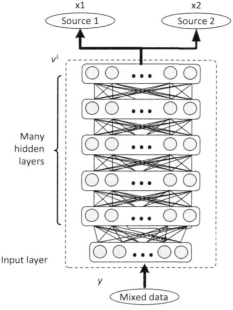

FIGURE 5.2

Speech separation with DNN.

should also be applicable to noise removal if we consider $\mathbf{x}_1$ and $\mathbf{x}_2$ are the clean speech and noise features, respectively.

Similar to the learning with SPLICE and empirical cepstral compensation, the supervised learning using DNN also needs stereo data which is hard to obtain in most real world scenarios. One solution is proposed in Du et al. (2014a), where pseudo clean data is generated with HMM-based synthesis. When the test noise is also available during the learning of the enhancement function $\mathcal{G}$, high performance can always be obtained by using DNN for noise removal. However, the challenge is the generalization to unseen conditions. This problem can be significantly alleviated by training the noise removal function on more acoustic conditions (Wang and Wang, 2013). In Xu et al. (2014), a set of more than 100 noise types is added when training the noise removal function in order to enrich the generalization of the DNN to unseen and non-stationary noise conditions. Although achieving satisfactory results in the area of speech separation, this noise enrichment method still degrades the recognition performance in the unseen test sets when the underlying acoustic model is a DNN (Du et al., 2014b).

5.2 LEARNING FROM MULTI-ENVIRONMENT DATA

This type of methods utilizes prior knowledge about the distortion by collecting and learning a set of models first, each corresponding to one specified environment in the training. These environment-specific models are then online combined to form a new model that fits the test environment best.

Usually, the acoustic model can be trained with a multi-condition training set to cover a wide range of application environments. However, there are two major problems with multi-style training. The first is that during training it is hard to enumerate all of the possible noise types and SNRs that may be present in future test environments. The second is that the distribution trained with multi-style training is too broad because it needs to model the data from all environments. Therefore, it is better to build environment-specific models, and use the model that best fits the test environment when doing runtime evaluation.

5.2.1 ONLINE MODEL COMBINATION

The model combination methods build a set of acoustic models, each modeling one specific environment. During testing, all the models are combined to construct a target model used to recognize the current test utterance. Denote the set of environment-dependent parameters as $\{\Lambda_1, \ldots, \Lambda_K\}$, where K is the total number of environments. Then the model parameters during testing can be obtained as

$$\hat{\Lambda} = \sum_{k=1}^{K} w_k \Lambda_k, \tag{5.24}$$

where w_k is the combination weight for the kth environment model. The model parameters can be Gaussian mean vectors or transforms when the underlying acoustic model is a GMM, and they can be weight matrices in the DNN case.

Online model combination for GMM

Assume that K environment-specific models share the same covariance matrix and only differ in mean parameters of GMMs. The mean parameters for each environment-specific model are concatenated together to form mean super-vectors ($\mathbf{s}_k, k = 1 \ldots K$), and the mean super-vector of the test utterance $\hat{\mathbf{s}}$ is obtained as a linear combination of K mean super-vectors of the environment-specific models

$$\hat{\mathbf{s}} = \sum_{k=1}^{K} w_k \mathbf{s}_k, \tag{5.25}$$

where w_k is the combination weight for the kth mean super-vector, and $\mathbf{w} = [w_1, w_2, \ldots, w_K]^T$. The combination weights $\mathbf{w}$ can be obtained with the maximum likelihood estimation (MLE) criterion as

$$\hat{\mathbf{w}} = \arg\max_{\mathbf{w}} \log p(\mathbf{Y}|\hat{\mathbf{s}}) \tag{5.26}$$

This is solved with the expectation-maximization (EM) algorithm which finds the solution of $\mathbf{w}$ iteratively. The auxiliary function is defined as the following by ignoring standard constants and terms independent of $\mathbf{w}$

$$Q(\mathbf{w}; \mathbf{w}_0) = -\frac{1}{2} \sum_{m,t} \gamma_t(m)(\mathbf{y}_t - \boldsymbol{\mu}(m))^T \boldsymbol{\Sigma}^{-1}(m)(\mathbf{y}_t - \boldsymbol{\mu}(m)), \qquad (5.27)$$

where $\mathbf{w}_0$ is the previous weight estimate, $\gamma_t(m)$ is the posterior of Gaussian component m at time t determined using the previous model parameters, and $\mathbf{y}_t$ is the feature vector of frame t. $\boldsymbol{\mu}(m)$ is the adapted mean of Gaussian component m, represented as

$$\boldsymbol{\mu}(m) = \sum_{k=1}^{K} w_k \mathbf{s}_k(m) = \mathbf{S}(m)\mathbf{w}, \qquad (5.28)$$

where $\mathbf{s}_k(m)$ is the subvector for Gaussian component m in super-vector $\mathbf{s}_k$ and $\mathbf{S}(m) = [\mathbf{s}_1(m), \ldots, \mathbf{s}_K(m)]$. $\boldsymbol{\Sigma}(m)$ is the variance of the Gaussian component m, shared by all the environment-specific models. By maximizing the auxiliary function, the combination weight $\mathbf{w}$ can be solved as

$$\mathbf{w} = \left[\sum_{m,t} \gamma_t(m) \mathbf{S}^T(m) \boldsymbol{\Sigma}^{-1}(m) \mathbf{S}(m) \right]^{-1} \sum_{m,t} \gamma_t(m) \mathbf{S}^T(m) \boldsymbol{\Sigma}^{-1}(m) \mathbf{y}_t. \qquad (5.29)$$

This model combination method is very similar to general speaker adaptation methods such as cluster adaptive training (CAT) (Gales, 2000) and eigenvoice (Kuhn et al., 2000). In the CAT approach, the speakers are clustered together and $\mathbf{s}_k$ stands for clusters instead of individual speakers. In the eigenvoice approach, a small number of eigenvectors are extracted from all the super-vectors and are used as $\mathbf{s}_k$. These eigenvectors are orthogonal to each other and guaranteed to represent the most important information. Although originally developed for speaker adaptation, both CAT and eigenvoice methods can be used for robust speech recognition. Storing K super-vectors in memory during online model combination may be too demanding. One way to reduce the cost is to use methods such as eigenMLLR (Chen et al., 2000; Wang et al., 2001) and transform-based CAT (Gales, 2000) by adapting the mean vector with environment dependent transforms. In this way, only K transforms are stored in memory. Moreover, adaptive training can be used to find the canonical mean as in CAT (Gales, 2000).

One potential problem of MLE model combination is that usually all combination weights are nonzero, that is, every environment-dependent model contributes to the final model. This is obviously not optimal if the test environment is exactly the same as one of the training environments. There is also a scenario where the test environment can be approximated well by interpolating only few training environments. Including unrelated models into the construction brings unnecessary distortion to the target model. In ensemble speaker and speaking environment modeling (ESSEM) (Tsao and Lee, 2007, 2009; Tsao et al., 2009), environment clustering is first used to cluster environments into several groups, each of which consists of environments having similar acoustic properties. During online model combination, an online cluster selection is first used to locate the most relevant cluster and then only the super-vectors in this selected cluster contribute to the model combination in Equation 5.25. In this way, most weights of the super-vectors are set to 0 and the method is shown to have better accuracy than simply combining all the super-vectors. By suitably incorporating prior knowledge, ESSEM can estimate combination weights accurately with a limited amount of adaptation data and has been shown to achieve very high accuracy on the standard Aurora 2 task (Tsao et al., 2014).

Instead of first doing the online clustering as in ESSEM, weights can also be automatically set to 0 (Xiao et al., 2012) by using Lasso (least absolute shrinkage and selection operator) (Tibshirani, 1996) which imposes an L_1 regularization term in the weight estimation problem to shrink some weights to exactly zero. The auxiliary function in Equation 5.27 is modified with the L_1 regularization as

$$Q(\mathbf{w}; \mathbf{w}_0) = -\frac{1}{2} \sum_{m,t} \gamma_t(m)(\mathbf{y}_t - \boldsymbol{\mu}(m))^T \boldsymbol{\Sigma}^{-1}(m)(\mathbf{y}_t - \boldsymbol{\mu}(m)) - T\alpha \sum_{k=1}^{K} |w_k|, \quad (5.30)$$

where α is a tuning parameter that controls the weight of the L_1 constraint, T is the total number of frames in the current utterance, and $|w_k|$ denotes the absolute value of w_k. This can be solved iteratively using the method proposed in Li et al. (2011). In Xiao et al. (2012), it is shown that Lasso usually shrinks to zero the weights of those mean super-vectors not relevant to the test environment. By removing some irrelevant super-vectors, the obtained mean super-vectors are found to be more robust against noise distortions.

Note that the noisy speech feature variance changes with the introduction of noise, therefore simply adjusting the mean vector of the speech model cannot solve all of the problems. It is better to adjust the model variance as well. One way is to combine the pre-trained CMLLR matrices as in Cui et al. (2009b). However, this is not trivial, requiring numerical optimization methods, such as the gradient descent method or a Newton method (Cui et al., 2009b).

Online model combination for DNN

The realization of Equation 5.24 in a DNN is done in the weight matrix and bias level as shown in Figure 5.3 (Tan et al., 2015; Wu and Gales, 2015). Suppose in the lth layer, we have trained the weight matrix set $\{\mathbf{H}_1^l, \dots, \mathbf{H}_K^l\}$ and bias set $\{\mathbf{p}_1^l, \dots, \mathbf{p}_K^l\}$ for all the environments. Then at test time, the weight matrix $\hat{\mathbf{A}}^l$ and bias $\hat{\mathbf{b}}^l$ for the new environment can be obtained as a linear combination of the trained counter parts as

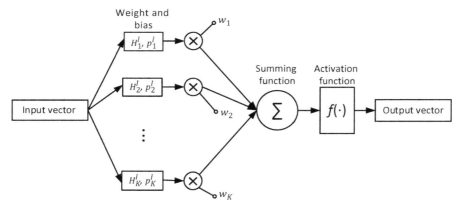

FIGURE 5.3

Linear model combination for DNN.

$$\hat{\mathbf{A}}^l = \sum_{k=1}^{K} w_k \mathbf{H}_k^l, \tag{5.31}$$

$$\hat{\mathbf{b}}^l = \sum_{k=1}^{K} w_k \mathbf{p}_k^l. \tag{5.32}$$

Because the number of environments observed during training is usually much less than the number of parameters in a DNN, the combination weight $\mathbf{w} = [w_1, w_2, \ldots, w_K]^T$ can be easily estimated online with only a few utterances with the standard error back propagation training.

5.2.2 NON-NEGATIVE MATRIX FACTORIZATION

In Section 5.2.1, the acoustic model for the current test utterance is obtained by combining the pre-learned acoustic models. Recently, there is increasing interest to use exemplar-based methods for general ASR (Demuynck et al., 2011; Sainath et al., 2011) and noise-robust ASR (Gemmeke and Virtanen, 2010; Gemmeke et al., 2011; Raj et al., 2010). Exemplar refers to an example speech segment from the training corpus. In exemplar-based noise-robust ASR (Gemmeke and Virtanen, 2010; Gemmeke et al., 2011; Raj et al., 2010), noisy speech is modeled by a linear combination of speech and noise (Gemmeke and Virtanen, 2010; Gemmeke et al., 2011) (or other interfering factors, such as music (Raj et al., 2010)) exemplars. If the reconstructed speech consists of only the exemplars of clean speech, the impact of noise is removed. This is a source separation approach, and non-negative matrix factorization (NMF) (Lee and Seung, 2000) has been shown to be a very successful method (Schmidt and Olsson, 2007; Smaragdis and Brown, 2003; Virtanen, 2007), and can directly benefit noise-robust ASR (Gemmeke and Virtanen, 2010; Gemmeke et al., 2011; Mohammadiha et al., 2013; Raj et al., 2010). An advantage of the exemplar-based approach is that it can deal with highly non-stationary noise, such as speech recognition in the presence of background music. The source separation process with NMF is described below.

First the training corpus is used to create a dictionary $x_l (1 \leq l \leq L)$ of clean speech exemplars and a matrix $\mathbf{X}$ is formed as $\mathbf{X} = [\mathbf{x}_1 \mathbf{x}_2 \ldots \mathbf{x}_L]$. The exemplars are drawn randomly from a collection of magnitude spectral vectors in a training set. Similarly, the noise matrix $\mathbf{N}$ is formed with noise exemplars. Then speech and noise exemplars are concatenated together to form a single matrix $\mathbf{A} = [\mathbf{XN}]$, with a total of K exemplars. The exemplars of $\mathbf{A}$ are denoted $\mathbf{a}_k, 1 \leq k \leq K$. The reconstruction feature is

$$\hat{\mathbf{y}} = \sum_{k=1}^{K} w_k \mathbf{a}_k = \mathbf{Aw}, \quad \text{s.t.} \quad w_k \geq 0 \tag{5.33}$$

with $\mathbf{w}$ as the K-dimensional activation vector. All exemplars and activation weights are required to be non-negative. The objective is to minimize the reconstruction error $d(\mathbf{y}, \mathbf{Aw})$ between the observation $\mathbf{y}$ and the reconstruction feature $\hat{\mathbf{y}}$ while constraining the matrices to be element-wise non-negative. It is also good to embed sparsity into the objective function so that the noisy speech can be represented as a combination of a small set of exemplars, similar to the concept of online GMM model combination with Lasso regularization in Section 5.2.1. This is done by penalizing the nonzero entries of $\mathbf{w}$ with the L_1 norm of the activation vector $\mathbf{w}$, weighted by element-wise multiplication (operation .*) of a non-negative vector $\boldsymbol{\lambda}$. Therefore the objective function is

$$d(\mathbf{y}, \mathbf{Aw}) + \|\boldsymbol{\lambda} . * \mathbf{w}\|_1 \quad \text{s.t.} \quad w_k \geq 0 \tag{5.34}$$

If all the elements of $\boldsymbol{\lambda}$ are zero, there is no enforced sparsity (Raj et al., 2010). Otherwise, sparsity is enforced (Gemmeke and Virtanen, 2010; Gemmeke et al., 2011). In Lee and Seung (2000), two measures are used for the reconstruction error $d(\mathbf{y}, \hat{\mathbf{y}})$, namely Euclidean distance and divergence. In most speech-related work (Gemmeke and Virtanen, 2010; Gemmeke et al., 2011; Raj et al., 2010), Kullback-Leibler (KL) divergence is used to measure the reconstruction error.

$$d(\mathbf{y}, \hat{\mathbf{y}}) = \sum_{e=1}^{E} y_e \log \left(\frac{y_e}{\hat{y}_e} \right) - y_e + \hat{y}_e, \tag{5.35}$$

where E is the vector dimension.

To solve Equation 5.34, the entries of the vector $\mathbf{w}$ are initialized to unity. Then Equation 5.34 can be minimized by iteratively applying the multiplicative update rule (Gemmeke et al., 2011)

$$\mathbf{w} \leftarrow \mathbf{w} . * (\mathbf{A}(\mathbf{y}./(\mathbf{Aw}))) ./ (\mathbf{A1} + \boldsymbol{\lambda}) \tag{5.36}$$

with .∗ and ./ denoting element-wise multiplication and division, respectively. **1** is a vector with all elements set to 1.

After getting **w**, the clean speech feature can be reconstructed by simply combining all the speech exemplars with nonzero weights (Schmidt and Olsson, 2007). Good recognition performance has been observed particularly at very low SNR (below 0 dB). Better results are reported using the following filtering (Gemmeke and Van hamme, 2012; Gemmeke et al., 2011; Raj et al., 2010) as

$$\hat{\mathbf{x}} = \mathbf{y}. * \mathbf{A}^x \mathbf{w}^x . / (\mathbf{A}^x \mathbf{w}^x + \mathbf{A}^n \mathbf{w}^n), \qquad (5.37)$$

where $\mathbf{A}^x$ and $\mathbf{w}^x$ denote the exemplars and activation vector for clean speech, respectively, and $\mathbf{A}^n$ and $\mathbf{w}^n$ denote the exemplars and activation vector for noise, respectively. This procedure can be viewed as filtering the noisy speech spectrum with a time-varying filter defined by $\mathbf{A}^x \mathbf{w}^x . / (\mathbf{A}^x \mathbf{w}^x + \mathbf{A}^n \mathbf{w}^n)$, similar to Wiener filtering in Equation 4.28. This is referred as feature enhancement (FE) in Gemmeke et al. (2011) and Gemmeke and Van hamme (2012).

Instead of cleaning the noisy speech magnitude spectrum, a sparse classification (SC) method is proposed in Gemmeke and Virtanen (2010) to directly use the activation weights to estimate the state or word likelihood. Since each frame of each speech exemplar in the speech dictionary has state or word labels obtained from the alignment with conventional HMMs, the weights of the exemplars in the sparse representation $\mathbf{w}^x$ can be used to calculate the state or word likelihood. Then, these activation-based likelihoods are used in a Viterbi search to obtain the state sequence with the maximum likelihood criterion.

Although the root methodology of FE and SC are the same, that is, NMF source separation, it is shown in Weninger et al. (2012) and Gemmeke and Van hamme (2012) that they are complementary. If combined together, more gain can be achieved. There are also variations of standard NMF source separation. For example, a sliding time window approach (Gemmeke et al., 2009) that allows the exemplars to span multiple frames is used for decoding utterances of arbitrary length. Convolutive extension of NMF is proposed to handle potential dependencies across successive input columns (Smaragdis, 2007; Weninger et al., 2012). Prior knowledge of the co-occurrence statistics of the basis functions for each source can also be employed to improve the performance of NMF (Wilson et al., 2008). In Grais and Erdogan (2013), by minimizing cross-coherence between the dictionaries of all sources in the mixed signal, the bases set of one source dictionary can be prevented from representing the other source signals. This clearly gives better separation results than the traditional NMF. Superior digit recognition accuracy has been reported in Gemmeke and Van hamme (2012) with the exemplar-based method by increasing the number of update iterations and exemplars, designing artificial noise dictionary, doing noise sniffing, and combining SC with FE.

Although the objective of NMF is to accurately recover clean features from noisy features, most NMF approaches have not directly optimized this objective. This problem is addressed in Weninger et al. (2014c), where discriminative training of the NMF bases are performed so that, given the weight coefficients obtained on a noisy feature, the desired clean feature is optimally recovered. This is done by minimizing the distance between the recovered and reference clean feature. However, this objective becomes a bi-level optimization problem because the recovered clean feature also depends on the bases. Therefore, this involves very complicated iterative inference. In Hershey et al. (2014) a concept of deep unfolding is proposed to address this issue by unfolding the inference iterations as layers in a DNN. Rather than optimizing the original model, the method unties the model parameters across layers to create a more powerful DNN. Then this DNN can be optimized with the back-propagation algorithm. This deep unfolding method gives superior performance to discriminative NMF than the solution in Weninger et al. (2014c).

There are still plenty of challenges, for example, how to deal with convolutive channel distortions (Gemmeke et al., 2013), how to most effectively deal with noise types in testing that have not been previously seen in the development of the noise dictionary (Gemmeke and Van hamme, 2012), and how to generalize to LVCSR tasks although there are recent improvements on Aurora 4 tasks (Geiger et al., 2014). Finally, although it is challenging to a noise-robust front-end to improve over the performance of a DNN back-end fed with raw features, it is reported in (Geiger et al., 2014) that NMF enhancement improves the recognition accuracy substantially when the training data is clean, and it still brings improvement even with multi-condition training data.

5.2.3 VARIABLE-PARAMETER MODELING

We have seen two broad classes of variables that affect the observation of speech signals: discrete (e.g., speaker and speaker classes, types of noises) and continuous (e.g., SNR, speaking rate, distance to the microphone). The variability of speech signals as a function of continuous variables can be explicitly modeled in the acoustic models. The concept of variable-parameter modeling is that speech model parameters in a specific test environment can be obtained as a function of environment variables. There are three advantages with this modeling techniques.

- When the test environment is unseen during training, the model parameters can still be extrapolated very well with the learned function. Therefore, this method generalizes very well to unseen test environments.
- With the introduction of a continuous parameterization of the model, any data sample contributes to the training of all the model parameters of the variable-parameter model. This improves training data effectiveness compared to multi-condition training where there is no design to leverage data across conditions.
- Another advantage is that the model is sharper than the model trained with standard multi-style training because it can fit the underlying individual test environments better by adjusting its parameters according to the test environment variable. This concept is first proposed to dynamically adjust GMM

parameters (Cui and Gong, 2003, 2006, 2007), and then extended to DNN modeling (Zhao et al., 2014a,b).

Variable-parameter modeling for GMM

As shown in Cui and Gong (2007), the mean and variance of the Gaussian distribution of the observed speech acoustic feature are functions of SNR. Pooling such distributions together and training SNR-independent models, as multi-style training does, inevitably yields relatively flat distributions. Apparently, the standard GMM-HMM which employs a constant set of model parameters to describe the acoustics under all different environments is imperfect and inadequate to deal with the phenomena.

To improve the modeling accuracy and performance, it is better to make the parameters of the acoustic model change according to the environment. This is the motivation of variable-parameter HMM (VPHMM) (Cui and Gong, 2003, 2006, 2007) which models the speech Gaussian mean and variance parameters as a set of polynomial functions of an environment variable u. A popular environment variable is SNR (Cui and Gong, 2003, 2006, 2007). Hence, the Gaussian component m is now modeled as $\mathcal{N}(\mathbf{y}; \boldsymbol{\mu}(m, u), \boldsymbol{\Sigma}(m, u))$. $\boldsymbol{\mu}(m, u)$ and $\boldsymbol{\Sigma}(m, u)$ are polynomial functions of environment variable u. For example, $\boldsymbol{\mu}(m, u)$ can be denoted by

$$\boldsymbol{\mu}(m, u) = \sum_{j=0}^{J} \mathbf{c}_j(m) u^j, \tag{5.38}$$

where $\mathbf{c}_j(m)$ is a vector with the same dimension as the input feature vectors. The choice of polynomial function is based on its good approximation to continuous functions, its simple derivation operations, and the fact that the change of means and variances in terms of the environment is smooth and can be modeled by low order polynomials. Note that strictly speaking, variable parameter modeling using Equation 5.38 should be called variable-parameter Gaussian mixture model instead of VPHMM because only Gaussian parameters are modeled with polynomial functions, although transition probabilities can also be modeled using variable parameter techniques (e.g., speaking rate changes). Here we still use the term VPHMM to follow the literature in which it was proposed (Cui and Gong, 2003, 2006, 2007).

Other functions can also be used for a VPHMM. For example, in Yu et al. (2009), piecewise spline interpolation is used to represent the dependency of the HMM parameters on the environment parameters. To reduce the total number of parameters for a VPHMM, parameter clustering can be employed (Yu et al., 2008b). The VPHMM parameters can be trained either with the MLE criterion (Cui and Gong, 2007) or a discriminative criterion (Yu et al., 2008a). In addition to Gaussian mean and variance parameters, other model parameters can also be modeled. In Cheng et al. (2011) and Li et al. (2013), a more generalized form of VPHMM is investigated by modeling tied linear transforms as a function of environment variables. In addition to using the standard MFCC as the input feature for a GMM, Xie et al. (2014) shows the effectiveness of using bottle-neck features generated from a DNN as the features of a VPHMM.

During testing, the actual set of speech model parameters can be calculated by evaluating the parametric function with the estimated environment variable. Even if the estimated environment is not seen during training, the curve fitting optimization naturally uses the information on articulation/context from neighboring environments. Therefore, VPHMM can work well in unseen environment instances modeled by the environment variable.

Variable-component DNN

Usually multi-style data is used to train a DNN (Seltzer et al., 2013) and good accuracies can be obtained. However, as shown in Section 3.4, speech samples from different environments cannot be well aligned even with the DNN's high-level feature extraction. Therefore, if a single DNN is used to model the multi-style speech data, it is possible to end up with "flat" distributions. So for the test speech produced in a particular environment, such a "flat" model would not be the optimal matched model. Actually, a flat model does not represent any of the training environments. It is also difficult to collect training data to cover all possible types of environments, so the performance on unseen noisy environments remains unpredictable. Therefore, it is desirable that DNN components can be modeled as a function of a continuous environment-dependent variable. At the recognition time, a set of DNN components specific to the given value of the environment variable is instantiated and used for recognition. Even if the test environment is not seen in the training, the estimated DNN components can still work well because the change of DNN components in terms of the environment variable can be predicted. Variable-component DNN (VCDNN) (Zhao et al., 2014a,b) is proposed for this purpose.

In the VCDNN method, any component in the DNN can be modeled as a set of polynomial functions of an environment variable. To that end, four types of variation can be defined for VCDNN: variable-parameter DNN (VPDNN) in which the weight matrix and bias are variable dependent, variable-output DNN (VODNN) in which the output of each hidden layer is variable dependent, variable-activation DNN (VADNN) in which the activation function is variable dependent, and variable-input DNN (VIDNN) in which the input feature is variable dependent.

Figure 5.4 shows the flow chart of one layer of a VPDNN, in which the weight matrix $\mathbf{A}$ and bias $\mathbf{b}$ of layer l is modeled as a function of the environment variable u:

$$\mathbf{A}^l = \sum_{j=0}^{J} \mathbf{H}_j^l u^j \qquad 0 < l \leq L \tag{5.39}$$

$$\mathbf{b}^l = \sum_{j=0}^{J} \mathbf{p}_j^l u^j \qquad 0 < l \leq L \tag{5.40}$$

J is the polynomial function order. $\mathbf{H}_j^l$ is a matrix with the same dimensions as $\mathbf{A}^l$ and $\mathbf{p}_j^l$ is a vector with the same dimension as $\mathbf{b}^l$.

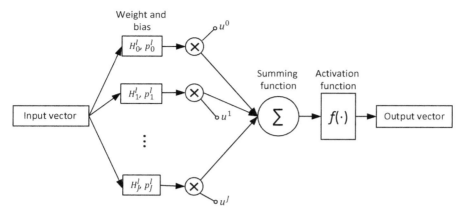

FIGURE 5.4

Variable-parameter DNN.

Then the relation between the input $\mathbf{v}^l$ and the output $\mathbf{v}^{l+1}$ of the lth layer at a VPDNN is

$$\mathbf{v}^{l+1} = \sigma(\mathbf{z}^l), \tag{5.41}$$

where

$$\mathbf{z}^l = \mathbf{A}^l \mathbf{v}^l + \mathbf{b}^l \tag{5.42}$$

and $\sigma(\cdot)$ is the sigmoid function.

Combining Equations 5.39 and 5.40 with the error back propagation algorithm introduced in Section 2.4, the update formulas for $\mathbf{H}_j^l$ and $\mathbf{p}_j^l$ can be obtained as:

$$\hat{\mathbf{H}}_j^l = \mathbf{H}_j^l + \alpha \mathbf{v}^l (\mathbf{e}^l)^T u^j, \tag{5.43}$$

$$\hat{\mathbf{p}}_j^l = \mathbf{p}_j^l + \alpha \mathbf{e}^l u^j, \tag{5.44}$$

where $\mathbf{v}^l$ is the input to the lth layer, α is the learning rate, and $\mathbf{e}^l$ is the error signal at the lth layer, defined in Equation 2.38.

In the recognition stage, the weight matrix $\mathbf{A}$ and bias $\mathbf{b}$ of each layer are instantiated according to Equations 5.39 and 5.40 with the estimated environment variable of the test data. Then the senone posterior can be calculated in the same way as in the standard DNN.

Comparing Equations 5.39 and 5.40 in VPDNN with Equations 5.31 and 5.32 in linear DNN model combination, we can see that they are very similar—both methods linearly combine a set of basis matrix and bias at test time. However, they also have several different aspects as shown in Table 5.1. Similar comparison can also be applied to VPHMM and linear GMM model combination.

Table 5.1 Difference Between VPDNN and Linear DNN Model Combination

	VPDNN	**Linear DNN Model Combination**
DNN weight matrix and bias	A learned polynomial function of environment variables	Linear combination of a set of weight matrix and bias trained from different environments
Combination coefficients	directly calculated with environment variables	Online estimated
Environment variables	can be continuous such as SNR	Discrete, each associated with a weight matrix and bias

In a VODNN, it is assumed that the output of each hidden layer could be described by a polynomial function of the environment variable u:

$$\mathbf{v}^{l+1} = \sum_{j=0}^{J} \sigma(\mathbf{z}_j^l) u^j \qquad 0 < l < L, \qquad (5.45)$$

where

$$\mathbf{z}_j^l = (\mathbf{H}_j^l)^T \mathbf{v}^l + \mathbf{p}_j^l. \qquad (5.46)$$

The framework of one layer in a VODNN is shown in Figure 5.5. Similarly, the update formulas can be obtained by combining Equations 5.45 and 5.46 with the error back propagation algorithm:

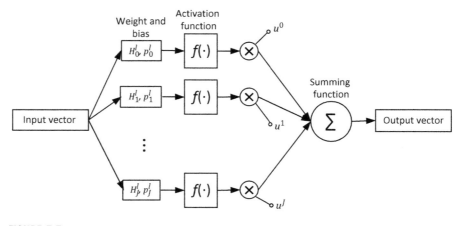

FIGURE 5.5

Variable-output DNN.

$$\hat{\mathbf{H}}_j^l = \mathbf{H}_j^l + \alpha \mathbf{v}^l (\mathbf{e}_j^l)^T u^j \tag{5.47}$$

$$\hat{\mathbf{p}}_j^l = \mathbf{p}_j^l + \alpha \mathbf{e}_j^l u^j. \tag{5.48}$$

The difference between the update formulas of VPDNN and VODNN parameters is that the order-independent error signal $\mathbf{e}^l$ is used in Equations 5.43 and 5.44 while the error signal $\mathbf{e}_j^l$ used in Equations 5.47 and 5.48 depends on the polynomial order j as

$$e_{i(j)}^l = \left[\sum_{n=0}^{J} \sum_{k=1}^{N_{l+1}} h_{ik(n)}^{l+1} e_{k(n)}^{l+1} \right] \sigma'(z_{ij}^l), \tag{5.49}$$

where $e_{i(j)}^l$ is the ith element of the error signal vector $\mathbf{e}_j^l$ at the lth layer, z_{ij}^l is the ith element of $\mathbf{z}_j^l$, and $h_{ik(n)}^{l+1}$ is the element of matrix $\mathbf{H}_n^{l+1}$ in the ith row and kth column at the layer $l+1$. $\sigma'(\cdot)$ is the derivative of the sigmoid function.

In a VADNN, the activation function of hidden layers has environment-variable-dependent parameters as

$$\mathbf{v}^{l+1} = \sigma \left(\mathbf{a}^l . * \mathbf{z}^l + \mathbf{m}^l \right), \tag{5.50}$$

where $\mathbf{z}^l$ is defined in Equation 3.21 and $.*$ means the element-wise product. $\mathbf{a}^l$ and $\mathbf{m}^l$ are defined as the polynomial functions of the environment variable u

$$\mathbf{a}^l = \sum_{j=0}^{J} \mathbf{h}_j^l u^j \qquad 0 < l < L, \tag{5.51}$$

$$\mathbf{m}^l = \sum_{j=0}^{J} \mathbf{p}_j^l u^j \qquad 0 < l < L. \tag{5.52}$$

Figure 5.6 shows one layer of a VADNN. The additional variable-dependent parameters $\mathbf{h}_j^l$ and $\mathbf{p}_j^l$ in a VADNN for each hidden layer are vectors with dimension N_l, which is the number of nodes of the lth layer. Hence its number of parameters is much smaller than that in a VPDNN or a VODNN. In the training of a VADNN, $\mathbf{h}_j^l$ and $\mathbf{p}_j^l$ as well as the DNN parameters $\mathbf{A}^l$ and $\mathbf{b}^l$ need to be updated with the error back propagation algorithm as

$$\hat{\mathbf{A}}^l = \mathbf{A}^l + \alpha \mathbf{v}^l (\mathbf{e}^l . * \mathbf{a}^l)^T, \tag{5.53}$$

$$\hat{\mathbf{b}}^l = \mathbf{b}^l + \alpha (\mathbf{e}^l . * \mathbf{a}^l), \tag{5.54}$$

$$\hat{\mathbf{h}}_j^l = \mathbf{h}_j^l + \alpha (\mathbf{e}^l . * \mathbf{z}^l) u^j, \tag{5.55}$$

$$\hat{\mathbf{p}}_j^l = \mathbf{p}_j^l + \alpha \mathbf{e}^l u^j. \tag{5.56}$$

Finally, the simplest DNN structure to use environment variables is VIDNN, which concatenates environment variables with the original input feature. Even with

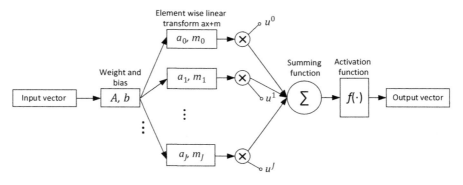

FIGURE 5.6

Variable-activation DNN.

the first-order polynomial, a VPDNN or a VODNN doubles the number of parameters from the standard DNN. If a large amount of training data is available, these two models may give better accuracy. In contrast, a VADNN or a VIDNN only increases negligibly the number of parameters, but still achieves satisfactory robustness.

The advantage of VCDNNs is shown in Zhao et al. (2014b) where VCDNNs achieved better relative WER reduction from the standard DNN under unseen SNR conditions than under the seen SNR conditions. This indicates that a standard DNN has a strong power to model the various environments it has observed, but for the unseen environments, there is more room for improvement from the standard DNN. With the polynomial function, VCDNNs can very well predict the DNN components used for unseen condition by extrapolation. Therefore, VCDNNs can generalize very well to unseen environments.

5.3 SUMMARY

To provide a better view of the development trend of robust methods from the GMM era to the DNN era, we summarize the representative methods described in this chapter for robust ASR exploiting prior knowledge originally proposed for GMM and DNN in Tables 5.2 and 5.3, respectively, in a chronological order. Further comments and summary of these methods, are made below:

- If stereo data is available, a mapping from noisy feature to clean feature can be learned. Empirical cepstral compensation is widely used to address all kinds of factors (SNR, VQ cluster, phoneme identity, etc.) with a bias, and it is improved by SPLICE which uses piecewise linear compensation. With the layer-by-layer nonlinear modeling power of a DNN, a much better learning of the noisy-to-clean feature mapping can be obtained. This is further improved by introducing

Table 5.2 Compensation with Prior Knowledge Methods Originally Proposed for GMMs in Chapter 5, Arranged Chronologically

Method	Proposed Around	Characteristics
Empirical cepstral compensation (Acero, 1993; Acero and Stern, 1990; Droppo et al., 2001a; Liu et al., 1994; Stern et al., 1996)	1990	Calculates all kinds of factor-dependent (including SNR, VQ cluster, phoneme identity, etc.) bias using stereo training data, and removes that bias during testing
Online GMM model combination (Gales, 2000; Kuhn et al., 2000)	2000	Online combines a set of environment-specific GMM models, representative methods are eigenvoice and cluster adaptive training
Stereo piecewise linear compensation for environment (SPLICE) (Deng et al., 2000)	2000	The additive correction vector is piecewise linear between the noisy speech observation and the clean speech of the stereo training data
Variable-parameter HMM (VPHMM) (Cui and Gong, 2003)	2003	Models the Gaussian mean and variance parameters as a set of polynomial functions of the environment variable
Ensemble speaker and speaking environment modeling (ESSEM) (Tsao and Lee, 2007; Tsao et al., 2009)	2007	To remove unrelated models into construction, an online cluster selection is first used to locate the most relevant cluster and then only the super-vectors in this selected cluster contribute to the model combination
Exemplar-based reconstruction with non-negative matrix factorization (NMF) (Gemmeke and Virtanen, 2010; Raj et al., 2010)	2010	NMF is used to reconstruct speech with only clean speech exemplars extracted from the training dictionary
Lasso model combination (Xiao et al., 2012)	2012	Imposes an L_1 regularization term in the weight estimation problem of online model combination to shrink some weights to exactly zero
Discriminative NMF (Weninger et al., 2014c)	2014	Discriminative training of the NMF bases is performed so that the desired clean feature is optimally recovered given the weight coefficients obtained on a noisy feature

Table 5.3 Compensation with Prior Knowledge Methods Originally Proposed for DNNs in Chapter 5, Arranged Chronologically

Method	Proposed Around	Characteristics
RNN for noise removal (Maas et al., 2012; Weninger et al., 2014a,b; Wöllmer et al., 2013)	2012	Uses a RNN which better models temporal sequence to learn the mapping from noisy feature to clean feature, and it is extended with advanced structure such as LSTM and BLSTM
DNN for noise removal (Du et al., 2014a; Feng et al., 2014; Lu et al., 2013; Narayanan and Wang, 2014a)	2013	Uses a DNN to learn the mapping from noisy feature to clean feature
online DNN model combination (Tan et al., 2015; Wu and Gales, 2015)	2015	Online combines a set of environment-specific DNN models
variable-component DNN (Zhao et al., 2014a,b)	2014	Any component in the DNN can be modeled as a set of polynomial functions of an environment variable so that better modeling of test environments can be achieved.
Deep unfolding (Hershey et al., 2014)	2014	Solves the complicated bi-level optimization problem in discriminative NMF by unfolding the inference iterations as layers in a DNN

the recurrent structure and (B)LSTM units which better model the temporal sequence of speech signals.

- Online model combination is one way to fast adapt acoustic models to environments with limited adaptation data because only combination coefficients need to be computed online. For GMM models, eigenvoice and cluster adaptive training are representative methods. The combination coefficients can be made sparse with either clustering or L_1 regularization. Similar idea can be easily extended to DNN by online combining weight matrices.
- VPHMM is another way to online construction of an adapted GMM model with a set of polynomial functions of the environment variable. It is extended to VCDNN in which any component in the DNN (parameter in VPDNN, output in VODNN, activation in VADNN, and input in VIDNN, respectively) can be modeled as a set of polynomial functions of an environment variable. With the polynomial functions, the model can be instantiated even in the unseen case by extrapolation, thus enjoying good generalization property.
- NMF is used to reconstruct the clean speech spectrum from the noisy speech spectrum using pre-constructed clean speech and noise exemplars. While no

stereo data is required, examples of the corrupting noise are nevertheless required to form the noise dictionary. There are plenty of extensions of NMF, including discriminative NMF in which discriminative training of the NMF bases is performed so that the desired clean feature is optimally recovered given the weight coefficients obtained on a noisy feature.

- Last, deep unfolding was proposed to solve the complicated bi-level optimization problem in discriminative NMF. It builds a bridge between DNN modeling and model-based approaches. As will be described in detail in Chapter 6, model-based approaches are very powerful because of the use of explicit distortion models between the clean and distorted speech. However, the inference is sometimes very complicated and may rely on the underlying Gaussian model assumption. On the other hand, it is straightforward to optimize parameters in DNN modeling with the back propagation algorithm. A well-known disadvantage of the DNN is that it is closer to mechanisms than problem-level formulation, and is usually considered as a "black-box". Deep unfolding may be a potential framework that allows model-based approaches to guide the exploration of the space of DNNs, which is important but missing in current literature.

REFERENCES

Acero, A., 1993. Acoustical and Environmental Robustness in Automatic Speech Recognition. Cambridge University Press, Cambridge, UK.

Acero, A., Stern, R., 1990. Environmental robustness in automatic speech recognition. In: Proc. International Conference on Acoustics, Speech and Signal Processing (ICASSP), vol. 2, pp. 849-852.

Afify, M., Cui, X., Gao, Y., 2007. Stereo-based stochastic mapping for robust speech recognition. In: Proc. International Conference on Acoustics, Speech and Signal Processing (ICASSP), vol. IV, pp. 377-380.

Afify, M., Cui, X., Gao, Y., 2009. Stereo-based stochastic mapping for robust speech recognition. IEEE Trans. Audio Speech Lang. Process. 17 (7), 1325-1334.

Bahl, L.R., Brown, P.F., Souza, P.V.D., Mercer, R.L., 1997. Maximum mutual information estimation of hidden Markov model parameters for speech recognition. In: Proc. International Conference on Acoustics, Speech and Signal Processing (ICASSP), vol. 11, pp. 49-52.

Chen, K.T., Liau, W.W., Wang, H.M., Lee, L.S., 2000. Fast speaker adaptation using eigenspace-based maximum likelihood linear regression. In: Proc. International Conference on Spoken Language Processing (ICSLP), vol. 3, pp. 742-745.

Cheng, N., Liu, X., Wang, L., 2011. Generalized variable parameter HMMs for noise robust speech recognition. In: Proc. Interspeech, pp. 481-484.

Cui, X., Afify, M., Gao, Y., 2008a. MMSE-based stereo feature stochastic mapping for noise robust speech recognition. In: Proc. International Conference on Acoustics, Speech and Signal Processing (ICASSP), pp. 4077-4080.

Cui, X., Afify, M., Gao, Y., 2008b. N-best based stochastic mapping on stereo HMM for noise robust speech recognition. In: Proc. Interspeech, pp. 1261-1264.

Cui, X., Afify, M., Gao, Y., 2009a. Stereo-based stochastic mapping with discriminative training for noise robust speech recognition. In: Proc. International Conference on Acoustics, Speech and Signal Processing (ICASSP), pp. 3933-3936.

Cui, X., Gong, Y., 2003. Variable parameter Gaussian mixture hidden Markov modeling for speech recognition. In: Proc. International Conference on Acoustics, Speech and Signal Processing (ICASSP), vol. I, pp. 12-15.

Cui, X., Gong, Y., 2006. Modeling variance variation in a variable parameter HMM framework for noise robust speech recognition. In: Proc. International Conference on Acoustics, Speech and Signal Processing (ICASSP), vol. I.

Cui, X., Gong, Y., 2007. A study of variable-parameter Gaussian mixture hidden Markov modeling for noisy speech recognition. IEEE Trans. Audio Speech Lang. Process. 15 (4), 1366-1376.

Cui, X., Xue, J., Zhou, B., 2009b. Improving online incremental speaker adaptation with eigen feature space MLLR. In: Proc. IEEE Workshop on Automatic Speech Recognition and Understanding (ASRU), pp. 136-140.

Demuynck, K., Seppi, D., Compernolle, D.V., Nguyen, P., Zweig, G., 2011. Integrating meta-information into exemplar-based speech recognition with segmental conditional random fields. In: Proc. International Conference on Acoustics, Speech and Signal Processing (ICASSP), pp. 5048-5051.

Deng, L., Acero, A., Jiang, L., Droppo, J., Huang, X.D., 2001. High-performance robust speech recognition using stereo training data. In: Proc. International Conference on Acoustics, Speech and Signal Processing (ICASSP), pp. 301-304.

Deng, L., Acero, A., Plumpe, M., Huang, X., 2000. Large vocabulary speech recognition under adverse acoustic environment. In: Proc. International Conference on Spoken Language Processing (ICSLP), vol. 3, pp. 806-809.

Deng, L., Droppo, J., Acero, A., 2003. Recursive estimation of nonstationary noise using iterative stochastic approximation for robust speech recognition. IEEE Trans. Audio Speech Lang. Process. 11 (6), 568-580.

Deng, L., Wu, J., Droppo, J., Acero, A., 2005. Analysis and comparison of two speech feature extraction/compensation algorithms. IEEE Signal Process. Lett. 12 (6), 477-480.

Droppo, J., Acero, A., 2005. Maximum mutual information SPLICE transform for seen and unseen conditions. In: Proc. Interspeech, pp. 989-992.

Droppo, J., Acero, A., Deng, L., 2001a. Efficient on-line acoustic environment estimation for FCDCN in a continuous speech recognition system. In: Proc. International Conference on Acoustics, Speech and Signal Processing (ICASSP), vol. 1, pp. 209-212.

Droppo, J., Deng, L., Acero, A., 2001b. Evaluation of the SPLICE algorithm on the Aurora2 database. In: Proc. European Conference on Speech Communication and Technology (EUROSPEECH), pp. 217-220.

Droppo, J., Deng, L., Acero, A., 2002. Uncertainty decoding with SPLICE for noise robust speech recognition. In: Proc. International Conference on Acoustics, Speech and Signal Processing (ICASSP), vol. 1, pp. 57-60.

Du, J., Dai, L., Huo, Q., 2014a. Synthesized stereo mapping via deep neural networks for noisy speech recognition. In: Proc. International Conference on Acoustics, Speech and Signal Processing (ICASSP), pp. 1764-1768.

Du, J., Hu, Y., Dai, L.R., Wang, R.H., 2010. HMM-based pseudo-clean speech synthesis for splice algorithm. In: Proc. International Conference on Acoustics, Speech and Signal Processing (ICASSP), pp. 4570-4573.

Du, J., Wang, Q., Gao, T., Xu, Y., Dai, L., Lee, C.H., 2014b. Robust speech recognition with speech enhanced deep neural networks. In: Proc. Interspeech.

ETSI, 2002. Speech processing, transmission and quality aspects (STQ); distributed speech recognition; advanced front-end feature extraction algorithm; compression algorithms. In: ETSI.

Feng, X., Zhang, Y., Glass, J., 2014. Speech feature denoising and dereverberation via deep autoencoders for noisy reverberant speech recognition. In: Proc. International Conference on Acoustics, Speech and Signal Processing (ICASSP).

Gales, M.J.F., 2000. Cluster adaptive training of hidden Markov models. IEEE Trans. Speech Audio Process. 8 (4), 417-428.

Gao, T., Du, J., Dai, L.R., Lee, C.H., 2015. Joint training of front-end and back-end deep neural networks for robust speech recognition. In: Proc. International Conference on Acoustics, Speech and Signal Processing (ICASSP).

Geiger, J., Gemmeke, J., Schuller, B., Rigoll, G., 2014. Investigating NMF speech enhancement for neural network based acoustic models. Proc. Interspeech.

Gemmeke, J.F., ten Bosch, L., Boves, L., Cranen, B., 2009. Using sparse representations for exemplar based continuous digit recognition. In: Proc. EUSIPCO., pp. 1755-1759.

Gemmeke, J.F., Van hamme, H., 2012. Advances in noise robust digit recognition using hybrid exemplar-based techniques. In: Proc. Interspeech, pp. 2134-2137.

Gemmeke, J.F., Virtanen, T., 2010. Noise robust exemplar-based connected digit recognition. In: Proc. International Conference on Acoustics, Speech and Signal Processing (ICASSP), pp. 4546-4549.

Gemmeke, J.F., Virtanen, T., Demuynck, K., 2013. Exemplar-based joint channel and noise compensation. In: Proc. International Conference on Acoustics, Speech and Signal Processing (ICASSP), pp. 868-872.

Gemmeke, J.F., Virtanen, T., Hurmalainen, A., 2011. Exemplar-based sparse representations for noise robust automatic speech recognition. IEEE Trans. Audio Speech Lang. Process. 19 (7), 2067-2080.

Grais, E.M., Erdogan, H., 2013. Discriminative nonnegative dictionary learning using cross-coherence penalties for single channel source separation. In: Proc. Interspeech, pp. 808-812.

Hershey, J., Le Roux, J., Weninger, F., 2014. Deep unfolding: Model-based inspiration of novel deep architectures. arXiv preprint arXiv:1409.2574.

Huang, P.S., Kim, M., Hasegawa-Johnson, M., Smaragdis, P., 2014. Deep learning for monaural speech separation. In: Proc. International Conference on Acoustics, Speech and Signal Processing (ICASSP).

Kuhn, R., Junqua, J.C., Nguyen, P., Niedzielski, N., 2000. Rapid speaker adaptation in eigenvoice space. IEEE Trans. Speech Audio Process. 8 (6), 695-707.

Lee, D.D., Seung, H.S., 2000. Algorithms for non-negative matrix factorization. In: Proc. Neural Information Processing Systems (NIPS), pp. 556-562.

Li, J., Yuan, M., Lee, C.H., 2011. Lasso model adaptation for automatic speech recognition. In: ICML Workshop on Learning Architectures, Representations, and Optimization for Speech and Visual Information Processing.

Li, Y., Liu, X., Wang, L., 2013. Feature space generalized variable parameter HMMs for noise robust recognition. In: Proc. Interspeech, pp. 2968-2972.

Liu, B., Dai, L., Li, J., Wang, R.H., 2004. Double Gaussian based feature normalization for robust speech. In: International Symposium on Chinese Spoken Language Processing, pp. 705-708.

Liu, F.H., Stern, R.M., Acero, A., Moreno, P., 1994. Environment normalization for robust speech recognition using direct cepstral comparison. In: Proc. International Conference on Acoustics, Speech and Signal Processing (ICASSP), vol. I, pp. 61-64.

Lu, X., Tsao, Y., Matsuda, S., Hori, C., 2013. Speech enhancement based on deep denoising autoencoder. In: Proc. Interspeech, pp. 436-440.

Maas, A.L., Le, Q.V., O'Neil, T.M., Vinyals, O., Nguyen, P., Ng, A.Y., 2012. Recurrent neural networks for noise reduction in robust ASR. In: Proc. Interspeech, pp. 22-25.

Mohammadiha, N., Smaragdis, P., Leijon, A., 2013. Supervised and unsupervised speech enhancement using nonnegative matrix factorization. IEEE Trans. Audio Speech Lang. Process. 21 (10), 2140-2151.

Narayanan, A., Wang, D., 2013. Ideal ratio mask estimation using deep neural networks. In: Proc. International Conference on Acoustics, Speech and Signal Processing (ICASSP), pp. 7092-7096.

Narayanan, A., Wang, D., 2014a. Investigation of speech separation as a front-end for noise robust speech recognition. IEEE/ACM Trans. Audio Speech Lang. Process. 22 (4), 826-835.

Narayanan, A., Wang, D., 2014b. Joint noise adaptive training for robust automatic speech recognition. In: Proc. International Conference on Acoustics, Speech and Signal Processing (ICASSP).

Povey, D., Kingsbury, B., Mangu, L., Saon, G., Soltau, H., Zweig, G., 2005. fMPE: Discriminatively trained features for speech recognition. In: Proc. International Conference on Acoustics, Speech and Signal Processing (ICASSP), vol. 1, pp. 961-964.

Raj, B., Virtanen, T., Chaudhuri, S., Singh, R., 2010. Non-negative matrix factorization based compensation of music for automatic speech recognition. In: Proc. Interspeech, pp. 717-720.

Sainath, T.N., Ramabhadran, B., Picheny, M., Nahamoo, D., Kanevsky, D., 2011. Exemplar-based sparse representation features: from TIMIT to LVCSR. IEEE Trans. Audio Speech Lang. Process. 19 (8), 2598-2613.

Schmidt, M.N., Olsson, R.K., 2007. Linear regression on sparse features for single-channel speech separation. In: IEEE Workshop on Applications of Signal Processing to Audio and Acoustics, pp. 26-29.

Seltzer, M.L., Yu, D., Wang, Y., 2013. An investigation of deep neural networks for noise robust speech recognition. In: Proc. International Conference on Acoustics, Speech and Signal Processing (ICASSP), pp. 7398-7402.

Smaragdis, P., 2007. Convolutive speech bases and their application to supervised speech separation. IEEE Trans. Audio Speech Lang. Process. 15 (1), 1-12.

Smaragdis, P., Brown, J.C., 2003. Non-negative matrix factorization for polyphonic music transcription. In: IEEE Workshop on Applications of Signal Processing to Audio and Acoustics, pp. 177-180.

Stern, R., Acero, A., Liu, F.H., Ohshima, Y., 1996. Signal processing for robust speech recognition. In: Lee, C.H., Soong, F.K., Paliwal, K.K. (Eds.), Automatic Speech and Speaker recognition: Advanced Topics. Kluwer Academic Publishers, Boston, MA, pp. 357-384.

Tan, T., Qian, Y., Yin, M., Zhuang, Y., Yu, K., 2015. Cluster adaptive training for deep neural network. In: Proc. International Conference on Acoustics, Speech and Signal Processing (ICASSP).

Tibshirani, R., 1996. Regression shrinkage and selection via the lasso. J. Royal. Statist. Soc B. 58 (1), 267-288.

Tokuda, K., Yoshimura, T., Masuko, T., Kobayashi, T., Kitamura, T., 2000. Speech parameter generation algorithms for HMM-based speech synthesis. In: Proc. International Conference on Acoustics, Speech and Signal Processing (ICASSP), vol. 3, pp. 1315-1318.

Tsao, Y., Lee, C.H., 2007. An ensemble modeling approach to joint characterization of speaker and speaking environments. In: Proc. Interspeech, pp. 1050-1053.

Tsao, Y., Lee, C.H., 2009. An ensemble speaker and speaking environment modeling approach to robust speech recognition. IEEE Trans. Speech Audio Process. 17 (5), 1025-1037.

Tsao, Y., Li, J., Lee, C.H., 2009. Ensemble speaker and speaking environment modeling approach with advanced online estimation process. In: Proc. International Conference on Acoustics, Speech and Signal Processing (ICASSP), pp. 3833-3836.

Tsao, Y., Matsuda, S., Hori, C., Kashioka, H., Lee, C.H., 2014. A map-based online estimation approach to ensemble speaker and speaking environment. IEEE/ACM Trans. Audio Speech Lang. Process.

Tu, Y., Du, J., Dai, L.R., Lee, C.H., 2015. Speech separation based on signal-noise-dependent deep neural networks for robust speech recognition. In: Proc. International Conference on Acoustics, Speech and Signal Processing (ICASSP).

Virtanen, T., 2007. Monaural sound source separation by nonnegative matrix factorization with temporal continuity and sparseness criteria. IEEE Trans. Audio Speech Lang. Process. 15 (3), 1066-1074.

Wang, N.C., Lee, S.M., Seide, F., Lee, L.S., 2001. Rapid speaker adaptation using a priori knowledge by eigenspace analysis of MLLR parameters. In: Proc. International Conference on Acoustics, Speech and Signal Processing (ICASSP), vol. 1, pp. 345-348.

Wang, Y., Narayanan, A., Wang, D., 2014. On training targets for supervised speech separation. IEEE/ACM Trans. Audio Speech Lang. Process. 22 (12), 1849-1858.

Wang, Y., Wang, D., 2013. Towards scaling up classification-based speech separation. IEEE Trans. Audio Speech Lang. Process. 21 (7), 1381-1390.

Weng, C, Yu, D., Seltzer, M., Droppo, J., 2014. Single-channel mixed speech recognition using deep neural networks. In: Proc. IEEE Intl. Conf. on Acoustic, Speech and Signal Processing.

Weninger, F., Eyben, F., Schuller, B., 2014a. Single-channel speech separation with memory-enhanced recurrent neural networks. In: Proc. International Conference on Acoustics, Speech and Signal Processing (ICASSP).

Weninger, F., Hershey, J., Le Roux, J., Schuller, B., 2014b. Discriminatively trained recurrent neural networks for single-channel speech separation. In: Proc. IEEE GlobalSIP Symposium on Machine Learning Applications in Speech Processing.

Weninger, F., Le Roux, J., Hershey, J., Watanabe, S., 2014c. Discriminative NMF and its application to single-channel source separation. In: Proc. Interspeech.

Weninger, F., Wöllmer, M., Geiger, J., Schuller, B., Gemmeke, J., Hurmalainen, A., et al., 2012. Non-negative matrix factorization for highly noise-robust ASR: to enhance or to recognize? In: Proc. International Conference on Acoustics, Speech and Signal Processing (ICASSP), pp. 4681-4684.

Wilson, K.W., Raj, B., Smaragdis, P., Divakaran, A., 2008. speech denoising using nonnegative matrix factorization with priors. In: Proc. International Conference on Acoustics, Speech and Signal Processing (ICASSP), pp. 4029-4032.

Wöllmer, M., Zhang, Z., Weninger, F., Schuller, B., Rigoll, G., 2013. Feature enhancement by bidirectional lstm networks for conversational speech recognition in highly non-stationary noise. In: Proc. International Conference on Acoustics, Speech and Signal Processing (ICASSP), pp. 6822-6826.

Wu, C., Gales, M.J.F., 2015. Multi-basis adaptive neural network for rapid adaptation in speech recognition. In: Proc. International Conference on Acoustics, Speech and Signal Processing (ICASSP).

Xiao, X., Li, J., Chng, E.S., Li, H., 2012. Lasso environment model combination for robust speech recognition. In: Proc. International Conference on Acoustics, Speech and Signal Processing (ICASSP), pp. 4305-4308.

Xie, X., Su, R., Liu, X., Wang, L., 2014. Deep neural network bottleneck features for generalized variable parameter HMMs. In: Proc. Interspeech.

Xu, Y., Du, J., Dai, L.R., Lee, C.H., 2014. Dynamic noise aware training for speech enhancement based on deep neural networks. In: Proc. Interspeech, pp. 2670-2674.

Yu, D., Deng, L., Gong, Y., Acero, A., 2008a. Discriminative training of variable-parameter HMMs for noise robust speech recognition. In: Proc. Interspeech, pp. 285-288.

Yu, D., Deng, L., Gong, Y., Acero, A., 2008b. Parameter clustering and sharing in variable-parameter HMMs for noise robust speech recognition. In: Proc. Interspeech, pp. 1253-1256.

Yu, D., Deng, L., Gong, Y., Acero, A., 2009. A novel framework and training algorithm for variable-parameter hidden Markov models. IEEE Trans. Audio Speech Lang. Process. 17 (7), 1348-1360.

Zhao, R., Li, J., Gong, Y., 2014a. Variable-activation and variable-input deep neural network for robust speech recognition. In: Proc. IEEE Spoken Language Technology Workshop.

Zhao, R., Li, J., Gong, Y., 2014b. Variable-component deep neural network for robust speech recognition. In: Proc. Interspeech.

Explicit distortion modeling

CHAPTER OUTLINE

In Chapters 4 and 5, we use two separate attributes to classify and analyze many popular techniques developed for noise-robust ASR: feature- versus model-domain processing, and the use of prior knowledge in the algorithm or otherwise. In this chapter, we use the third attribute to perform the analysis and categorization: the use of explicit models of acoustic distortion or otherwise.

In the absence of explicit models of acoustic distortion, general adaptation techniques for acoustic models of ASR usually involve either changing all model parameters or using a generic set of linear transformations to compensate for the mismatch between training and test conditions. Either case necessitates the use of many free parameters and thus typically requires a large amount of adaptation data for parameter estimation. This difficulty can be overcome when one exploits explicit distortion models which take into account the way in which distorted speech features are generated (Acero, 1993; Deng et al., 2002, 2004b; Frey et al., 2001b). In this approach, the distorted speech features are represented explicitly as a nonlinear function of clean speech features, additive noise, and convolutive distortion, as in Equation 3.13 or Equation 3.14. This type of physical model enables structured transforms to be used, which are generally nonlinear and involve only a parsimonious set of free parameters to be estimated. An additional advantage of using such a physically meaningful generative model of distortion is that it can be naturally integrated to more comprehensive probabilistic generative models for the observed distorted speech feature sequences (Baker et al., 2009a,b; Deng, 2006; Deng and Li, 2013; Deng and O'Shaughnessy, 2003).

In this chapter, we unify the terminology and refer to the set of methods as explicit distortion modeling when a physical model for the generation of distorted speech features is employed either in the recognizer-model-domain or in feature-domain processing. If no physical model is explicitly used, the methods will be referred to as implicit distortion modeling.

It is noted that there has been inconsistency in the current literature when naming and discussing the methods with explicit distortion modeling. Some researchers call explicit distortion modeling methods as structured methods (Deng, 2011) while others call them predictive modeling (Gales, 1998). Further, the name involving "structure" does not explicitly show that a physical model is used and this causes confusions in the literature. For example, some adaptation techniques make use of a hierarchical tree as a structure (e.g., structural MAP (Shinoda and Lee, 2001)) without using an explicit distortion model. The name of predictive model is not self-explanatory either and it causes confusions with other predictive techniques in ASR. One task of this chapter is to clear up all these confusions, and to distinguish the use of the term "model" in "distortion modeling" from its use in "acoustic modeling" in the recognizer. We intend to categorize major noise-robust ASR techniques using the criterion of whether or not an explicit distortion model is exploited in representing acoustic distortion of speech.

Feature-space approaches can also be classified into explicit and implicit distortion modeling methods. Spectral subtraction and Wiener filtering discussed in Section 4.1.3 belong to explicit modeling methods, since they model the noise as an additive component and explicitly remove it from the noisy observation in the spectral domain. Two-stage Mel-warped Wiener filtering, the major component of the ETSI advanced front-end (AFE) (Macho et al., 2002) which has shown excellent performance on standard noise robustness tasks, also belongs to this category. Feature vector Taylor series (VTS) techniques are the most prominent examples in the explicit distortion modeling category, with a more complicated nonlinear distortion model that can work in either the log-spectral or the cepstral domain. In contrast, the methods (e.g., statistical moment normalization) without using any physical model can be considered as implicit feature processing methods. Importantly, because the physical constraints are explicitly modeled, the explicit distortion modeling methods require only a relatively small number of distortion parameters to be estimated. They also exhibit high performance due to the explicit modeling of the distorted speech generation process. Although the most representative explicit distortion modeling method, VTS, was proposed in 1996 (Moreno, 1996), it has been extended considerably in recent years. Notable examples of extended work with impact include phase-sensitive modeling (Deng et al., 2004a; Leutnant and Haeb-Umbach, 2009;

Li et al., 2008), higher-order VTS (Du and Huo, 2011), and unscented transform (Hu and Huo, 2006; Li et al., 2010) which provide a more precise physical modeling of the speech distortion, and joint uncertainty decoding (Liao and Gales, 2006) which aims at an efficient computation.

In this chapter, we will also include a section with several acoustic factorization methods, which handle each acoustic factor with an individual transform. On the one hand, acoustic factorization can be considered as a general form of explicit distortion modeling in which underlying acoustic factors are embedded into a single distortion function. However, on the other hand, strictly speaking, most acoustic factorization methods cannot be considered as explicit distortion modeling methods because they do not explicitly use any physical model for representing the distortion.

6.1 PARALLEL MODEL COMBINATION

Parallel model combination (PMC) uses the explicit distortion model to adapt the clean speech model. Figure 6.1 gives the framework of a basic PMC, which is

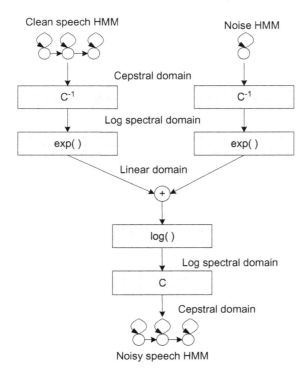

FIGURE 6.1

Parallel model combination.

described in the following. The model parameters of noise in the cepstral domain are first transformed to the log-Mel-filter-bank domain with

$$\boldsymbol{\mu}_{\tilde{n}} = \mathbf{C}^{-1}\boldsymbol{\mu}_n, \tag{6.1}$$

$$\boldsymbol{\Sigma}_{\tilde{n}} = \mathbf{C}^{-1}\boldsymbol{\Sigma}_n(\mathbf{C}^{-1})^T. \tag{6.2}$$

Then in the Mel-filter-bank domain, if the distribution is assumed log-normal, the elements of the mean and covariance of noise in the Mel-filter-bank domain are (Gales, 1995):

$$\boldsymbol{\mu}_{\check{n}}[i] = \exp\left(\boldsymbol{\mu}_{\tilde{n}}[i] + \frac{1}{2}\boldsymbol{\Sigma}_{\tilde{n}}[i,i]\right), \tag{6.3}$$

$$\boldsymbol{\Sigma}_{\check{n}}[i,j] = \boldsymbol{\mu}_{\check{n}}[i]\boldsymbol{\mu}_{\check{n}}[j]\left(\exp(\boldsymbol{\Sigma}_{\tilde{n}}[i,j]) - 1\right). \tag{6.4}$$

Similar expressions are also applied to the mean and covariance of clean speech in the Mel-filter-bank domain. Then, the explicit distortion model assumes that noise and clean speech are independent and additive in the Mel-filter-bank domain.

$$\check{\boldsymbol{y}} = \check{\boldsymbol{x}} + \check{\boldsymbol{n}}. \tag{6.5}$$

Therefore, the mean and covariance of distorted speech in the Mel-filter-bank domain are:

$$\boldsymbol{\mu}_{\check{y}} = \boldsymbol{\mu}_{\check{x}} + \boldsymbol{\mu}_{\check{n}}, \tag{6.6}$$

$$\boldsymbol{\Sigma}_{\check{y}} = \boldsymbol{\Sigma}_{\check{x}} + \boldsymbol{\Sigma}_{\check{n}}. \tag{6.7}$$

With the log-normal approximation (Gales, 1995) on $\check{\boldsymbol{y}}$, the elements of the mean and covariance of distorted speech in the log-Mel-filter-bank domain are

$$\boldsymbol{\mu}_{\tilde{y}}[i] = \log(\boldsymbol{\mu}_{\check{y}}[i]) - \frac{1}{2}\log\left(\frac{\boldsymbol{\Sigma}_{\check{y}}[i,i]}{\boldsymbol{\mu}_{\check{y}}^2[i]} + 1\right), \tag{6.8}$$

$$\boldsymbol{\Sigma}_{\tilde{y}}[i,j] = \log\left(\frac{\boldsymbol{\Sigma}_{\check{y}}[i,j]}{\boldsymbol{\mu}_{\check{y}}[i]\boldsymbol{\mu}_{\check{y}}[j]} + 1\right). \tag{6.9}$$

Finally, the parameters of distorted speech in the cepstral domain are obtained with

$$\boldsymbol{\mu}_y = \mathbf{C}\boldsymbol{\mu}_{\tilde{y}}, \tag{6.10}$$

$$\boldsymbol{\Sigma}_y = \mathbf{C}\boldsymbol{\Sigma}_{\tilde{y}}\mathbf{C}^T. \tag{6.11}$$

The basic PMC method cannot be directly used for adapting the dynamic cepstrum. Instead of the log-normal approximation used above, a simple log-add approximation is also proposed in Gales (1995) by directly working on the log-Mel-filter-bank domain without transforming back to the Mel-filter-bank domain:

$$\boldsymbol{\mu}_{\tilde{y}}[i] = \log\left(\exp(\boldsymbol{\mu}_{\tilde{x}}[i]) + \exp(\boldsymbol{\mu}_{\tilde{n}}[i])\right). \tag{6.12}$$

The log-add approximation is simpler but rougher than the log-normal approximation. Another advantage of the log-add approximation is that it can be applied to the dynamic parameters as well (Gales, 1995).

The basic PMC method can also be extended for situations where there is convolutive channel distortion as well as additive noise (Gales, 1995). A simple technique presented in Gales (1995) uses a one-state single-Gaussian speech model to calculate the convolutive component. An approximate solution of the convolutive component by steepest descent methods has also been reported (Minami and Furui, 1995), which relies on the Viterbi approximation and does not handle Gaussian mixture distributions. The method in Minami and Furui (1996) uses an additional universal speech Gaussian mixture model and incorporates an existing bias estimation procedure (Sankar and Lee, 1995) for channel estimation. In Gong (2005), a joint estimation of log-spectral noise and channel component is developed using an EM algorithm.

As shown in Acero et al. (2000), the vector Taylor series (VTS) approximation appears to be more accurate than the log-normal approximation in PMC. Therefore, many studies of explicit distortion modeling have been switched to the VTS direction over the last decade.

6.2 VECTOR TAYLOR SERIES

In recent years, model-domain approaches that jointly compensate for additive and convolutive (JAC) distortions (Acero et al., 2000; Gong, 2005; Kim et al., 1998; Li et al., 2007, 2009; Moreno, 1996; Sagayama et al., 1997; Stouten, 2006)) have yielded promising results. The various methods proposed so far use a parsimonious nonlinear physical model to describe the environmental distortion and use the vector Taylor series (VTS) approximation technique to find closed-form HMM adaptation and noise/channel parameter estimation formulas. Although some methods are referred to with different names, such as Jacobian adaptation (Sagayama et al., 1997) and JAC (Li et al., 2007, 2009), they are in essence the VTS methods since VTS is used to linearize the involved nonlinearity, from which the solutions are derived.

Equation 3.14 is a popular nonlinear distortion model between clean and distorted speech in the cepstral domain. We recap it in the following:

$$y = x + h + C \log(1 + \exp(C^{-1}(n - x - h))), \tag{6.13}$$

where C denotes the DCT matrix, y, x, n, and h are the distorted speech, clean speech, additive noise, and convolutive channel in the cepstral domain, respectively. Equation 6.13 can be re-written as

$$y = x + h + g(x, h, n), \tag{6.14}$$

where

$$g(\mathbf{x}, \mathbf{h}, \mathbf{n}) = \mathbf{C} \log(1 + \exp(\mathbf{C}^{-1}(\mathbf{n} - \mathbf{x} - \mathbf{h})))). \tag{6.15}$$

In standard VTS adaptation (Moreno, 1996), the nonlinear function in Equation 6.15 is approximated using a first-order VTS expansion at point $(\boldsymbol{\mu}_x(m), \boldsymbol{\mu}_h, \boldsymbol{\mu}_n)$. $\boldsymbol{\mu}_x(m)$, $\boldsymbol{\mu}_n$, and $\boldsymbol{\mu}_h$ are the static cepstral means of the clean speech Gaussian component m, noise, and channel, respectively.

Denoting

$$\frac{\partial \mathbf{y}}{\partial \mathbf{x}}\bigg|_{\boldsymbol{\mu}_x(m), \boldsymbol{\mu}_n, \boldsymbol{\mu}_h} = \frac{\partial \mathbf{y}}{\partial \mathbf{h}}\big|_{\boldsymbol{\mu}_x(m), \boldsymbol{\mu}_n, \boldsymbol{\mu}_h} = \mathbf{G}(m), \tag{6.16}$$

$$\frac{\partial \mathbf{y}}{\partial \mathbf{n}}\bigg|_{\boldsymbol{\mu}_x(m), \boldsymbol{\mu}_n, \boldsymbol{\mu}_h} = \mathbf{I} - \mathbf{G}(m) = \mathbf{F}(m), \tag{6.17}$$

where $\mathbf{G}(m)$ is the Jacobian matrix for Gaussian component m, defined as

$$\mathbf{G}(m) = \mathbf{C} \text{diag} \left(\frac{1}{1 + \exp(\mathbf{C}^{-1}(\boldsymbol{\mu}_n - \boldsymbol{\mu}_x(m) - \boldsymbol{\mu}_h))} \right) \mathbf{C}^{-1}, \tag{6.18}$$

Equation 6.14 can then be approximated by a VTS expansion, truncated after the linear term:

$$\mathbf{y}|_{\boldsymbol{\mu}_x(m), \boldsymbol{\mu}_h, \boldsymbol{\mu}_n} \approx \boldsymbol{\mu}_x(m) + \boldsymbol{\mu}_h + g(\boldsymbol{\mu}_x(m), \boldsymbol{\mu}_h, \boldsymbol{\mu}_n) + \mathbf{G}(m)(\mathbf{x} - \boldsymbol{\mu}_x(m))$$
$$+ \mathbf{G}(m)(\mathbf{h} - \boldsymbol{\mu}_h) + \mathbf{F}(m)(\mathbf{n} - \boldsymbol{\mu}_n), \tag{6.19}$$

With Equation 6.19, the distorted speech $\mathbf{y}$ is now a linear function of the clean speech $\mathbf{x}$, the noise $\mathbf{n}$, and the channel $\mathbf{h}$, in the cepstral domain. This linearity facilitates the HMM model adaptation and distortion parameter estimation by providing possible closed-form solutions, because if a linear operation is applied to a Gaussian distributed random variable the resulting random variable still follows a Gaussian distribution.

6.2.1 VTS MODEL ADAPTATION

By taking the expectation on both sides of Equation 6.19, the static mean of the distorted speech signal $\boldsymbol{\mu}_y$ for Gaussian component m can be written as

$$\boldsymbol{\mu}_y(m) \approx \boldsymbol{\mu}_x(m) + \boldsymbol{\mu}_h + g(\boldsymbol{\mu}_x(m), \boldsymbol{\mu}_h, \boldsymbol{\mu}_n), \tag{6.20}$$

and the static variance of the distorted speech signal $\boldsymbol{\mu}_y$ for Gaussian component m can be obtained by taking the variance operation on both sides of Equation 6.19:

$$\boldsymbol{\Sigma}_y(m) \approx \text{diag}(\mathbf{G}(m)\boldsymbol{\Sigma}_x(m)\mathbf{G}(m)^T + \mathbf{F}(m)\boldsymbol{\Sigma}_n\mathbf{F}(m)^T), \tag{6.21}$$

The delta parameters can be updated (Acero et al., 2000) with the continuous time approximation (Gopinath et al., 1995), which makes the assumption that the dynamic cepstral coefficients are the time derivatives of the static cepstral coefficients.

$$\Delta \mathbf{y}_t \approx \frac{\partial \mathbf{y}}{\partial \mathbf{x}} \frac{\partial \mathbf{x}}{\partial t} + \frac{\partial \mathbf{y}}{\partial \mathbf{n}} \frac{\partial \mathbf{n}}{\partial t}. \tag{6.22}$$

The delta model parameters adaptation formulas in Equations 6.23 and 6.25 are obtained by plugging Equations 6.16 and 6.17, and taking the mean and variance operation to both sides of Equation 6.22. It is easy to extend this to obtain the delta-delta model parameters adaptation formulas in Equations 6.24 and 6.26.

$$\boldsymbol{\mu}_{\Delta y}(m) \approx \mathbf{G}(m)\boldsymbol{\mu}_{\Delta x}(m), \tag{6.23}$$

$$\boldsymbol{\mu}_{\Delta \Delta y}(m) \approx \mathbf{G}(m)\boldsymbol{\mu}_{\Delta \Delta x}(m), \tag{6.24}$$

$$\boldsymbol{\Sigma}_{\Delta y}(m) \approx \text{diag}\left(\mathbf{G}(m)\boldsymbol{\Sigma}_{\Delta x}(m)\mathbf{G}(m)^T + \mathbf{F}(m)\boldsymbol{\Sigma}_{\Delta n}\mathbf{F}(m)^T\right), \tag{6.25}$$

$$\boldsymbol{\Sigma}_{\Delta \Delta y}(m) \approx \text{diag}(\mathbf{G}(m)\boldsymbol{\Sigma}_{\Delta \Delta x}(m)\mathbf{G}(m)^T + \mathbf{F}(m)\boldsymbol{\Sigma}_{\Delta \Delta n}\mathbf{F}(m)^T). \tag{6.26}$$

Note that although the cepstral distortion formulation that Equation 6.13, is widely used in VTS studies, the log spectral distortion formulation, that is, Equation 3.13, can also be used (Gong, 2005). Some studies (Zhao, 2000) even work in the linear frequency domain. However, this brings a large computational cost and the diagonal covariance assumption used in Zhao (2000) may not be valid for linear frequency. Therefore, there are only a small number of VTS methods working in the linear frequency domain.

6.2.2 DISTORTION ESTIMATION IN VTS

Although proposed in 1996 (Moreno, 1996), VTS model adaptation has shown great accuracy advantages over other noise-robustness methods only recently (Li et al., 2007, 2009) when the distortion model parameters are re-estimated based on the first-pass decoding result with the expectation-maximization (EM) algorithm (Li et al., 2007, 2009). First, an auxiliary Q function for an utterance is

$$Q(\hat{\Lambda}; \Lambda) = \sum_{t,m} \gamma_t(m) \log p_{\hat{\Lambda}}(\mathbf{y}_t|m), \tag{6.27}$$

where $\hat{\Lambda}$ denotes the adapted model, and $\gamma_t(m)$ is the posterior probability for the Gaussian component m of the HMM at time t, that is,

$$\gamma_t(m) = p_\Lambda(m|\mathbf{y}_t), \tag{6.28}$$

where Λ denotes the previous model.

To maximize the auxiliary function in the M-step of the EM algorithm, the derivatives of Q are taken with respect to $\boldsymbol{\mu}_n$ and $\boldsymbol{\mu}_h$, and are set to 0. Then, the mean of the noise $\boldsymbol{\mu}_n$ can be updated according to (Li et al., 2009):

$$\boldsymbol{\mu}_n = \boldsymbol{\mu}_{n,0} + \mathbf{A}_n^{-1}\mathbf{b}_n, \tag{6.29}$$

with

$$\mathbf{A}_n = \sum_{t,m} \gamma_t(m)\mathbf{F}(m)^T \mathbf{\Sigma}_y^{-1}(m)\mathbf{F}(m), \tag{6.30}$$

$$\mathbf{b}_n = \sum_{t,m} \gamma_t(m)\mathbf{F}(m)^T \mathbf{\Sigma}_y^{-1}(m)(\mathbf{y}_t - \boldsymbol{\mu}_{y,0}(m)), \tag{6.31}$$

$$\boldsymbol{\mu}_{y,0}(m) = \boldsymbol{\mu}_x(m) - \boldsymbol{\mu}_{h,0} - \mathbf{g}(\boldsymbol{\mu}_x(m), \boldsymbol{\mu}_{h,0}, \boldsymbol{\mu}_{n,0}). \tag{6.32}$$

where $\boldsymbol{\mu}_{n,0}$ and $\boldsymbol{\mu}_{h,0}$ are the VTS expansion points for $\boldsymbol{\mu}_n$ and $\boldsymbol{\mu}_h$, respectively. The channel mean $\boldsymbol{\mu}_h$ can be updated as in Li et al. (2009).

$$\boldsymbol{\mu}_h = \boldsymbol{\mu}_{h,0} + \mathbf{A}_h^{-1}\mathbf{b}_h, \tag{6.33}$$

with

$$\mathbf{A}_h = \sum_{t,m} \gamma_t(m)\mathbf{G}(m)^T \mathbf{\Sigma}_y^{-1}(m)\mathbf{G}(m), \tag{6.34}$$

$$\mathbf{b}_h = \sum_{t,m} \gamma_t(m)\mathbf{G}(m)^T \mathbf{\Sigma}_y^{-1}(m)(\mathbf{y}_t - \boldsymbol{\mu}_{y,0}(m)). \tag{6.35}$$

To estimate the D-dimensional static noise variance vector $\mathbf{\Sigma}_n = \text{diag}(\boldsymbol{\sigma}_n^2)$ with $\boldsymbol{\sigma}_n^2 = [\sigma_{n,1}^2, \sigma_{n,2}^2, \ldots \sigma_{n,D}^2]^T$, a Newton's method, an iterative second-order approach, can be used (Li et al., 2009). To guarantee that the adapted noise variance is positive, one solution is to transform the static noise variance first with

$$\tilde{\boldsymbol{\sigma}}_n^2 = \log(\boldsymbol{\sigma}_n^2), \tag{6.36}$$

and then update $\tilde{\boldsymbol{\sigma}}_n^2$ with the Newton's method as

$$\tilde{\boldsymbol{\sigma}}_n^2 = \tilde{\boldsymbol{\sigma}}_{n,0}^2 - \left(\frac{\partial^2 Q}{\partial^2 \tilde{\boldsymbol{\sigma}}_n^2}\right)^{-1} \frac{\partial Q}{\partial \tilde{\boldsymbol{\sigma}}_n^2}, \tag{6.37}$$

where $\tilde{\boldsymbol{\sigma}}_{n,0}^2$ is the expansion point for $\tilde{\boldsymbol{\sigma}}_n^2$. Finally, the static noise variance in the linear scale can be obtained with

$$\boldsymbol{\sigma}_n^2 = \exp(\tilde{\boldsymbol{\sigma}}_n^2). \tag{6.38}$$

The cth element of $\frac{\partial Q}{\partial \tilde{\boldsymbol{\sigma}}_n^2}$ and the (c, e)–th element of $\frac{\partial^2 Q}{\partial^2 \tilde{\boldsymbol{\sigma}}_n^2}$ in Equation 6.37 can be obtained with Equations 6.39 and 6.40, respectively.

$$\frac{\partial Q}{\partial \tilde{\sigma}_{n,c}^2} = -\frac{1}{2}\sum_{t,m} \gamma_t(m) \sum_{d=1}^{D} \left(\frac{\sigma_{n,c}^2 F_{dc}^2(m)}{\tau_d(m)}\left(1 - \frac{(y_{t,d} - \mu_{y,d}(m))^2}{\tau_d(m)}\right)\right), \tag{6.39}$$

$$\frac{\partial^2 Q}{\partial \tilde{\sigma}_{n,c}^2 \partial \tilde{\sigma}_{n,e}^2} = -\frac{1}{2} \sum_{t,m} \gamma_t(m) \sum_{d=1}^{D} \left(\frac{\sigma_{n,c}^2 \mathbf{F}_{dc}^2(m)}{\tau_d(m)} \left(1 - \frac{(\mathbf{y}_{t,d} - \boldsymbol{\mu}_{y,d}(m))^2}{\tau_d(m)} \right) \delta(c - e) \right.$$
$$\left. + \frac{\sigma_{n,c}^2 \mathbf{F}_{dc}^2(m) \sigma_{n,e}^2 \mathbf{F}_{de}^2(m)}{\tau_d^2(m)} \left(-1 + 2 \frac{(\mathbf{y}_{t,d} - \boldsymbol{\mu}_{y,d}(m))^2}{\tau_d(m)} \right) \right), \tag{6.40}$$

where

$$\tau_d(m) = \sum_{i=1}^{D} (\sigma_{x,i}^2(m) \mathbf{G}_{di}^2(m) + \sigma_{n,i}^2 \mathbf{F}_{di}^2(m)). \tag{6.41}$$

$\mathbf{G}_{di}(m)$ and $\mathbf{F}_{di}(m)$ are the dth row and ith column elements of $\mathbf{G}(m)$ and $\mathbf{F}(m)$, respectively.

The delta and 2nd delta noise variance can be estimated in a similar way. Full derivations of the update formula for the distortion model parameters can be found in Li et al. (2009).

Distortion parameter estimation can also be done in other ways. In Bu et al. (2014a,b), $\mathbf{C}^{-1}(\mathbf{n} - \mathbf{x})$ in Equation 6.15 is replaced with a single variable $\mathbf{z}$ and channel distortion $\mathbf{h}$ is ignored, and the optimization becomes easier for dynamic parameter calculation and second order statistics calculation. In Liao and Gales (2006), a gradient-descent method is used to obtain the noise variance estimate. Since there is no guarantee that the auxiliary function will increase, a back-off step is needed. In Zhao and Juang (2012), a Gauss-Newton method is used by discarding the second derivative of the residual with respect to the distortion parameters when calculating the Hessian. In Kim et al. (1998), both the static mean and variance parameters in the cepstral domain are adjusted using the VTS approximation technique. In that work, however, noise was estimated on a frame-by-frame basis, which is complex and computationally costly. It is shown in Zhao and Juang (2010) that the estimation method used in this section (Li et al., 2007, 2009) is clearly better than the estimation method in Kim et al. (1998).

The implementation steps of the VTS model adaptation algorithm are described as follows and plotted in Figure 6.2:

1. read in a distorted speech utterance;
2. set the channel mean vector to all zeros;
3. initialize the noise mean vector and diagonal covariance matrix using the first and last N frames (speech-free) from the utterance using sample estimates;
4. compute the Gaussian-dependent $\mathbf{G}(m)$ with Equation 6.18, and adapt the HMM parameters with Equations 6.20–6.26;
5. decode the utterance with the adapted HMM parameters;
6. re-estimate noise and channel distortions using the above-decoded transcription with formulations in Section 6.2.2;
7. adapt the HMM parameters again with Equation 6.20–6.26;
8. use the final adapted HMM model to decode the distorted speech feature and get output transcription.

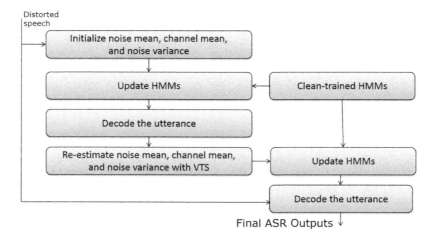

FIGURE 6.2

VTS model adaptation.

If multiple-pass decoding is desired, there would be a loop between steps 5 and 7.

6.2.3 VTS FEATURE ENHANCEMENT

As shown in Li et al. (2009), VTS model adaptation achieves much better accuracy than several popular model adaptation technologies. Although VTS model adaptation can achieve high accuracy, the computational cost is very high as all the Gaussian parameters in the recognizer need to be updated every time the environmental parameters change. This time-consuming requirement prevents VTS model adaptation from being widely used, especially in LVCSR tasks where the number of model parameters is large.

On the other hand, VTS feature enhancement has been proposed as a lower-cost alternative to VTS model adaptation. For example, a number of techniques have been proposed that can be categorized as model-based feature enhancement schemes (Moreno, 1996; Stouten et al., 2003; Droppo et al., 2003; Li et al., 2011). These methods use a small GMM in the front-end and the same methodology used in VTS model adaptation to derive a minimum-mean-square-error (MMSE) estimate of the clean speech features given the noisy observations. In addition to the advantage of a low runtime cost, VTS feature enhancement can be easily combined with other popular feature-based technologies, such as CMN, HLDA, fMPE, etc., which are challenging to VTS model adaptation.

In general, the MMSE method can be used to get the estimate of clean speech

$$\hat{\mathbf{x}} = E(\mathbf{x}|\mathbf{y}) = \int \mathbf{x} p(\mathbf{x}|\mathbf{y}) \, d\mathbf{x}. \tag{6.42}$$

Denote the clean-trained GMM as

$$p_\Lambda(x) = \sum_{k=1}^{K} \mathbf{c}(k)\mathcal{N}(\mathbf{x}; \boldsymbol{\mu}_x(k), \boldsymbol{\Sigma}_x(k)), \tag{6.43}$$

along with Equation 6.13, the MMSE estimate of clean speech becomes

$$\hat{\mathbf{x}} = \mathbf{y} - \mathbf{h} - \int \mathbf{C} \log(1 + \exp(\mathbf{C}^{-1}(\mathbf{n} - \mathbf{x} - \mathbf{h})))p(\mathbf{x}|\mathbf{y})d\mathbf{x},$$

$$= \mathbf{y} - \mathbf{h} - \int \mathbf{C} \log(1 + \exp(\mathbf{C}^{-1}(\mathbf{n} - \mathbf{x} - \mathbf{h}))) \sum_{k=1}^{K} P(k|\mathbf{y})p(\mathbf{x}|\mathbf{y},k) \, d\mathbf{x},$$

$$= \mathbf{y} - \mathbf{h} - \sum_{k=1}^{K} P(k|\mathbf{y}) \int \mathbf{C} \log(1 + \exp(\mathbf{C}^{-1}(\mathbf{n} - \mathbf{x} - \mathbf{h})))p(\mathbf{x}|\mathbf{y},k) \, d\mathbf{x}, \tag{6.44}$$

where $P(k|\mathbf{y})$ is the Gaussian posterior probability, calculated as

$$P(k|\mathbf{y}) = \frac{\mathbf{c}(k)\mathcal{N}(\mathbf{y}; \boldsymbol{\mu}_y(k), \boldsymbol{\Sigma}_y(k))}{\sum_{k=1}^{K} \mathbf{c}(k)\mathcal{N}(\mathbf{y}; \boldsymbol{\mu}_y(k), \boldsymbol{\Sigma}_y(k))}. \tag{6.45}$$

If the 0th-order VTS approximation is used for the nonlinear term in Equation 6.44, the MMSE estimate of cleaned speech $\mathbf{x}$ is obtained as

$$\hat{\mathbf{x}} = \mathbf{y} - \mathbf{h} - \sum_{k=1}^{K} P(k|\mathbf{y})\mathbf{C} \log(1 + \exp(\mathbf{C}^{-1}(\boldsymbol{\mu}_n - \boldsymbol{\mu}_x(k) - \boldsymbol{\mu}_h))). \tag{6.46}$$

This formulation was first proposed in Moreno (1996). In Stouten et al. (2003), another solution was proposed when expanding Equation 6.13 with the 1st-order VTS. For the kth GMM component, the joint distribution of $\mathbf{x}$ and $\mathbf{y}$ is modeled as

$$\mathcal{N}\left(\begin{bmatrix} \mathbf{x} \\ \mathbf{y} \end{bmatrix}; \begin{bmatrix} \boldsymbol{\mu}_x(k) \\ \boldsymbol{\mu}_y(k) \end{bmatrix}, \begin{bmatrix} \boldsymbol{\Sigma}_x(k) & \boldsymbol{\Sigma}_{xy}(k) \\ \boldsymbol{\Sigma}_{yx}(k) & \boldsymbol{\Sigma}_y(k) \end{bmatrix}\right). \tag{6.47}$$

The following can be derived (Stouten et al., 2003)

$$E(\mathbf{x}|\mathbf{y},k) = \boldsymbol{\mu}_{\mathbf{x}|\mathbf{y}}(k) = \boldsymbol{\mu}_x(k) + \boldsymbol{\Sigma}_{xy}(k)\boldsymbol{\Sigma}_y^{-1}(k)(\mathbf{y} - \boldsymbol{\mu}_y(k)). \tag{6.48}$$

The covariance between $\mathbf{x}$ and $\mathbf{y}$ can be derived as

$$\boldsymbol{\Sigma}_{xy}(k) = E\left[(\mathbf{x} - \boldsymbol{\mu}_x(k))(\mathbf{y} - \boldsymbol{\mu}_y(k))^T\right] \tag{6.49}$$

$$= E\left[(\mathbf{x} - \boldsymbol{\mu}_x(k))(\mathbf{G}(k)(\mathbf{x} - \boldsymbol{\mu}_x(k))\right.$$

$$\left. + \mathbf{G}(k)(\mathbf{h} - \boldsymbol{\mu}_h) + \mathbf{F}(k)(\mathbf{n} - \boldsymbol{\mu}_n))^T\right] \tag{6.50}$$

$$= E\left[(\mathbf{x} - \boldsymbol{\mu}_x(k))(\mathbf{G}(k)(\mathbf{x} - \boldsymbol{\mu}_x(k)))^T\right] \tag{6.51}$$

$$= \boldsymbol{\Sigma}_x(k)\mathbf{G}(k)^T \tag{6.52}$$

Equation 6.50 is obtained by subtracting Equation 6.20 from the 1st-order VTS expansion in Equation 6.19. Equation 6.51 is obtained by using the property that speech, channel, and noise are independent.

Then the MMSE estimate of clean speech is (Stouten et al., 2003)

$$\hat{\mathbf{x}} = \sum_{k=1}^{K} P(k|\mathbf{y}) \left(\boldsymbol{\mu}_x(k) + \boldsymbol{\Sigma}_x(k)\mathbf{G}(k)^T \boldsymbol{\Sigma}_y^{-1}(k)(\mathbf{y} - \boldsymbol{\mu}_y(k)) \right). \tag{6.53}$$

The flowchart of feature VTS is in Figure 6.3. The following are the detailed implementation steps (Li et al., 2011):

1. read in a distorted speech utterance;
2. set the channel mean vector to all zeros;
3. initialize the noise mean vector and diagonal covariance matrix using the first and last N frames (speech-free) from the utterance using sample estimates;
4. compute the Gaussian-dependent $\mathbf{G}(k)$ with Equation 6.18, and adapt the GMM parameters with Equations 6.20–6.26;
5. re-estimate noise and channel distortions with formulations in Section 6.2.2;
6. adapt the GMM parameters with Equations 6.20–6.26;
7. use the final adapted GMM model to clean the distorted speech feature with Equation 6.46 or Equation 6.53;
8. use the clean-trained HMM model to decode the cleaned speech feature obtained in step 7 and get output transcription.

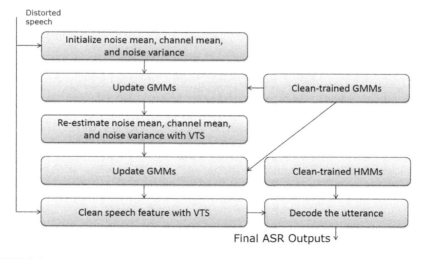

FIGURE 6.3

VTS feature enhancement.

Two key aspects of VTS feature enhancement are how to obtain reliable estimates of the noise and channel distortion parameters and how to accurately calculate the Gaussian occupancy probability. In contrast to using static features alone to calculate the Gaussian occupancy probability (Stouten, 2006), both static and dynamic features are used to obtain more reliable Gaussian occupancy probabilities. Then, these probabilities are plugged into Equation 6.46 or Equation 6.53. In Li et al. (2012b), it is shown that recent improvements in VTS model adaptation can be incorporated into VTS feature enhancement to improve the algorithm performance: updating all of the environment distortion parameters (Li et al., 2007) and subsequently carrying out noise adaptive training (Kalinli et al., 2010).

Figures 6.4(a) and (b) show the distribution of the C1 and C0 of word *oh* in Aurora 2 test set A with noise type 1. Comparing with Figure 3.2a and b, it is clear that after VTS feature enhancement (fVTS), the distributions of cleaned signals in all SNR conditions are now very close to the original distribution of clean signal. This benefits both training and testing. In training, only small numbers of Gaussians are now needed to model the distributions of all data. In testing, even clean-trained model can well deal with the cleaned low SNR signals because the distributions are similar now.

A common concern of feature enhancement is that after the enhancement, the clean speech signal is distorted and the accuracy on clean test sets will drop. As shown in Li et al. (2012a), VTS feature enhancement enjoys the nice property that it significantly improves accuracy in noisy test conditions without degrading accuracy in clean test conditions. This can also be visualized in Figure 6.4(c) and (d) which shows the distribution of the C1 and C0 of word "oh" in Aurora 2 test set A clean condition. The solid line denotes the distribution of raw clean data, and the dotted line denotes the distribution of clean data after feature VTS enhancement. As shown in the figures, they are very close.

By incorporating the recent advances in VTS model adaptation, VTS feature enhancement can obtain very high accuracy on some noisy tasks (Li et al., 2012b). However, it is shown that there is still a small accuracy gap between VTS feature enhancement and VTS model adaptation (Li et al., 2012b). Regarding the runtime cost, VTS model adaptation needs to adapt the back-end HMM parameters twice, while VTS feature enhancement needs to adapt the front-end GMM parameters twice. Usually, the number of parameters in a front-end GMM is much smaller than that in the back-end HMM. Furthermore, two rounds of decoding are needed in VTS model adaptation while only one round of decoding is performed in VTS feature enhancement. As a consequence, VTS feature enhancement has a much lower computational cost than VTS model adaptation. Therefore, the tradeoff between accuracy and computational cost will determine which technology is more suitable in a real world deployment scenario if the underlying acoustic model is a GMM. However, if the underlying acoustic model is a DNN, VTS feature enhancement is a more natural choice. In Section 6.2.5, we will discuss how to combine the VTS technology with DNN.

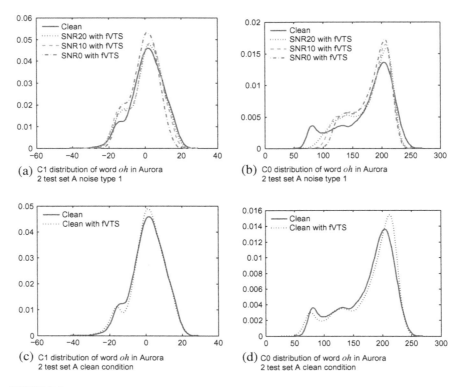

FIGURE 6.4

Cepstral distribution of word *oh* in Aurora 2 after VTS feature enhancement (fVTS).

6.2.4 IMPROVEMENTS OVER VTS

Recently, there has been a series of studies focusing on how to improve the performance of VTS. A natural extension to the VTS methods described in the previous section is to use the high-order VTS expansion instead of the first-order VTS expansion. That way, the nonlinear relation found in Equation 6.13 can be well modeled. There are several studies (Ding and Xu, 2004; Du and Huo, 2011; Kim, 1998; Stouten et al., 2005a; Xu and Chin, 2009b) along this line. As shown in Du and Huo (2011), the second-order VTS is shown to achieve a noticeable performance gain over the first-order VTS, although the accuracy gap between the third-order VTS and second-order VTS is small. Another way to address the inaccuracy problem of the first-order expansion in the VTS is to use piecewise functions to model the nonlinearity, such as a piecewise linear approximation (Du and Huo, 2008) or linear spline interpolation (Kalgaonkar et al., 2009; Seltzer et al., 2010).

Equation 6.13 is simplified from Equation 3.12, the phase-sensitive model, by approximating $\alpha[l]$ as 0 (Deng et al., 2004a). There is some work (Stouten et al.,

2005a) that uses the phase-sensitive model for better modeling of the distortion. $\boldsymbol{\alpha}$ can be estimated from the training set (Stouten et al., 2005a). Due to its physical interpretation, the value of elements in $\boldsymbol{\alpha}$ ranges from -1 to 1. In Li et al. (2008), this value constraint is broken by assigning a constant value α to the elements of $\boldsymbol{\alpha}$. If α is set to 0 and 1, VTS can be considered to work with MFCCs extracted from the power spectrum and magnitude spectrum, respectively (Li et al., 2009). It is shown in Li et al. (2008) that the best accuracy is obtained on the Aurora 2 task when the value is set to around 2.5, which is larger than 1, the theoretical maximum value. Therefore, the phase-sensitive model with a constant α value can also be considered to be a generalization function of the distortion model. The phase-sensitive model with a large α value may be considered as a way to compensate for the loss brought by the inaccurate approximation in VTS. In Gales and Flego (2008), it is shown that this phase-sensitive model has its counter part when using the domain-based (Gales, 1995) distortion function:

$$\mathbf{y} = \mathbf{x} + \mathbf{h} + \frac{1}{\gamma}\mathbf{C}\log(1 + \exp(\gamma\mathbf{C}^{-1}(\mathbf{n} - \mathbf{x} - \mathbf{h}))), \tag{6.54}$$

where γ with values 2, 1, and 0.75, corresponds to α with values 0, 1, and 2.5, in phase-sensitive model. Another way to handle the phase term is to use the ALGONQUIN algorithm (Frey et al., 2001a), which models the phase term as the modeling error with Equation 6.14. Therefore, the observation likelihood is

$$p(\mathbf{y}|\mathbf{x}, \mathbf{h}, \mathbf{n}) = \mathcal{N}(\mathbf{y}; \mathbf{x} + \mathbf{h} + \mathbf{g}(\mathbf{x}, \mathbf{h}, \mathbf{n}), \Psi), \tag{6.55}$$

where Ψ is the covariance matrix of the modeling error. The MMSE estimate of $\mathbf{x}$ is

$$\hat{\mathbf{x}} = \int \mathbf{x}p(\mathbf{x}|\mathbf{y})\,d\mathbf{x}, \tag{6.56}$$

with

$$p(\mathbf{x}|\mathbf{y}) = \int p(\mathbf{x}, \mathbf{h}, \mathbf{n})p(\mathbf{y}|\mathbf{x}, \mathbf{h}, \mathbf{n})\,d\mathbf{n}\,d\mathbf{h} \tag{6.57}$$

In standard VTS, the delta and delta-delta model parameters are updated (Acero et al., 2000) with the continuous time approximation (Gopinath et al., 1995). In van Dalen and Gales (2009) and van Dalen and Gales (2011a), extended VTS is proposed to provide a more accurate form to adapt dynamic model parameters. For the purpose of illustration, suppose the delta coefficient is calculated as

$$\begin{bmatrix} \mathbf{y}_t \\ \Delta\mathbf{y}_t \end{bmatrix} = \begin{bmatrix} \mathbf{0} & \mathbf{I} & \mathbf{0} \\ -\frac{\mathbf{I}}{2} & \mathbf{0} & \frac{\mathbf{I}}{2} \end{bmatrix} \begin{bmatrix} \mathbf{y}_{t-1} \\ \mathbf{y}_t \\ \mathbf{y}_{t+1} \end{bmatrix} = \mathbf{D}\mathbf{y}_t^e \tag{6.58}$$

with

$$\mathbf{D} = \begin{bmatrix} \mathbf{0} & \mathbf{I} & \mathbf{0} \\ -\frac{\mathbf{I}}{2} & \mathbf{0} & \frac{\mathbf{I}}{2} \end{bmatrix}, \tag{6.59}$$

$$\mathbf{y}_t^e = \begin{bmatrix} \mathbf{y}_{t-1} \\ \mathbf{y}_t \\ \mathbf{y}_{t+1} \end{bmatrix}. \tag{6.60}$$

Equation 6.58 can be generalized to have a large context window and can be extended for delta-delta coefficients also. The linear transform $\mathbf{D}$ applies to the extended distorted speech, $\mathbf{y}_t^e$, to generate the joint vector of the static and delta coefficients. Then, the mean and covariances of joint vectors are calculated as

$$\begin{bmatrix} \mu_y \\ \mu_{\Delta y} \end{bmatrix} = \mathbf{D}\mu_y^e, \tag{6.61}$$

$$\begin{bmatrix} \Sigma_y & \Sigma_{y,\Delta y} \\ \Sigma_{\Delta y,y} & \Sigma_{\Delta y} \end{bmatrix} = \mathbf{D}\Sigma_y^e\mathbf{D}^T, \tag{6.62}$$

The key is how to get the distribution of the extended distorted speech, $\mathbf{y}_t^e$. This is solved in van Dalen and Gales (2009, 2011a). Extended VTS has been shown to outperform standard VTS at the cost of more expensive computation (van Dalen and Gales, 2011a).

Recently, research on VTS-based methods has been extended to broader areas such as joint modeling speaker and noise effects (Chin et al., 2011; Wang and Gales, 2011b, 2012), and handling reverberation effects (Wang and Gales, 2011a).

As mentioned before, high computational cost is a concern for VTS model adaptation. VTS feature enhancement uses a small GMM on the front-end and the same methodology used in VTS model adaptation to derive a MMSE estimate of the clean speech features given the noisy observations. However, even after employing the recent advanced methods in VTS feature adaptation, VTS feature enhancement still shows an obvious accuracy gap compared to VTS model adaptation (Li et al., 2012b). In Li et al. (2012a), VTS model adaptation with diagonal Jacobian approximation method is proposed, which gives a relatively small accuracy loss at a drastic savings in computational cost over all three major components in standard VTS model adaptation. The computational cost reduction is on a scale of $\mathcal{O}(D)$ for the Jacobian calculation and most parts of parameter adaptation, and $\mathcal{O}(D^2)$ for online distortion estimation, where D is the dimension of static cepstral feature. There is also a family of joint uncertainty decoding (JUD) methods (Gales and van Dalen, 2007; Liao and Gales, 2008; Xu et al., 2011) that can reduce the computational cost of VTS by changing the Jacobian in Equation 6.18 from being Gaussian-dependent to regression-class-dependent. We will discuss the JUD methods in Section 7.3.2.

6.2.5 **VTS FOR THE DNN-BASED ACOUSTIC MODEL**

Last, all the VTS methods are built on top of the underlying GMM models parameterized by Gaussian mean and variance. The parameter relation between the clean speech and distorted speech can be derived by taking the mean and variance operation on both sides of the distortion function in Equation 6.14. However, currently there is no clear way to apply VTS to DNN models in which model parameters are weight

matrices because there is no simple and analytic relationship between the values of
weight matrices and feature vectors. Therefore, it is not straightforward to adapt the
DNN weight matrices according to the distortion function in Equation 6.14. It is
also shown in Li et al. (2014) that approximation is needed to derive the relationship
between the clean and distorted output vector of a DNN using the distortion function.
Some fundamental reasons for the difficulty of embedding generative models such as
the VTS into the overall discriminative model such as the DNN have been discussed
and analyzed in Yu and Deng (2014) and Deng and Yu (2014).

One natural way to employ the power of VTS for the DNN-based acoustic model
is to use VTS feature enhancement to clean up or denoise the distorted features,
which has been shown to be effective as reported in Li and Sim (2013). Another way
is to use VTS to compensate for the global feature normalization of the DNN. In
DNN training, the global mean $\boldsymbol{\mu}_x$ and standard deviation $\boldsymbol{\sigma}_x$ are calculated from the
training set. Then all samples in both training and test sets can be normalized with

$$\tilde{x}_i = \frac{x_i - \mu_{x,i}}{\sigma_{x,i}}, \tag{6.63}$$

where x_i, $\boldsymbol{\mu}_{x,i}$, and $\sigma_{x,i}$ are the ith dimension feature value, mean value, and standard
deviation value, respectively. $\tilde{\mathbf{x}}$ is then used as the input to the DNN's first hidden
layer. This global feature normalization is useful because the later processing steps
often perform better when each dimension is scaled to a similar range of numerical
values (LeCun et al., 1998). However, if the test data $\mathbf{y}$ are distorted with noise and
channel factors, simply applying the feature normalization learned from training set
to the test data is not optimal. Therefore, a transformed feature normalization method
is proposed in Li and Sim (2013), where the global mean and variance of the clean
training data are transformed to the distorted mean and variance of the current test
utterance with

$$\boldsymbol{\mu}_y \approx \boldsymbol{\mu}_x + \boldsymbol{\mu}_h + \mathbf{g}(\boldsymbol{\mu}_x, \boldsymbol{\mu}_h, \boldsymbol{\mu}_n), \tag{6.64}$$

$$\boldsymbol{\Sigma}_y \approx \text{diag}(\mathbf{G}\boldsymbol{\Sigma}_x\mathbf{G}^T + \mathbf{F}\boldsymbol{\Sigma}_n\mathbf{F}^T). \tag{6.65}$$

$\mathbf{g}$, $\mathbf{G}$, and $\mathbf{F}$ are defined in Equations 6.15, 6.16, and 6.17, respectively. Note here
only the global mean and variance are concerned. Therefore all the state indices
are removed, compared to Equations 6.20 and 6.21. Finally, $\boldsymbol{\mu}_y$ and $\boldsymbol{\sigma}_y$ are used
to transform the test feature $\mathbf{y}$ with

$$\tilde{y}_i = \frac{y_i - \mu_{y,i}}{\sigma_{y,i}}, \tag{6.66}$$

for each dimension i, and $\tilde{\mathbf{y}}$ is then used as the input to the DNN's first hidden layer
for the current test utterance.

6.3 SAMPLING-BASED METHODS

The PMC methods in Section 6.1 rely on either the log-normal or the log-add approximation while the VTS methods in Section 6.2 rely on the first-order or higher-order VTS approximation. These approximations inevitably cause loss in model adaptation or feature enhancement. To improve the implementation accuracy of explicit distortion modeling, sampling-based methods can be used.

6.3.1 DATA-DRIVEN PMC

Data-driven parallel model combination (DPMC) (Gales, 1995) can be used to improve the modeling accuracy of PMC. This method is based on Monte-Carlo (MC) sampling by drawing random samples from the clean speech and noise distributions. In a non-iterative DPMC, the frame/state component alignment within a state does not change, and the clean speech samples are drawn from each Gaussian of the clean speech distributions.

$$\mathbf{x}(m) \sim \mathcal{N}(\boldsymbol{\mu}_x(m), \boldsymbol{\Sigma}_x(m))$$
$$\mathbf{n} \sim \mathcal{N}(\boldsymbol{\mu}_n, \boldsymbol{\Sigma}_n) \tag{6.67}$$

Then, the distorted speech samples can be obtained with Equation 6.13,

$$\mathbf{y}_i(m) = \mathbf{x}_i(m) + \mathbf{h} + \mathbf{C}\log(1 + \exp(\mathbf{C}^{-1}(\mathbf{n}_i - \mathbf{x}_i(m) - \mathbf{h}))), \tag{6.68}$$

where $\mathbf{y}_i(m)$, $\mathbf{x}_i(m)$, and $\mathbf{n}_i$ are the ith sample of the distorted speech and clean speech for Gaussian component m, and the ith sample of noise, respectively. The static mean and variance of the distorted speech for Gaussian component m are estimated as the sample mean and variance of N distorted speech samples.

$$\boldsymbol{\mu}_y(m) = \frac{1}{N}\sum_{i=0}^{N-1}\mathbf{y}_i(m), \tag{6.69}$$

$$\boldsymbol{\Sigma}_y(m) = \frac{1}{N}\sum_{i=0}^{N-1}(\mathbf{y}_i(m) - \boldsymbol{\mu}_y(m))(\mathbf{y}_i(m) - \boldsymbol{\mu}_y(m))^T, \tag{6.70}$$

where N is the number of samples.

As $N \to \infty$, the sample mean and covariance are approaching the true values. However, due to the nature of random sampling, N needs to be very large to guarantee the approximation accuracy. Hence, the biggest disadvantage of DPMC is the computational cost. As a solution, a model adaptation method based on the unscented transform (Julier and Uhlmann, 2004) is proposed in Stouten et al. (2005b) and Hu and Huo (2006).

6.3.2 UNSCENTED TRANSFORM

Originally developed to improve the extended Kalman filter and introduced to the field of robust ASR in Stouten et al. (2005b) and Hu and Huo (2006), the unscented transform (UT) (Julier and Uhlmann, 2004) gives an accurate estimate of the mean and variance parameters of a Gaussian distribution under a nonlinear transform by drawing only a limited number of samples. This is achieved by systematically drawing samples jointly from the clean speech and noise distributions, described in the following.

An augmented signal $\mathbf{s} = [\mathbf{x}^T, \mathbf{n}^T]^T$ is formed with a D-dimensional clean speech cepstral vector $\mathbf{x}$ and a noise cepstral vector $\mathbf{n}$, with dimensionality $D_s = D_x + D_n = 2D$. The UT algorithm samples the Gaussian-dependent augmented signal with $4D + 1$ sigma points $\mathbf{s}_i(m)$.

$$\mathbf{s}_0(m) = \boldsymbol{\mu}_s(m), w_0 = \frac{\kappa}{2D + \kappa},$$

$$\mathbf{s}_i(m) = \boldsymbol{\mu}_s(m) + \left(\sqrt{2D\boldsymbol{\Sigma}_s(m)}\right)_i, \quad w_i = \frac{1}{2(2D + \kappa)}, i = 1 \cdots 2D,$$

$$\mathbf{s}_i(m) = \boldsymbol{\mu}_s(m) - \left(\sqrt{2D\boldsymbol{\Sigma}_s(m)}\right)_{i-2D}, \quad w_i = \frac{1}{2(2D + \kappa)}, i = 1 + 2D \cdots 4D, \quad (6.71)$$

where κ is a control parameter whose effect is explained in Julier and Uhlmann (2004), $\boldsymbol{\mu}_s(m)$ and $\boldsymbol{\Sigma}_s(m)$ are the mean and covariance of the augmented signal, and $\left(\sqrt{\boldsymbol{\Sigma}_s(m)}\right)_i$ denotes the ith column of the square root matrix of $\boldsymbol{\Sigma}_s(m)$.

In the feature space, the ith transformed sample $\mathbf{y}_i(m)$ from the sigma point $\mathbf{s}_i(m) = [\mathbf{x}_i(m)^T, \mathbf{n}^T]^T$ is obtained with the mapping function of Equation 6.13. Then, the static mean and variance of the distorted speech are estimated as the sample mean and variance of these $4D + 1$ transformed samples.

$$\boldsymbol{\mu}_y(m) = \sum_{i=0}^{4D} w_i \mathbf{y}_i(m), \quad (6.72)$$

$$\boldsymbol{\Sigma}_y(m) = \sum_{i=0}^{4D} w_i (\mathbf{y}_i(m) - \boldsymbol{\mu}_y(m))(\mathbf{y}_i(m) - \boldsymbol{\mu}_y(m))^T. \quad (6.73)$$

It is shown in Julier and Uhlmann (2004) that the UT accurately matches the mean and covariance of the true distribution. Due to the special sampling strategy of the UT, the number of samples to be computed, $4D + 1$, is much smaller than N. Therefore model adaptation with the UT is more affordable than the MC method, although Equations 6.69 and 6.70 are very similar to Equations 6.72 and 6.73.

In Hu and Huo (2006), the static mean and variance of nonlinearly distorted speech signals are estimated using the UT, but the static noise mean and variance are estimated from a simple average of the beginning and ending frames of the current utterance. This technique was improved in Xu and Chin (2009a), where the static noise parameters were estimated online with MLE using the VTS approximation and the estimates were subsequently plugged into the UT formulation to obtain the estimate of the mean and variance of the static distorted speech features. In Faubel et al. (2010), a robust feature extraction technique is proposed to estimate the parameters of the conditional noise and channel distribution using the UT and embed the estimated parameters into the EM framework. In all of these approaches (Faubel et al., 2010; Hu and Huo, 2006; Xu and Chin, 2009a), sufficient statistics of only the static features or model parameters are estimated using the UT although adaptation of the dynamic model parameters with reliable noise and channel estimations has shown to be important (Li et al., 2007, 2009). As a solution, an approach is proposed in Li et al. (2010) to unify static and dynamic model parameter adaptation with online estimation of noise and channel parameters in the UT framework.

6.3.3 METHODS BEYOND THE GAUSSIAN ASSUMPTION

As shown in Figure 3.3, with the introduction of noise, the distorted speech is no longer Gaussian distributed. Therefore, the popular one-to-one Gaussian mapping used in the aforementioned methods has a theoretic flaw. The iterative DPMC method (Gales, 1995) solves this problem by sampling from GMMs instead of Gaussians and then uses the Baum-Welch algorithm to re-estimate the distorted speech parameters. This is extended in van Dalen and Gales (2011b), where a variational method is used to remove the constraint that the samples must be used to model the Gaussians they are originally drawn from. This is also extended to variational PCMLLR (van Dalen and Gales, 2011b), which is shown to be better than PCMLLR (Gales and van Dalen, 2007) and has a much lower computational cost than variational DPMC. In Leutnant et al. (2011) the Gaussian at the input of the nonlinearity is approximated by a GMM whose individual components have a smaller variance than the original Gaussian. A VTS linearization of the individual GMM components then incurs smaller approximation error than the linearization of the original Gaussian. Thus the overall modeling accuracy could be improved.

6.4 ACOUSTIC FACTORIZATION

The explicit distortion modeling methods discussed so far separate the clean speech feature from the environment (noise and channel) factors. This can be further extended to an acoustic factorization (Gales, 2001) problem: separate the clean speech feature/model from the multiple speaker and environmental factors irrelevant to the phonetic classification. In Gales (2001), it is done by using MLLR adaptation to capture speaker variability and using cluster adaptive training to capture environment variability. This work can be considered as source separation with an implicit physical model. To achieve factorization, it is crucial that each factor transform only models the impact of its associated factor, and keeps it independent from others.

Acoustic factorization can be considered as a general form of explicit distortion modeling in which underlying acoustic factors are embedded into a single distortion function. Strictly speaking, not all acoustic factorization methods can be considered as explicit distortion modeling methods. Only the methods that embed all the factors into a mismatch function that describes the physical model of distortion fall into the category of explicit distortion modeling.

6.4.1 ACOUSTIC FACTORIZATION FRAMEWORK

Wang (2015) describes the acoustic factorization framework based on the assumption that acoustic factors impact speech signals independently. Without loss of generality, we consider a complex acoustic environment, simultaneously affected by two acoustic factors s_i and n_j respectively, where **s** and **n** are the types of acoustic factors and i, j are the indices within each acoustic factor. The factor-independent model Λ is adapted to represent this condition by the transform $\mathcal{T}^{(i,j)}$:

$$\hat{\Lambda}^{(i,j)} = \mathcal{F}(\Lambda, \mathcal{T}^{(i,j)}), \tag{6.74}$$

where $\hat{\Lambda}^{(i,j)}$ is the adapted acoustic model for the acoustic condition (s_i, n_j), $\mathcal{T}^{(i,j)}$ is the transform for the target condition, and $\mathcal{F}$ is a mapping function.

In the traditional approaches, the transform is usually estimated as

$$\mathcal{T}^{(i,j)} = \arg\max_{\mathcal{T}} p(Y^{(i,j)}|\Lambda, \mathcal{T}), \tag{6.75}$$

where $Y^{(i,j)}$ is the feature sequence in the acoustic condition (s_i, n_j). However, there may be not enough adaptation data for every combined acoustic condition from acoustic factor space **s** and **n**. Acoustic factorization (Gales, 2001) is proposed to deal with complex acoustic environments by decomposing the target transform into a few factor transforms, each of which is only related to an individual acoustic factor. For example, $\mathcal{T}^{(i,j)}$ in the above example can be decomposed as

$$\mathcal{T}^{(i,j)} = \mathcal{T}_s^{(i)} \otimes \mathcal{T}_n^{(j)} \qquad i \in 1, \dots, I, and \qquad j \in 1, \dots, J, \tag{6.76}$$

where $\otimes$ denotes transform composition and its exact form depends on the nature of factor transforms; $\mathcal{T}_s^{(i)}$ and $\mathcal{T}_n^{(j)}$ are the factor transforms associated with s_i and n_j, respectively. Now instead of estimating $I * J$ transforms, it only needs to estimate I transforms for factor **s** and J transforms for factor **n**. Figure 6.5 (Wang, 2015) shows the concepts of acoustic factorization with speaker factor s and noise factor n. Utterances from the ith speaker can be pooled to estimate the transform $\mathcal{T}_s^{(i)}$ while utterances recorded in the jth noise condition can be pooled to estimate the transform $\mathcal{T}_n^{(j)}$. For the data produced by speaker s_i in unseen noise condition $n_{j'}$, it only needs to estimate the noise transform $\mathcal{T}_n^{(j')}$ and combine this transform with the existing speaker transform $\mathcal{T}_s^{(i)}$ using Equation 6.76 to compose the target transform $\mathcal{T}^{(i,j')}$.

6.4.2 ACOUSTIC FACTORIZATION FOR GMM

The most popular acoustic factors addressed in literature are speaker and noise. There are several categories of methods that perform acoustic factorization for the underlying GMM models (Wang, 2015). The simplest category is called data-constrained approach which relies on balanced data to force a factor transform to learn the impact caused by its associated factor, while keeping other factors invariant. The representative method is proposed in Seltzer and Acero (2011a,b), where the impacts

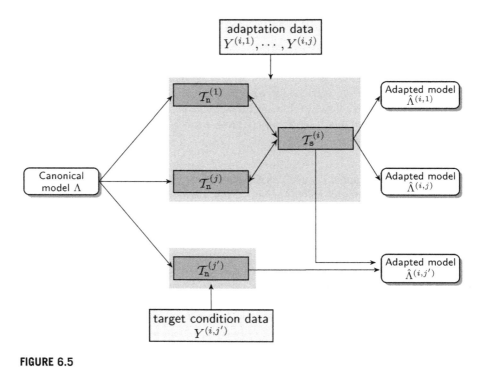

FIGURE 6.5

Acoustic factorization framework.

of speaker and noise factors are represented by two sets of CMLLR transforms in a cascaded fashion, that is, the likelihood of a particular observation vector $\mathbf{y}_t$ produced by speaker i under environment j emitted by the mth Gaussian component is:

$$p(\mathbf{y}_t|i,j,m) = |\mathbf{A}_s^{(i)}||\mathbf{A}_n^{(j)}|\mathcal{N}(\hat{\mathbf{x}}_t; \mu_x(m), \Sigma_x(m)) \qquad (6.77)$$

$$\hat{\mathbf{x}}_t = \mathbf{A}_s^{(i)}(\mathbf{A}_n^{(j)}\mathbf{y}_t + \mathbf{b}_n^{(j)}) + \mathbf{b}_s^{(i)}$$

where $\mu_x(m)$ and $\Sigma_x(m)$ are the factor-independent model parameters, and $W_s^{(i)} = [\mathbf{A}_s^{(i)}, \mathbf{b}_s^{(i)}]$ and $W_n^{(j)} = [\mathbf{A}_n^{(j)}, \mathbf{b}_n^{(j)}]$ are the transforms for acoustic factors s_i and n_j, respectively. In Seltzer and Acero (2011a,b), $\mathbf{s}$ is the speaker factor and $\mathbf{n}$ is the noise factor. Using this cascaded CMLLR transform, the noise and speaker distorted speech vector $\mathbf{y}_t$ is normalized to $\hat{\mathbf{x}}_t$, and the transforms are estimated from the corresponding speaker or noise data.

The second category embeds all the factors into a mismatch function. The mismatch function in Equation 6.14 takes the noise and channel factors into consideration. To address the speaker and noise factorization problem, the following distortion function is used for the ith speaker and jth noise condition

$$\mathbf{y}_t^{(i,j)} = \mathbf{C} \log \left(\exp(\mathbf{C}^{-1}(\mathbf{x}_t + \mathbf{b}_s^{(i)})) + \exp(\mathbf{C}^{-1}\mathbf{n}_t^{(j)}) \right), \tag{6.78}$$

where $\mathbf{y}_t^{(i,j)}$ is the observation in the ith speaker and jth noise condition, $\mathbf{b}_s^{(i)}$ is the ith speaker bias, $\mathbf{n}_t^{(j)}$ is the additive noise generated in the jth noise condition, which is assumed to follow a Gaussian distribution $\mathcal{N}(\mu_n^{(j)}, \Sigma_n^{(j)})$, and $\mathbf{C}$ is the DCT matrix. Note this mismatch function is similar to Equation 6.14, while the channel distortion is ignored here to concentrate on the effect of speaker and additive noise.

By combining the VTS transform and the speaker bias transform, a model transform which maps the factor-independent model $\Lambda = \mathcal{N}(\mu_x, \Sigma_x)$ to the adapted model $\hat{\Lambda}^{(i,j)} = \mathcal{N}(\mu_y^{(i,j)}, \Sigma_y^{(i,j)})$ can be derived:

$$\mu_y^{(i,j)} = \mathbf{C} \log \left(\exp(\mathbf{C}^{-1}(\mu_x + \mathbf{b}_s^{(i)})) + \exp(\mathbf{C}^{-1}\mu_n^{(j)}) \right), \tag{6.79}$$

$$\Sigma_y^{(i,j)} = \mathbf{J}_x \Sigma_x \mathbf{J}_x^T + \mathbf{J}_n \Sigma_n^{(j)} \mathbf{J}_n^T, \tag{6.80}$$

where $\mathbf{J}_x$ and $\mathbf{J}_n$ are the Jacobian matrices,

$$\mathbf{J}_x = \left. \frac{\partial \mathbf{y}_t^{(i,j)}}{\partial \mathbf{x}_t} \right|_{(\mu_x + \mathbf{b}_s^{(i)}, \mu_n^{(j)})} \qquad \mathbf{J}_n = \left. \frac{\partial \mathbf{y}_t^{(i,j)}}{\partial \mathbf{n}_t} \right|_{(\mu_x + \mathbf{b}_s^{(i)}, \mu_n^{(j)})}. \tag{6.81}$$

It can be seen from the compensation formula that the speaker transform $\mathcal{T}_s^{(i)} = \mathbf{b}_s^{(i)}$ linearly transforms the model parameters, while a nonlinear transform $\mathcal{T}_n^{(j)}$ parameterized by $(\mu_n^{(j)}, \Sigma_n^{(j)})$ is used for the noise factor. It can also be seen that these transforms are constrained in the mismatch function. Hence, this approach is called transform-constrained approach. In Rigazio et al. (2001), Jacobian adaptation (a form of VTS adaptation) for noise compensation is combined with MLLR for speaker adaptation. In Wang and Gales (2011b), Chin et al. (2011), and Wang and Gales (2012), VTS is used for noisy environment adaptation and MLLR or vocal tract length normalization (VTLN) (Lee and Rose, 1998) is used for speaker adaptation.

The first two factorization schemes rely on various assumptions to implicitly maintain the independence between factors: the data-constrained approach assumes that data is distributed over acoustic factors in a balanced manner, while the transform-constrained approach assumes the form of the relationship between acoustic factors and the observed speech feature, usually summarized in a so called mismatch function, is known. The third category, explicit-independence-constrained approach, enforces an explicit independence constraint (Wang and Gales, 2013) to assure the independence of acoustic factors. This is done by maximizing the log-likelihood function via the EM algorithm using the following auxiliary function

$$Q(\hat{\Lambda}; \Lambda) = \sum_{m,t} \gamma_t^{(m)} \log p(\mathbf{y}_t; \mu(m), \Sigma(m), \mathcal{T}_s \otimes \mathcal{T}_n) \tag{6.82}$$

with the constraint

$$\frac{\partial^2 Q(\hat{\Lambda}; \Lambda)}{\partial \mathcal{T}_s \partial \mathcal{T}_n} = \mathbf{0}. \tag{6.83}$$

There are also other studies for acoustic factorization. In García et al. (2011) and Joshi et al. (2011), VTLN used for speaker normalization is combined with HEQ used for environment normalization.

In Rouvier et al. (2011), factor analysis is used to remove the impact of speaker and channel variability. In Karafiát et al. (2011), the speaker and acoustic environment information is represented by i-vector (Dehak et al., 2011) which is a low-dimensional fixed length vector extracted from the feature space. Then, the compensated feature based on i-vector can well address the mismatch of speaker and environment between training and testing.

6.4.3 ACOUSTIC FACTORIZATION FOR DNN

Although the aforementioned methods use GMM as the back-end, acoustic factorization can also be applied to DNN models. In Li et al. (2014), a DNN adaptation method was proposed by taking into account the underlying factors that contribute to the distorted speech signal. Denote $\mathbf{r} = z(\mathbf{v}^L)$ as the output vector right before the softmax activation in Equation 2.32. Now we consider the case that the clean input feature, $\mathbf{x}$, has been distorted by environment factors to become $\mathbf{y}$. Merging the layer-by-layer nonlinear functions, we can denote $\mathbf{r} = R(\mathbf{y})$, where $R(\cdot)$ represents the overall nonlinear function in a DNN. To adapt a clean-trained DNN to a new environment, vector $\mathbf{r}$ is compensated by removing those unwanted parts in the network outputs caused by acoustic factors (Li et al., 2014), as shown in Figure 6.6. Specifically, the modified vector $\mathbf{r}'$ is obtained by

$$\mathbf{r}' = R(\mathbf{y}) + \sum_{n=1}^{N} \mathbf{Q}_n \mathbf{f}_n, \tag{6.84}$$

where $\mathbf{f}_n$ is the underlying nth acoustic factor and $\mathbf{Q}_n$ is the corresponding loading matrix. Then $\mathbf{r}'$ is used to calculate the posterior probability as

$$P(o = s|\mathbf{x}) = \text{softmax}(\mathbf{r}'). \tag{6.85}$$

When adapting the clean-trained DNN to a new environment, we extract the factors $[\mathbf{f}_1, \ldots \mathbf{f}_N]$ from adaptation utterances, and then train the loading matrices $[\mathbf{Q}_1, \ldots \mathbf{Q}_N]$ using the standard back-propagation algorithm while keeping the original DNN parameters intact. Then this adapted DNN can be used to decode the utterance from the new environment. The advantage of the factorized adaptation method is that the number of parameters in the factor loading matrices is much

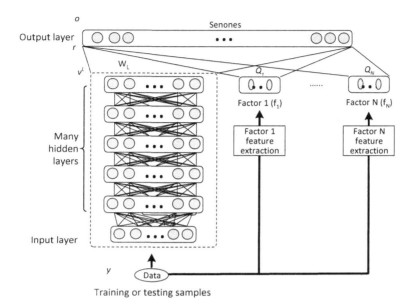

FIGURE 6.6

The flow chart of factorized adaptation for a DNN at the output layer.

smaller than that of the original DNN, therefore it only requires very few adaptation utterances.

Although factors can be used in the output layer of a DNN as in Figure 6.6, they can also be used in any layer of a DNN. For example, as in Figure 6.7, the factors can be used to augment the original feature as the new input feature.

Again, only the loading matrices $[\mathbf{Q}_1, \ldots, \mathbf{Q}_N]$ are trained using standard back-propagation, given few adaptation utterances. It is arguable as to which method is better: factorizing in the input or output layer. The argument of factorizing in the output layer is that it directly changes the output with factors. The argument of factorizing in the input layer is that the distortion caused by factors may be absorbed by the layer-by-layer nonlinearity, which indeed may not be effective because the original DNN parameters are not changed for adaptation scenarios with limited amount of training data.

However, if the original DNN parameters can be changed with large amount of adaptation utterances, it is clear that factorizing in the input layer for adaptation is better by leveraging the layer-by-layer nonlinearity until the output layer as in Figure 6.7. It also makes sense to factorize in the input layer for DNN training with large amount of data and can be benefited from optimizing the layer-by-layer nonlinearity.

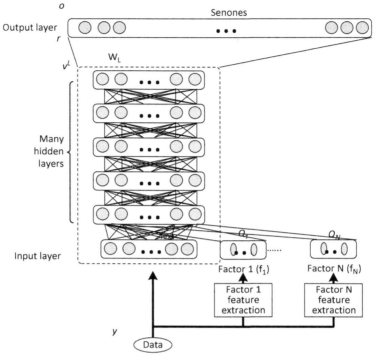

FIGURE 6.7

The flow chart of factorized training or adaptation for a DNN at the input layer.

This type of methods can be called factorized DNN training. In Seltzer et al. (2013), the factor is chosen as the noise estimate of the current utterance to train a DNN. It is extended in Narayanan and Wang (2014), where residual noise estimate or speech estimate is used as the factor in addition to the noise estimate. Other factors, such as i-vector used to characterize speakers, can also be used in the input layer for DNN training (Miao et al., 2014; Saon et al., 2013; Senior and Lopez-Moreno, 2014).

6.5 SUMMARY

We now summarize the major representative distortion modeling methods for noise-robust ASR discussed in this chapter in Table 6.1 in a chronological order to facilitate clear understanding of the technology development. Other methods, such as Wiener

Table 6.1 Distortion Modeling Methods in Chapter 6, Arranged Chronologically

Method	Proposed Around	Characteristics
Parallel model combination (PMC) (Gales, 1995)	1995	Combines noise and speech models to form a model for distorted speech signal in the log-Mel-filter-bank domain and then transfer back to the cepstral domain
Data-driven parallel model combination (DPMC) (Gales, 1995)	1995	Improves the modeling accuracy of PMC by drawing random samples from the clean speech and noise distribution to construct distorted speech samples for model parameter estimation
Vector Taylor series (VTS) model adaptation (Moreno, 1996; Acero et al., 2000)	1996	VTS expansion is used to linearize the nonlinear physical distortion model to find closed-form solutions to HMM parameter adaptation
VTS feature enhancement (Moreno, 1996; Stouten et al., 2003)	1996	Uses a small GMM in the front-end to derive a minimum mean square error estimate of the clean speech features given the noisy observations
high-order VTS (Kim, 1998; Stouten et al., 2005a)	1998	Uses the high-order instead of the first-order VTS expansion to better approximate the nonlinear physical distortion model
data-constrained acoustic factorization (Gales, 2001; Seltzer and Acero, 2011b)	2001	Forces a factor transform to learn the impact caused by its associated factor, while keeping other factors invariant
Jointly compensate for additive and convolutive (JAC) (Gong, 2005)	2005	Jointly compensates for additive and convolutive distortion using the physical distortion model and provides joint-estimation EM of channel and noise components
Unscented transform (UT) (Hu and Huo, 2006)	2006	Uses the UT as a better sampling strategy to draw only limited clean speech and noise samples and provide an accurate match to the mean and covariance of the true distribution
VTS model adaptation with all distortion re-estimation (Li et al., 2007, 2009)	2007	Re-estimates the static and dynamic mean and variance of all the distortion parameters
phase-sensitive VTS (Li et al., 2008; Li et al., 2009)	2008	A phase term is introduced into the nonlinear physical distortion model for better modeling

Continued

Table 6.1 Distortion Modeling Methods in Chapter 6, Arranged Chronologically—cont'd

Method	Proposed Around	Characteristics
extended VTS (van Dalen and Gales, 2009)	2009	Provides a more accurate form to adapt dynamic model parameters instead of using the continuous time approximation in VTS
UT with full model parameter adaptation (Li et al., 2010)	2010	Unifies the static and dynamic model parameter adaptation with online estimation of noise and channel parameters in the UT framework
Transform-constrained acoustic factorization (Wang and Gales, 2011b, 2012)	2011	Embeds all the acoustic factors into a mismatch function
VTS feature enhancement with all distortion re-estimation (Li et al., 2012b)	2012	Re-estimates the static and dynamic mean and variance of all the distortion parameters for VTS feature enhancement
Explicit-independence-constrained acoustic factorization (Wang and Gales, 2013)	2013	Enforces an explicit independence constraint into an objective function to assure the independence of acoustic factors
Transformed feature normalization for DNN (Li and Sim, 2013)	2013	The global mean and variance of the clean training data are transformed to the distorted mean and variance of the current test utterance, which are used to preform feature normalization for the DNN
Acoustic factorization for DNN (Li et al., 2014)	2014	Extracts different acoustic factors as augmented features and connects these factor features either at the input layer or at the output layer of the DNN

filtering and spectral subtraction described in Section 4.1.3 and the joint uncertainty decoding in Section 7.3, are not listed in the table, although they can also be viewed as distortion modeling methods.

Two main themes of the methods presented in this chapter can be highlighted as follows:

- re-estimation of all distortion parameters improves the performance of the related algorithms (e.g., VTS and UT);
- better distortion modeling results in better algorithm performance

From the chronology of the studies on developing explicit acoustic distortion models of speech represented in the table, we can see the trend of building the increasingly sophisticated models that characterize the impact of different mismatch acoustic sources, including noises, channels, and interfering speakers. Importantly, these better and better distortion models and the related robust techniques are already providing outstanding performance which is superior to other methods exploiting less powerful distortion models. Further, distortion-robust approaches based on the explicit distortion models require far less data to estimate parameters than the approaches without an explicit distortion model. A greater performance gap is expected in the future as more advanced distortion models are being developed and incorporated into robust ASR methods.

On the other hand, explicit distortion modeling is very hard to apply to DNN models in which model parameters are weight matrices because there is no simple and analytic relationship between the values of weight matrices and feature vectors. Therefore, it is far from straightforward to adapt the DNN weight matrices according to the distortion function. It was recently shown in Li et al. (2014) that complex approximations are needed to obtain the relationship between the clean and distorted output vectors of a DNN using the distortion function. As a result, we can appreciate why there are limited studies currently in the literature on the use of explicit distortion models in DNN-based ASR. Given the huge success of explicit distortion modeling in the GMM-based ASR, how to integrate explicit distortion modeling into DNN modeling is a promising but challenging topic. Nevertheless, we believe that DNN technology for ASR will be significantly advanced after this challenge is overcome in the future.

REFERENCES

Acero, A., 1993. Acoustical and Environmental Robustness in Automatic Speech Recognition. Cambridge University Press, Cambridge, UK.

Acero, A., Deng, L., Kristjansson, T., Zhang, J., 2000. HMM adaptation using vector Taylor series for noisy speech recognition. In: Proc. International Conference on Spoken Language Processing (ICSLP), pp. 869-872.

Baker, J., Deng, L., Glass, J., Khudanpur, S., Lee, C.H., Morgan, N., et al., 2009a. Research developments and directions in speech recognition and understanding, part i. IEEE Signal Process. Mag. 26 (3), 75-80.

Baker, J., Deng, L., Glass, J., Khudanpur, S., Lee, C.H., Morgan, N., et al., 2009b. Updated MINDS report on speech recognition and understanding. IEEE Signal Process. Mag. 26 (4), 78-85.

Bu, S., Qian, Y., Sim, K.C., You, Y., Yu, K., 2014a. Second order vector Taylor series based robust speech recognition. In: Proc. International Conference on Acoustics, Speech and Signal Processing (ICASSP), pp. 1769-1773.

Bu, S., Qian, Y., Yu, K., 2014b. A novel dynamic parameters calculation approach for model compensation. In: Proc. Interspeech.

Chin, K.K., Xu, H., Gales, M.J.F., Breslin, C., Knill, K., 2011. Rapid joint speaker and noise compensation for robust speech recognition. In: Proc. International Conference on Acoustics, Speech and Signal Processing (ICASSP), pp. 5500-5503.

Dehak, N., Kenny, P., Dehak, R., Dumouchel, P., Ouellet, P., 2011. Front-end factor analysis for speaker verification. IEEE Trans. Audio Speech Lang. Process. 19 (4), 788-798.

Deng, L., 2006. Dynamic Speech Models—Theory, Algorithm, and Applications. Morgan and Claypool, San Rafael, CA.

Deng, L., 2011. Front-end, back-end, and hybrid techniques for noise-robust speech recognition. In: Robust Speech Recognition of Uncertain or Missing Data: Theory and Application. Springer, New York, pp. 67-99.

Deng, L., Droppo, J., Acero, A., 2002. A Bayesian approach to speech feature enhancement using the dynamic cepstral prior. In: Proc. International Conference on Acoustics, Speech and Signal Processing (ICASSP), vol. 1, pp. I-829-I-832.

Deng, L., Droppo, J., Acero, A., 2004a. Enhancement of Log Mel power spectra of speech using a phase-sensitive model of the acoustic environment and sequential estimation of the corrupting noise. IEEE Trans. Speech Audio Process. 12 (2), 133-143.

Deng, L., Droppo, J., Acero, A., 2004b. Estimating cepstrum of speech under the presence of noise using a joint prior of static and dynamic features. IEEE Trans. Audio Speech Lang. Process. 12 (3), 218-233.

Deng, L., Li, X., 2013. Machine learning paradigms in speech recognition: An overview. IEEE Trans. Audio Speech and Lang. Process. 21 (5), 1060-1089.

Deng, L., O'Shaughnessy, D., 2003. Speech Processing—A Dynamic and Optimization-Oriented Approach. Marcel Dekker Inc., New York.

Deng, L., Yu, D., 2014. Deep Learning: Methods and Applications. Now Publishers, Hanover, MA.

Ding, G.H., Xu, B., 2004. Exploring high-performance speech recognition in noisy environments using high-order Taylor series expansion. In: Proc. International Conference on Spoken Language Processing (ICSLP), pp. 149-152.

Droppo, J., Deng, L., Acero, A., 2003. A comparison of three non-linear observation models for noisy speech features. In: Proc. European Conference on Speech Communication and Technology (EUROSPEECH), pp. 681-684.

Du, J., Huo, Q., 2008. A speech enhancement approach using piecewise linear approximation of an explicit model of environmental distortions. In: Proc. Interspeech, pp. 569-572.

Du, J., Huo, Q., 2011. A feature compensation approach using high-order vector Taylor series approximation of an explicit distortion model for noisy speech recognition. IEEE Trans. Audio Speech Lang. Process. 19 (8), 2285-2293.

Faubel, F., McDonough, J., Klakow, D., 2010. On expectation maximization based channel and noise estimation beyond the vector Taylor series expansion. In: Proc. International Conference on Acoustics, Speech and Signal Processing (ICASSP), pp. 4294-4297.

Frey, B., Deng, L., Acero, A., Kristjansson, T., 2001a. ALGONQUIN: iterating Laplace's method to remove multiple types of acoustic distortion for robust speech recognition. In: Proc. Interspeech, pp. 901-904.

Frey, B., Kristjansson, T., Deng, L., Acero, A., 2001b. ALGONQUIN—learning dynamic noise models from noisy speech for robust speech recognition. In: NIPS, pp. 1165-1171.

Gales, M.J.F., 1995. Model-based techniques for noise robust speech recognition. Ph.D. thesis, University of Cambridge.

Gales, M.J.F., 1998. Predictive model-based compensation schemes for robust speech recognition. Speech Commun. 25, 49-74.

Gales, M.J.F., 2001. Acoustic factorisation. In: Proc. IEEE Workshop on Automatic Speech Recognition and Understanding (ASRU), pp. 77-80.

Gales, M.J.F., Flego, F., 2008. Discriminative classifiers and generative kernels for noise robust speech recognition. University of Cambridge.

Gales, M.J.F., van Dalen, R.C., 2007. Predictive linear transforms for noise robust speech recognition. In: Proc. IEEE Workshop on Automatic Speech Recognition and Understanding (ASRU), pp. 59-64.

García, L., Benítez, C., Segura, J.C., Umesh, S., 2011. Combining speaker and noise feature normalization techniques for automatic speech recognition. In: Proc. International Conference on Acoustics, Speech and Signal Processing (ICASSP), pp. 5496-5499.

Gong, Y., 2005. A method of joint compensation of additive and convolutive distortions for speaker-independent speech recognition. IEEE Trans. Speech Audio Process. 13 (5), 975-983.

Gopinath, R.A., Gales, M., Gopalakrishnan, P.S., Balakrishnan-Aiyer, S., Picheny, M.A., 1995. Robust speech recognition in noise—performance of the IBM continuous speech recognizer on the ARPA noise spoken task. In: Proc. of DARPA Workshop on Spoken Language System and Technology, pp. 127-130.

Hu, Y., Huo, Q., 2006. An HMM compensation approach using unscented transformation for noisy speech recognition. In: ISCSLP.

Joshi, V., Bilgi, R., Umesh, S., Benitez, C., Garcia, L., 2011. Efficient speaker and noise normalization for robust speech recognition. In: Proc. Interspeech, pp. 2601-2604.

Julier, S.J., Uhlmann, J.K., 2004. Unscented filtering and nonlinear estimation. Proceed. IEEE 92 (3), 401-422.

Kalgaonkar, K., Seltzer, M.L., Acero, A., 2009. Noise robust model adaptation using linear spline interpolation. In: Proc. IEEE Workshop on Automatic Speech Recognition and Understanding (ASRU), pp. 199-204.

Kalinli, O., Seltzer, M.L., Droppo, J., Acero, A., 2010. Noise adaptive training for robust automatic speech recognition. IEEE Trans. Audio Speech Lang. Process. 18 (8), 1889-1901.

Karafiát, M., Burget, L., Matejka, P., Glembek, O., Cernocký, J., 2011. iVector-based discriminative adaptation for automatic speech recognition. In: Proc. IEEE Workshop on Automatic Speech Recognition and Understanding (ASRU), pp. 152-157.

Kim, D.Y., Un, C.K., Kim, N.S., 1998. Speech recognition in noisy environments using first-order vector Taylor series. Speech Commun. 24 (1), 39-49.

Kim, N.S., 1998. Statistical linear approximation for environment compensation. IEEE Signal Process. Lett. 5 (1), 8-10.

LeCun, Y., Bottou, L., Orr, G.B., Müller, K.R., 1998. Efficient backprop. In: Neural Networks: Tricks of the Trade. Springer, New York, pp. 9-50.

Lee, L., Rose, R., 1998. A frequency warping approach to speaker normalization. IEEE Trans. Speech Audio Process. 6 (1), 49-60.

Leutnant, V., Haeb-Umbach, R., 2009. An analytic derivation of a phase-sensitive observation model for noise-robust speech recognition. In: Proc. Interspeech, pp. 2395-2398.

Leutnant, V., Krueger, A., Haeb-Umbach, R., 2011. A versatile Gaussian splitting approach to non-linear state estimation and its application to noise-robust ASR. In: Proc. Interspeech, pp. 1641-1644.

Li, B., Sim, K.C., 2013. Noise adaptive front-end normalization based on vector Taylor series for deep neural networks in robust speech recognition. In: Proc. International Conference on Acoustics, Speech and Signal Processing (ICASSP), pp. 7408-7412.

Li, J., Deng, L., Yu, D., Gong, Y., 2011. Towards high-accuracy low-cost noisy robust speech recognition exploiting structured model. In: ICML Workshop on Learning Architectures, Representations, and Optimization for Speech and Visual Information Processing.

Li, J., Deng, L., Yu, D., Gong, Y., Acero, A., 2007. High-performance HMM adaptation with joint compensation of additive and convolutive distortions via vector Taylor series. In: Proc. IEEE Workshop on Automatic Speech Recognition and Understanding (ASRU), pp. 65-70.

Li, J., Deng, L., Yu, D., Gong, Y., Acero, A., 2008. HMM adaptation using a phase-sensitive acoustic distortion model for environment-robust speech recognition. In: Proc. International Conference on Acoustics, Speech and Signal Processing (ICASSP), pp. 4069-4072.

Li, J., Deng, L., Yu, D., Gong, Y., Acero, A., 2009. A unified framework of HMM adaptation with joint compensation of additive and convolutive distortions. Comput. Speech Lang. 23 (3), 389-405.

Li, J., Huang, J.T., Gong, Y., 2014. Factorized adaptation for deep neural network. In: Proc. International Conference on Acoustics, Speech and Signal Processing (ICASSP).

Li, J., Seltzer, M.L., Gong, Y., 2012a. Efficient VTS adaptation using Jacobian approximation. In: Proc. Interspeech, pp. 1906-1909.

Li, J., Seltzer, M.L., Gong, Y., 2012b. Improvements to VTS feature enhancement. In: Proc. International Conference on Acoustics, Speech and Signal Processing (ICASSP), pp. 4677-4680.

Li, J., Yu, D., Gong, Y., Deng, L., 2010. Unscented transform with online distortion estimation for HMM adaptation. In: Proc. Interspeech, pp. 1660-1663.

Liao, H., Gales, M.J.F., 2006. Joint uncertainty decoding for robust large vocabulary speech recognition. University of Cambridge.

Liao, H., Gales, M.J.F., 2008. Issues with uncertainty decoding for noise robust automatic speech recognition. Speech Commun. 50 (4), 265-277.

Macho, D., Mauuary, L., Noé., B., Cheng, Y.M., Ealey, D., Jouvet, D., et al., 2002. Evaluation of a noise-robust DSR front-end on Aurora databases. In: Proc. International Conference on Spoken Language Processing (ICSLP), pp. 17-20.

Miao, Y., Zhang, H., Metze, F., 2014. Towards speaker adaptive training of deep neural network acoustic models. In: Proc. Interspeech.

Minami, Y., Furui, S., 1995. A maximum likelihood procedure for a universal adaptation method based on HMM composition. In: Proc. International Conference on Acoustics, Speech and Signal Processing (ICASSP), pp. 129-132.

Minami, Y., Furui, S., 1996. Adaptation method based on HMM composition and EM algorithm. In: Proc. International Conference on Acoustics, Speech and Signal Processing (ICASSP), pp. 327-330.

Moreno, P.J., 1996, Speech recognition in noisy environments. PhD thesis, Carnegie Mellon University.

Narayanan, A., Wang, D., 2014. Joint noise adaptive training for robust automatic speech recognition. In: Proc. International Conference on Acoustics, Speech and Signal Processing (ICASSP).

Rigazio, L., Nguyen, P., Kryze, D., Junqua, J.C., 2001. Separating speaker and environment variabilities for improved recognition in non-stationary conditions. In: Proc. Interspeech, pp. 2347-2350.

Rouvier, M., Bouallegue, M., Matrouf, D., Linarès, G., 2011. Factor analysis based session variability compensation for automatic speech recognition. In: Proc. IEEE Workshop on Automatic Speech Recognition and Understanding (ASRU), pp. 141-145.

Sagayama, S., Yamaguchi, Y., Takahashi, S., Takahashi, J., 1997. Jacobian approach to fast acoustic model adaptation. In: Proc. International Conference on Acoustics, Speech and Signal Processing (ICASSP), vol. 2, pp. 835-838.

Sankar, A., Lee, C.H., 1995. Robust speech recognition based on stochastic matching. In: Proc. International Conference on Acoustics, Speech and Signal Processing (ICASSP), pp. 121-124.

Saon, G., Soltau, H., Nahamoo, D., Picheny, M., 2013. Speaker adaptation of neural network acoustic models using i-vectors. In: Proc. IEEE Workshop on Automatic Speech Recognition and Understanding (ASRU), pp. 55-59.

Seltzer, M.L., Acero, A., 2011a. Factored adaptation for separable compensation of speaker and environmental variability. In: Proc. IEEE Workshop on Automatic Speech Recognition and Understanding (ASRU), pp. 146-151.

Seltzer, M.L., Acero, A., 2011b. Separating speaker and environmental variability using factored transforms. In: Proc. Interspeech, pp. 1097-1100.

Seltzer, M.L., Kalgaonkar, K., Acero, A., 2010. Acoustic model adaptation via linear spline interpolation for robust speech recognition. In: Proc. International Conference on Acoustics, Speech and Signal Processing (ICASSP), pp. 4550-4553.

Seltzer, M.L., Yu, D., Wang, Y., 2013. An investigation of deep neural networks for noise robust speech recognition. In: Proc. International Conference on Acoustics, Speech and Signal Processing (ICASSP), pp. 7398-7402.

Senior, A., Lopez-Moreno, I., 2014. Improving DNN speaker independence with i-vector inputs. In: Proc. International Conference on Acoustics, Speech and Signal Processing (ICASSP), pp. 225-229.

Shinoda, K., Lee, C.H., 2001. A structural Bayes approach to speaker adaptation. IEEE Trans. Speech Audio Process. 9 (3), 276-287.

Stouten, V., 2006. Robust Automatic Speech Recognition in Time-varying Environments. Ph.D. thesis, K. U. Leuven.

Stouten, V., Hamme, H.V., Demuynck, K., Wambacq, P., 2003. Robust speech recognition using model-based feature enhancement. In: Proc. European Conference on Speech Communication and Technology (EUROSPEECH), pp. 17-20.

Stouten, V., Hamme, H.V., Wambacq, P., 2005a. Effect of phase-sensitive environment model and higher order VTS on noisy speech feature enhancement. In: Proc. International Conference on Acoustics, Speech and Signal Processing (ICASSP), vol. I, pp. 433-436.

Stouten, V., Hamme, H.V., Wambacq, P., 2005b. Kalman and unscented Kalman filter feature enhancement for noise robust ASR. In: Proc. Interspeech, pp. 953-956.

van Dalen, R.C., Gales, M.J.F., 2009. Extended VTS for noise-robust speech recognition. In: Proc. International Conference on Acoustics, Speech and Signal Processing (ICASSP), pp. 3829-3832.

van Dalen, R.C., Gales, M.J.F., 2011a. Extended VTS for noise-robust speech recognition. IEEE Trans. Audio Speech Lang. Process. 19 (4), 733-743.

van Dalen, R.C., Gales, M.J.F., 2011b. A variational perspective on noise-robust speech recognition. In: Proc. IEEE Workshop on Automatic Speech Recognition and Understanding (ASRU), pp. 125-130.

Wang, Y., 2015. Model-based Approaches to Robust Speech Recognition in Diverse Environments. Ph.D. thesis, University of Cambridge.

Wang, Y., Gales, M.J.F., 2011a. Improving reverberant vts for hands-free robust speech recognition. In: Proc. IEEE Workshop on Automatic Speech Recognition and Understanding (ASRU), pp. 113-118.

Wang, Y., Gales, M.J.F., 2011b. Speaker and noise factorisation on aurora4 task. In: Proc. International Conference on Acoustics, Speech and Signal Processing (ICASSP), pp. 4584-4587.

Wang, Y., Gales, M.J.F., 2012. Speaker and noise factorisation for robust speech recognition. IEEE Trans. Audio Speech Lang. Process. 20 (7), 2149-2158.

Wang, Y., Gales, M.J.F., 2013. An explicit independence constraint for factorised adaptation in speech recognition. In: Proc. Interspeech, pp. 1233-1237.

Xu, H., Chin, K.K., 2009a. Comparison of estimation techniques in joint uncertainty decoding for noise robust speech recognition. In: Proc. Interspeech, pp. 2403-2406.

Xu, H., Chin, K.K., 2009b. Joint uncertainty decoding with the second order approximation for noise robust speech recognition. In: Proc. International Conference on Acoustics, Speech and Signal Processing (ICASSP), vols. 3841-3844.

Xu, H., Gales, M.J.F., Chin, K.K., 2011. Joint uncertainty decoding with predictive methods for noise robust speech recognition. IEEE Trans. Audio Speech Lang. Process. 19 (6), 1665-1676.

Yu, D., Deng, L., 2014. Automatic Speech Recognition—A Deep Learning Approach. Springer, New York.

Zhao, Y., 2000. Frequency-domain maximum likelihood estimation for automatic speech recognition in additive and convolutive noises. IEEE Trans. Speech Audio Process. 8 (3), 255-266.

Zhao, Y., Juang, B.H., 2010. A comparative study of noise estimation algorithms for VTS-based robust speech recognition. In: Proc. Interspeech, pp. 2090-2093.

Zhao, Y., Juang, B.H., 2012. Nonlinear compensation using the Gauss-Newton method for noise-robust speech recognition. IEEE Trans. Audio Speech Lang. Process. 20 (8), 2191-2206.

Uncertainty processing

7

CHAPTER OUTLINE

Most noise-robust methods use a deterministic strategy; that is, the compensated feature is a deterministic estimate computed from the corrupted speech feature or the compensated model is a point estimate adapted from the clean speech model. This category of methods is referred to as deterministic processing methods. The effects of strong noise necessarily create inherent uncertainty, either in the feature or model space, which can be beneficially integrated into the MAP decoding in the ASR process. When a noise-robust method takes into consideration that uncertainty, we call it an uncertainty processing method. In this chapter, we use this new attribute—exploiting the uncertainty or otherwise—to categorize and analyze noise-robust ASR techniques.

In the feature space, the presence of noise brings uncertainty into the enhanced speech signal, which is modeled as a distribution instead of a deterministic value. This can be formulated in a computationally efficient framework called uncertainty decoding (Deng et al., 2002; Droppo et al., 2002), in which the model variance is increased with a bias to address the uncertainty brought in by unreliable feature estimation. It can then be extended to joint uncertainty decoding, which uses a feature transform derived from the joint distribution between the clean and noisy speech and an uncertainty variance bias to modify the decoder.

Uncertainty can also be introduced in the model space when assuming the true model parameters are in a neighborhood of the compensated or trained model

Robust Automatic Speech Recognition. http://dx.doi.org/10.1016/B978-0-12-802398-3.00007-6

parameters. Robust techniques that make use of minimax classification (Merhav and Lee, 1993) and Bayesian predictive classification (Huo et al., 1997; Huo and Lee, 2000; Jiang and Deng, 2002) in decoding belong to this category.

7.1 MODEL-DOMAIN UNCERTAINTY

Uncertainty in the GMM-HMM parameters has been represented by their statistical distribution (Jiang and Deng, 2002; Deng, 2011). In order to take advantage of the model parameter uncertainty, the decision rule for recognition can be improved from the conventional MAP decision rule in Equation 7.1

$$\hat{\mathbf{W}} = \underset{\mathbf{W}}{\operatorname{argmax}}\, p_\Lambda(\mathbf{Y}|\mathbf{W})P_\Gamma(\mathbf{W}) \tag{7.1}$$

to the minimax decision rule (Merhav and Lee, 1993)

$$\hat{\mathbf{W}} = \underset{\mathbf{W}}{\operatorname{argmax}} \left(P_\Gamma(\mathbf{W}) \max_{\Lambda \in \Omega} p_\Lambda(\mathbf{Y}|\mathbf{W}) \right), \tag{7.2}$$

or to the Bayesian prediction classification (BPC) rule (Huo et al., 1997; Huo and Lee, 2000)

$$\hat{\mathbf{W}} = \underset{\mathbf{W}}{\operatorname{argmax}} \left(\int_{\Lambda \in \Omega} p_\Lambda(\mathbf{Y}|\mathbf{W}) p(\Lambda|\phi, \mathbf{W})\, d\Lambda \right) P_\Gamma(\mathbf{W}), \tag{7.3}$$

where ϕ is the hyper-parameter characterizing the distribution of acoustic model parameter Λ, and Ω denotes the space that Λ lies in. Equation 7.2 is derived from minimizing the upper bound of the worse-case probability of classification error. Both minimax classification and BPC consider the uncertainty of the estimated model, reflected by Ω. They change the decision rule to address this uncertainty using two steps. In the first step, either the maximum value of $p_\Lambda(\mathbf{Y}|\mathbf{W})$ within the parameter neighborhood (as in minimax classification) or the integration of $p_\Lambda(\mathbf{Y}|\mathbf{W})$ in this parameter neighborhood (as in BPC) for word sequence $\mathbf{W}$ is obtained. In the second step, the value obtained in the first step is plugged into the MAP decision rule.

It is usually difficult to define the parameter neighborhood Ω. Moreover, with two-stage processing, the computational cost of model space uncertainty using the modified decision rule is very large. It usually involves a very complicated implementation, which prevents this type of methods from being widely used although there was some research into minimax classification (Merhav and Lee, 1993; Afify et al., 2002) and BPC (Huo et al., 1997; Huo and Lee, 2000; Jiang et al., 1999) until around 10 years ago. An alternative treatment of uncertainty is by integrating over the feature space instead of over the model space. This will offer a much simpler system implementation and lower computational cost. Therefore, research has switched to feature space uncertainty or joint uncertainty decoding as described in the following sections.

7.2 FEATURE-DOMAIN UNCERTAINTY
7.2.1 OBSERVATION UNCERTAINTY

Although it has been shown that more gain can be obtained with a context window (Hermansky et al., 2000; Du and Huo, 2012), it is still a very popular assumption in most robust ASR methods that the clean feature is only dependent on the distorted feature of current frame. In this way, $p(\mathbf{x}_t|\mathbf{Y})$ is replaced by $p(\mathbf{x}_t|\mathbf{y}_t)$ and Equation 3.29 becomes

$$P_{\Lambda,\Gamma}(\mathbf{W}|\mathbf{Y}) = P_{\Gamma}(\mathbf{W}) \sum_{\theta} \prod_{t=1}^{T} \int \frac{p_{\Lambda}(\mathbf{x}_t|\boldsymbol{\theta}_t)p(\mathbf{x}_t|\mathbf{y}_t)}{p(\mathbf{x}_t)} d\mathbf{x}_t P_{\Lambda}(\boldsymbol{\theta}_t|\boldsymbol{\theta}_{t-1}). \tag{7.4}$$

As in Equation 3.31, the most popular robustness techniques use a point estimate which means that the back-end recognizer considers the cleaned signal $\hat{\mathbf{x}}_t(\mathbf{Y})$ to be noise free. However, the de-noising process is not perfect and there may exist some residual uncertainty. Hence, in the observation uncertainty work (Arrowood and Clements, 2002), instead of using a point estimate of clean speech, a posterior is passed to the back-end recognizer. The prior $p(\mathbf{x}_t)$ always has a larger variance than the posterior $p(\mathbf{x}_t|\mathbf{y}_t)$. If it is much larger, the denominator $p(\mathbf{x}_t)$ in Equation 7.4 can be considered constant in the range of values around $\mathbf{x}_t$. As a consequence, the denominator $p(\mathbf{x}_t)$ is neglected in Deng et al. (2002, 2005) and Stouten et al. (2006), and Equation 7.4 becomes

$$P_{\Lambda,\Gamma}(\mathbf{W}|\mathbf{Y}) = P_{\Gamma}(\mathbf{W}) \sum_{\theta} \prod_{t=1}^{T} \int p_{\Lambda}(\mathbf{x}_t|\boldsymbol{\theta}_t)p(\mathbf{x}_t|\mathbf{y}_t) d\mathbf{x}_t P_{\Lambda}(\boldsymbol{\theta}_t|\boldsymbol{\theta}_{t-1}). \tag{7.5}$$

If the de-noised speech is $\hat{\mathbf{x}}_t(\mathbf{Y})$, the clean speech estimate can be considered as a Gaussian distribution $\mathbf{x}_t \sim \mathcal{N}(\mathbf{x}_t; \hat{\mathbf{x}}_t, \boldsymbol{\Sigma}_{\hat{x}})$. When the underlying acoustic model is a GMM, the integration in Equation 7.5 reduces to

$$\int p_{\Lambda}(\mathbf{x}_t|\boldsymbol{\theta}_t)p(\mathbf{x}_t|\mathbf{y}_t) d\mathbf{x}_t = \int \sum_{m} c(m)\mathcal{N}(\mathbf{x}_t; \boldsymbol{\mu}_x(m), \boldsymbol{\Sigma}_x(m))\mathcal{N}(\mathbf{x}_t; \hat{\mathbf{x}}_t, \boldsymbol{\Sigma}_{\hat{x}}) d\mathbf{x}_t$$
$$= \sum_{m} c(m)\mathcal{N}(\hat{\mathbf{x}}_t; \boldsymbol{\mu}_x(m), \boldsymbol{\Sigma}_x(m) + \boldsymbol{\Sigma}_{\hat{x}}). \tag{7.6}$$

This is the formulation of observation uncertainty. It is clear that during recognition, every Gaussian component in the GMM model has a variance bias $\boldsymbol{\Sigma}_{\hat{x}}$ in observation uncertainty methods. The key is to get this frame-dependent variance $\boldsymbol{\Sigma}_{\hat{x}}$, which depends on the noise-robustness method used to clean the noisy feature. If the MMSE-based VTS feature enhancement is used as in Equation 6.53, the variance bias is (Stouten et al., 2006)

$$\boldsymbol{\Sigma}_{\hat{x}}(k) = \boldsymbol{\Sigma}_x(k) - \boldsymbol{\Sigma}_{xy}(k)\boldsymbol{\Sigma}_y^{-1}(k)\boldsymbol{\Sigma}_{yx}(k), \tag{7.7}$$

where k is the component index of the front-end GMM. If SPLICE (Deng et al., 2000) is used as in Section 5.1.2, the bias variance is given in Equation 5.10. In Arrowood and Clements (2002), it is a polynomial function of SNR. An interesting alternative supervised approach to estimate uncertainties was proposed in Srinivasan and Wang (2007). A more recent study on propagating uncertainties from short-time Fourier transform into the nonlinear feature domain appeared in Astudillo and Orglmeister (2013) for noise-robust ASR.

Equation 7.4 is reduced to Equation 7.5 by omitting $p(\mathbf{x}_t)$ with the belief that it has a larger variance than the posterior $p(\mathbf{x}_t|\mathbf{y}_t)$. However, this assumption may not be always true. Another better variation of Equation 7.4 is to multiply both the numerator and denominator by $p(\mathbf{y}_t)$. By applying Bayes' rule, we get

$$\int \frac{p_\Lambda(\mathbf{x}_t|\boldsymbol{\theta}_t)p(\mathbf{x}_t|\mathbf{y}_t)}{p(\mathbf{x}_t)}d\mathbf{x}_t = \int \frac{p_\Lambda(\mathbf{x}_t|\boldsymbol{\theta}_t)p(\mathbf{x}_t|\mathbf{y}_t)p(\mathbf{y}_t)}{p(\mathbf{x}_t)p(\mathbf{y}_t)}d\mathbf{x}_t \tag{7.8}$$

$$= \int \frac{p_\Lambda(\mathbf{x}_t|\boldsymbol{\theta}_t)p(\mathbf{x}_t,\mathbf{y}_t)}{p(\mathbf{x}_t)}d\mathbf{x}_t\frac{1}{p(\mathbf{y}_t)} \tag{7.9}$$

$$= \int p_\Lambda(\mathbf{x}_t|\boldsymbol{\theta}_t)p(\mathbf{y}_t|\mathbf{x}_t)d\mathbf{x}_t\frac{1}{p(\mathbf{y}_t)}. \tag{7.10}$$

Then, Equation 7.8 is plugged into Equation 7.4 and $p(\mathbf{y}_t)$ is omitted since it does not affect the MAP decision rule, and we obtain

$$P_{\Lambda,\Gamma}(\mathbf{W}|\mathbf{Y}) = P_\Gamma(\mathbf{W})\sum_\theta \prod_{t=1}^{T} \int p_\Lambda(\mathbf{x}_t|\boldsymbol{\theta}_t)p(\mathbf{y}_t|\mathbf{x}_t)\,d\mathbf{x}_t P_\Lambda(\boldsymbol{\theta}_t|\boldsymbol{\theta}_{t-1}). \tag{7.11}$$

We can denote

$$p_\Lambda(\mathbf{y}_t|\boldsymbol{\theta}_t) = \int p_\Lambda(\mathbf{x}_t|\boldsymbol{\theta}_t)p(\mathbf{y}_t|\mathbf{x}_t)\,d\mathbf{x}_t. \tag{7.12}$$

The key to calculating $p_\Lambda(\mathbf{y}_t|\boldsymbol{\theta}_t)$ is to estimate the conditional distribution $p(\mathbf{y}_t|\mathbf{x}_t)$ because $p_\Lambda(\mathbf{x}_t|\boldsymbol{\theta}_t)$ has already been trained. It is better to denote $p(\mathbf{y}_t|\mathbf{x}_t)$ as a Gaussian or GMM so that the integration of Equation 7.12 is still a Gaussian or GMM. Equation 7.11 is used in uncertainty decoding with SPLICE (Droppo et al., 2002) and joint uncertainty decoding work (Liao and Gales, 2005, 2006b, 2007, 2008) which we will discuss in detail in Section 7.3.

Uncertainty propagation through multilayer perceptrons
The integration in Equation 7.5 reduced to Equation 7.6 when the underlying acoustic model is a GMM. Given that DNNs now give better performance than GMMs, a natural question is whether a solution to the integration in Equation 7.13 exists when the underlying acoustic model is a DNN:

$$\int p_\Lambda(\mathbf{x}_t|\boldsymbol{\theta}_t)p(\mathbf{x}_t|\mathbf{y}_t)\,d\mathbf{x}_t. \tag{7.13}$$

One possible solution is proposed by propagating the front-end uncertainty through the multilayer perceptrons (MLP-UD) (Astudillo et al., 2012; Astudillo and da Silva Neto, 2011). Plugged into the observation uncertainty framework, uncertainty propagation through the network gets significant improvement from its counterpart without uncertainty propagation (Astudillo and da Silva Neto, 2011).

Suppose the input to the network is Gaussian distributed as $\mathbf{x}_t \sim \mathcal{N}(\mathbf{x}_t; \hat{\mathbf{x}}_t, \boldsymbol{\Sigma}_{\hat{x}})$, then the outputs before and after the sigmoid operation in the first hidden layer are

$$\mathbf{z}^1 = \mathbf{A}^1\mathbf{v}^1 + \mathbf{b}^1, \tag{7.14}$$

$$\mathbf{v}^2 = \sigma(\mathbf{z}^1) \tag{7.15}$$

with $\mathbf{v}^1 = \mathbf{x}$. MLP-UD assumes that each node of $\mathbf{z}^1$ is Gaussian-distributed as $\mathcal{N}(z_n^1; \mu(z_n^1), \Sigma(z_n^1))$ due to the central limit theorem because it is a weighted sum of large number of inputs. Then the first and second moments of a Gaussian variable transformed through the sigmoid function via the piecewise exponential approximation as

$$\sigma(z_n^1) = \frac{1}{1 + e^{(-z_n^1)}} \approx 2^{z_n^1 - 1}u(-z_n^1) + (1 - 2^{-z_n^1 - 1})u(z_n^1), \tag{7.16}$$

where $u(\cdot)$ is the unit step function. A closed-form solution can be derived for both the first and second order moments as

$$E(\sigma(z_n^1)) \approx \frac{1}{2}\Omega\left(\log(2)\mu(z_n^1), \log^2(2)\Sigma(z_n^1)\right)$$
$$- \frac{1}{2}\Omega\left(-\log(2)\mu(z_n^1), \log^2(2)\Sigma(z_n^1)\right) + \phi(0, -\mu(z_n^1), \Sigma(z_n^1)), \tag{7.17}$$

$$E(\sigma^2(z_n^1)) \approx \frac{1}{4}\Omega\left(2\log(2)\mu(z_n^1), 4^2\log(2)\Sigma(z_n^1)\right) + \frac{1}{4}\Omega\left(-2\log(2)\mu(z_n^1), 4\log^2(2)\Sigma(z_n^1)\right)$$
$$- \frac{1}{2}\Omega\left(-\log(2)\mu(z_n^1), \log^2(2)\Sigma(z_n^1)\right) + \phi(0, -\mu(z_n^1), \Sigma(z_n^1)), \tag{7.18}$$

where $\phi(\cdot)$ is the cumulative density function (CDF) of the Gaussian variable and $\Omega(\cdot)$ is the partial expectation of the exponential of a Gaussian variable. As a result, the mean and variance of each node of $\mathbf{v}^2$, that is, the output of first layer or the input of the second layer, can be easily computed. Because each node of $\mathbf{v}^2$ is again Gaussian-distributed, it is easy to propagate through the upper level hidden layers with sigmoid units using the same procedure.

The final step is to estimate the mean and variance of the network output node

$$\mathbf{z}^L = \mathbf{A}^L\mathbf{v}^L + \mathbf{b}^L, \tag{7.19}$$

$$P(n|\mathbf{x}) = \text{softmax}(z_n^L) = \frac{e^{z_n^L}}{\sum_j e^{z_j^L}}. \tag{7.20}$$

Again, every node of $\mathbf{z}^L$ is Gaussian distributed due to the central limit theorem, and softmax operation is directly applied to $\mu(z_n^L)$ to approximate the output node mean. The variance is approximated as (Astudillo et al., 2012)

$$\Sigma\left(P(n|\mathbf{x})\right) \approx \left(\exp(\Sigma(z_n^L)) - 1\right) \exp\left(2\mu(z_n^L) - \log\left(\sum_j e^{z_j^L}\right) + \Sigma(z_n^L)\right). \qquad (7.21)$$

Note that in a recent work (Astudillo et al., 2014), an approximation of MLP inference is proposed to take into consideration the residual uncertainty of the mean field approximation. Although using a similar piecewise exponential sigmoid approximation, the work in Astudillo et al. (2014) deals with the uncertainty inside the MLP inference. This is in opposite with the method described in this section, which propagates uncertainty from the front-end while considering the MLP as a deterministic function.

Although the uncertainty propagation through network provides one possible research direction, it is too early to say whether this is the right method for uncertainty processing with DNNs because it has lots of approximation. Therefore, in the remaining of this chapter, we will still focus on uncertainty processing with GMMs.

7.3 JOINT UNCERTAINTY DECODING

Joint uncertainty decoding (JUD) uses a feature transform derived from the joint distribution between the clean and noisy speech and an uncertainty variance bias to modify the decoder. While the joint distribution can be estimated from stereo data as in SPLICE (Deng et al., 2001), the most popular way to obtain it is to use the physical distortion model as in Chapter 6. JUD has two implementation forms: front-end JUD and model JUD. In front-end JUD, a front-end GMM is built and one of its components is selected to pass one single transform and bias variance to the decoder. In contrast, model JUD is connected with the acoustic model and generates transform and uncertainty variance bias based on the regression class that the individual acoustic model component belongs to.

7.3.1 FRONT-END JUD

In front-end JUD, $p(\mathbf{y}_t|\mathbf{x}_t)$ is represented by a GMM

$$p(\mathbf{y}_t|\mathbf{x}_t) \approx \sum_k P(k|\mathbf{x}_t)\mathcal{N}\left(\mathbf{y}_t; f_\mu(\mathbf{x}_t, k), f_\Sigma(\mathbf{x}_t, k)\right), \qquad (7.22)$$

where $f_\mu(\mathbf{x}_t, k)$ and $f_\Sigma(\mathbf{x}_t, k)$ are functions used to calculate the mean vector and covariances matrix. The joint distribution of clean and noisy speech can be modeled in the same way as in Equation 6.47. It can be either obtained from stereo training data or derived with physical distortion modeling as in Chapter 6. In most JUD methods, the latter option is used given the difficulty to obtain stereo data.

The Gaussian conditional distribution $\mathcal{N}\left(\mathbf{y}_t; f_\mu(\mathbf{x}_t, k), f_\Sigma(\mathbf{x}_t, k)\right)$ in Equation 7.22 can be derived from the joint distribution as

$$\mathcal{N}\left(\mathbf{y}_t; \boldsymbol{\mu}_y(k) + \boldsymbol{\Sigma}_{yx}(k)\boldsymbol{\Sigma}_x^{-1}(k)(\mathbf{x}_t - \boldsymbol{\mu}_x(k)), \boldsymbol{\Sigma}_y(k) - \boldsymbol{\Sigma}_{yx}(k)\boldsymbol{\Sigma}_x^{-1}(k)\boldsymbol{\Sigma}_{xy}(k)\right). \qquad (7.23)$$

By applying the transformation on the observed feature $\mathbf{y}_t$ instead of $\mathbf{x}_t$ using $\boldsymbol{\Sigma}_x(k)\boldsymbol{\Sigma}_{yx}^{-1}(k)$, Equation 7.23 can be rewritten as (Liao and Gales, 2006b, 2008)

$$|\mathbf{A}(k)|\mathcal{N}\left(\mathbf{A}(k)\mathbf{y}_t + \mathbf{b}(k); \mathbf{x}_t, \boldsymbol{\Sigma}_b(k)\right), \tag{7.24}$$

with

$$\mathbf{A}(k) = \boldsymbol{\Sigma}_x(k)\boldsymbol{\Sigma}_{yx}^{-1}(k), \tag{7.25}$$

$$\mathbf{b}(k) = \boldsymbol{\mu}_x(k) - \mathbf{A}(k)\boldsymbol{\mu}_y(k), \tag{7.26}$$

$$\boldsymbol{\Sigma}_b(k) = \mathbf{A}(k)\boldsymbol{\Sigma}_y(k)\mathbf{A}^T(k) - \boldsymbol{\Sigma}_x(k). \tag{7.27}$$

These transforms can be obtained using VTS related schemes.

Using Equation 7.24 and the rough assumption that $P(k|\mathbf{x}_t) \approx P(k|\mathbf{y}_t)$, Equation 7.12 can be rewritten as (Liao and Gales, 2008)

$$\begin{aligned}
p_\Lambda(\mathbf{y}_t|\boldsymbol{\theta}_t) &= \int p_\Lambda(\mathbf{x}_t|\boldsymbol{\theta}_t)p(\mathbf{y}_t|\mathbf{x}_t)\,\mathrm{d}\mathbf{x}_t \\
&\approx \int \sum_k P(k|\mathbf{y}_t)|\mathbf{A}(k)|\mathcal{N}\left(\mathbf{A}(k)\mathbf{y}_t + \mathbf{b}(k); \mathbf{x}_t, \boldsymbol{\Sigma}_b(k)\right) \\
&\quad \sum_m c(m)\mathcal{N}(\mathbf{x}_t; \boldsymbol{\mu}_x(m), \boldsymbol{\Sigma}_x(m))\,\mathrm{d}\mathbf{x}_t \\
&= \sum_k \sum_m c(m)P(k|\mathbf{y}_t) \\
&\quad |\mathbf{A}(k)|\mathcal{N}\left(\mathbf{A}(k)\mathbf{y}_t + \mathbf{b}(k); \boldsymbol{\mu}_x(m), \boldsymbol{\Sigma}_x(m) + \boldsymbol{\Sigma}_b(k)\right). \tag{7.28}
\end{aligned}$$

The summation in Equation 7.28 is time consuming, involving the clean speech GMM component m and the front-end GMM component k. One popular approach is to select the most dominant front-end component k^*

$$k^* = \underset{k}{\mathrm{argmax}}\, P(k|\mathbf{y}_t) \tag{7.29}$$

and Equation 7.28 can be simplified as

$$p_\Lambda(\mathbf{y}_t|\boldsymbol{\theta}_t) \approx \sum_m c(m)|\mathbf{A}(k^*)|\mathcal{N}\left(\mathbf{A}(k^*)\mathbf{y}_t + \mathbf{b}(k^*); \boldsymbol{\mu}_x(m), \boldsymbol{\Sigma}_x(m) + \boldsymbol{\Sigma}_b(k^*)\right). \tag{7.30}$$

Comparing Equation 7.30 with Equation 7.6, we can see that front-end JUD transfers the distorted feature $\mathbf{y}_t$ in addition to adding a variance bias. The front-end JUD can be summarized in two steps. The first step is to select the front-end component k^* using Equation 7.29. The second step is to use Equation 7.30 to evaluate the likelihood of every state in acoustic model. $\mathbf{A}(k^*)$, $\mathbf{b}(k^*)$, and $\boldsymbol{\Sigma}_b(k^*)$ are the same for every acoustic model state. SPLICE with uncertainty decoding (Droppo et al., 2002) is similar to front-end JUD, but with a different format of $\mathbf{A}(k)$, $\mathbf{b}(k)$, and $\boldsymbol{\Sigma}_b(k)$.

As discussed in Liao and Gales (2006a, 2008), in low SNR conditions where noise dominates the speech signal, the conditional distribution $p(\mathbf{y}_t|\mathbf{x}_t)$ degenerates to the distribution of additive noise $\mathbf{n}_t$ as

$$p(\mathbf{y}_t|\mathbf{x}_t) \approx \mathcal{N}(\mathbf{n}_t; \boldsymbol{\mu}_n, \boldsymbol{\Sigma}_n). \tag{7.31}$$

Then, the distribution of distorted speech in Equation 7.12 also becomes the distribution of additive noise

$$p_\Lambda(\mathbf{y}_t|\boldsymbol{\theta}_t) = \int p_\Lambda(\mathbf{x}_t|\boldsymbol{\theta}_t)\mathcal{N}(\mathbf{n}_t; \boldsymbol{\mu}_n, \boldsymbol{\Sigma}_n)d\mathbf{x}_t$$
$$= \mathcal{N}(\mathbf{n}_t; \boldsymbol{\mu}_n, \boldsymbol{\Sigma}_n). \tag{7.32}$$

With Equation 7.32, the distribution of every state is the same. Therefore, the current frame cannot contribute to differentiating states using acoustic model scores. This is the biggest theoretical issue with front-end JUD, although SPLICE with uncertainty decoding can circumvent this issue with additional processing (Droppo et al., 2002).

7.3.2 MODEL JUD

In front-end JUD, its conditional distribution is completely decoupled from the acoustic model used for recognition. In contrast, model JUD (Liao and Gales, 2006b) links them together with

$$p(\mathbf{y}_t|\mathbf{x}_t) \approx \mathcal{N}\left(\mathbf{y}_t; f_\mu(\mathbf{x}_t, r_m), f_\Sigma(\mathbf{x}_t, r_m)\right), \tag{7.33}$$

where r_m is the regression class index of Gaussian component m in the acoustic model, generated with the method in Gales (1996). The joint distribution of clean and noisy speech can be modeled similarly as in Equation 6.47 by replacing the front-end component index k with the regression class index r_m. With a similar derivation as in front-end JUD, the likelihood of distorted speech can be denoted by

$$p_\Lambda(\mathbf{y}_t|\boldsymbol{\theta}_t) \approx \sum_m c(m)|\mathbf{A}(r_m)|\mathcal{N}\left(\mathbf{A}(r_m)\mathbf{y}_t + \mathbf{b}(r_m); \boldsymbol{\mu}_x(m), \boldsymbol{\Sigma}_x(m) + \boldsymbol{\Sigma}_b(r_m)\right). \tag{7.34}$$

Comparing Equation 7.34 with Equation 7.30, we can tell that the difference between model JUD and front-end JUD is that in front-end JUD only the best component k^* selected in front-end processing is passed to modify the likelihood evaluation during decoding while in model JUD every Gaussian component is associated with a regression-class-dependent transform. Therefore, in model JUD, the distorted feature $\mathbf{y}_t$ is transformed by multiple transforms, similar to CMLLR (Gales, 1998). However, it differs from CMLLR due to the regression-class-dependent variance term $\boldsymbol{\Sigma}_b(r_m)$.

There are several extensions of model JUD. $\boldsymbol{\Sigma}_b(r_m)$ in Equation 7.34 is a full covariance matrix which brings a large computational cost when evaluating the likelihood. One direct solution is to diagonalize it. However, this solution turns out to have poor performance (Liao and Gales, 2006a). Predictive CMLLR (PCMLLR) (Gales and van Dalen, 2007) can be used to avoid the full covariance matrix by applying a CMLLR-like transform in the feature space transformed by model JUD

$$\tilde{\mathbf{y}}_t = \mathbf{A}(r_m)\mathbf{y}_t + \mathbf{b}(r_m). \tag{7.35}$$

The likelihood for PCMLLR decoding is given by

$$p_\Lambda(\mathbf{y}_t|\boldsymbol{\theta}_t) \approx \sum_m c(m)|\mathbf{A}_p(r_m)||\mathbf{A}(r_m)|\mathcal{N}\left(\mathbf{A}_p(r_m)\tilde{\mathbf{y}}_t + \mathbf{b}_p(r_m); \boldsymbol{\mu}_x(m), \boldsymbol{\Sigma}_x(m)\right) \tag{7.36}$$

$\mathbf{A}_p(r_m)$ and $\mathbf{b}_p(r_m)$ are obtained by using CMLLR, with the statistics obtained from the model JUD transformed feature $\tilde{\mathbf{y}}_t$. With Equation 7.36, the clean acoustic model is unchanged.

Another alternative is with VTS-JUD (Xu et al., 2009, 2011) in which the likelihood is computed as

$$p_\Lambda(\mathbf{y}_t|\boldsymbol{\theta}_t) \approx \sum_m c(m)\mathcal{N}(\mathbf{y}_t; \mathbf{H}(r_m)(\boldsymbol{\mu}_x(m) - \mathbf{b}(r_m)),$$

$$\text{diag}(\mathbf{H}(r_m)(\boldsymbol{\Sigma}_x(m) + \boldsymbol{\Sigma}_b(r_m))\mathbf{H}^T(r_m))), \tag{7.37}$$

where $\mathbf{H}(r_m) = \mathbf{A}^{-1}(r_m)$. VTS-JUD can be considered as the model space implementation of model JUD, very similar to VTS but with less computational cost. In Kim and Gales (2011), noise CMLLR is also proposed to extend the conventional CMLLR in Section 4.2 to reflect additional uncertainty from noisy features by introducing a covariance bias with the same form as Equation 7.33. All of model JUD, VTS-JUD, and PCMLLR use VTS in Chapter 6 to calculate the regression-class-dependent transforms. If the number of regression classes is identical to the number of Gaussians, it can be proven that all of these methods are the same as VTS. By using regression classes, some computational cost can be saved. For example, in Equation 7.37 of VTS-JUD, although it still needs to apply transforms to every Gaussian mean and variance of the clean acoustic model, the cost of calculating transforms is reduced because they are now regression-class-dependent instead of Gaussian-dependent.

Recently, subspace Gaussian mixture models (SGMMs) are proposed in Povey et al. (2010) with better performance than GMMs. In Lu et al. (2013), an extension of JUD when using SGMMs is presented with good improvements in noisy conditions.

7.4 MISSING-FEATURE APPROACHES

Missing-feature approaches (Cooke et al., 1994; Lippmann and Carlson, 1997), also known as missing-data approaches, introduce the concept of uncertainty into feature processing. The methods are based on the inherent redundancy in the speech signal: one may still be able to recognize speech effectively even with only a fraction of the spectro-temporal information in the speech signal. They attempt to determine which time-frequency cells are unreliable due to the introduction of noise or other types of interference. These unreliable cells are either ignored or filled in by estimates of their putative values in subsequent processing (Barker et al., 2005; Cooke et al., 2001).

The components of the distorted feature vector $\mathbf{y}$ are divided into two mutually exclusive sets $\mathcal{U}$ and $\mathcal{R}$, denoting the unreliable and reliable components, respectively. $\mathbf{y}_u$ is a vector constructed from all elements of $\mathbf{y}$ lying in $\mathcal{U}$, while $\mathbf{y}_r$ is a vector formed by all elements of $\mathbf{y}$ lying in $\mathcal{R}$. $\mathbf{x}_u$ and $\mathbf{x}_r$ are the corresponding clean vectors. Usually, it is assumed that $\mathbf{y}_r$ is a close approximation to $\mathbf{x}_r$, that is, $\mathbf{y}_r \approx \mathbf{x}_r$, while $\mathbf{y}_u$ is an upper bound of $\mathbf{x}_u$, that is, $\mathbf{x}_u \leq \mathbf{y}_u$.

There are two major types of missing-feature approaches, namely, feature vector imputation and classifier modification (Raj and Stern, 2005). Feature imputation methods treat unreliable spectral components as missing components and attempt to reconstruct them by utilizing spectrum statistics. There are several typical methods. In correlation-based reconstruction (Raj et al., 2004), the spectral samples are considered to be the output of a Gaussian wide-sense stationary random process, which implies that the means of the spectral vectors and the covariance between spectral components are independent of their positions in the spectrogram. A joint distribution of unreliable components and reliable neighborhood components can be constructed and the reconstruction is then estimated using a bounded MAP estimation procedure. On the other hand, in cluster-based reconstruction (Raj et al., 2004), the unreliable components are reconstructed only based on the relationships among the components within individual vectors. Soft-mask-based MMSE estimation is similar to cluster-based reconstruction, but with soft masks (Raj and Singh, 2005).

In the second category of missing-feature approaches, classifier modification, one may discern between class-conditional imputation and marginalization. In class-conditional imputation, HMM state-specific estimates are derived for the missing components (Josifovski et al., 1999). Marginalization, on the other hand, directly performs optimal classification based on the observed reliable and unreliable components. One extreme and popularly used case is where only the reliable component is used during recognition.

While with feature vector imputation recognition can be done with features that may be different from the reconstructed log-spectral vectors, it was, until recently, common understanding that state-based imputation and marginalization precluded the use of cepstral features for recognition. This was a major drawback, since the log-spectral features to be used instead exhibit strong spatial correlations, which either resulted in a loss of recognition accuracy at comparable acoustic model size or required significantly more mixture components to achieve competitive recognition rates, compared to cepstral features. However, recently it has been demonstrated that the techniques can be applied in any domain that is a linear transform of log-spectra (Van Segbroeck and Van Hamme, 2011). In Hartmann et al. (2013), it has been shown that (cepstral) features directly computed from the masked spectrum can outperform imputation techniques, as long as variance normalization is applied on the resulting features. Given the ideal binary mask, recognition on masked speech has been shown to outperform recognition on reconstructed speech. However, mask estimation is never perfect, and as the quality of the mask estimation degrades, recognition on reconstructed speech begins to outperform recognition on masked speech (Hartmann and Fosler-Lussier, 2011).

The most difficult part of missing-feature methods is the accurate estimation of spectral masks which identify unreliable spectrum cells. The estimation can be performed in multiple ways: SNR-dependent estimation (El-Maliki and Drygajlo, 1999; van Hout and Alwan, 2012; Vizinho et al., 1999), Bayesian estimation (Renevey and Drygajlo, 1999; Seltzer et al., 2004), and with perceptual criteria (Barker et al., 2001; Palomäki et al., 2004). Also, deep neural networks have been employed for supervised learning of the mapping of the features to the desired soft mask target (Narayanan and Wang, 2013b).

However, it is impossible to estimate the mask perfectly. Unreliable mask estimation significantly reduces the recognition accuracy of missing-feature approaches (Seltzer et al., 2004). This problem can be remedied to some extent by using soft masks (Barker et al., 2000; Renevey and Drygajlo, 1999; Seltzer et al., 2004), which use a probability to represent the reliability of a spectrum cell. Strictly speaking, missing feature approaches using soft masks can be categorized as uncertainty processing methods, but not those that use binary masks. In Van Segbroeck and Van Hamme (2011), it was shown how soft masks can be used with imputation techniques. Further, the estimation of the ideal ratio mask, a soft mask version of the ideal binary mask, was shown to outperform the estimated ideal binary mask in Narayanan and Wang (2013b). There is a close link between missing data approaches employing a soft mask and optimal MMSE estimation of the clean speech features, as was shown, among others, in Gonzalez et al. (2013).

Instead of treating the mask estimation and the classification as two separate tasks, combining the two promises superior performance. A first approach in this direction was the speech fragment decoder of Barker et al. (2005). The fragment decoder simultaneously searches for the optimal mask and the optimal HMM state sequence. Its initial limitations, which were that ASR had to be carried out in the spectral domain and that the time-frequency fragments were formed prior to the ASR decoding stage and therefore could not benefit from the powerful ASR acoustic models, have been recently overcome (Hartmann and Fosler-Lussier, 2012a,b; Ma and Barker, 2012; Narayanan and Wang, 2013a). In Hartmann and Fosler-Lussier (2012a), ASR-driven mask estimation is proposed. Similarly, the bidirectional speech decoding of (Narayanan and Wang, 2013a) also exploits the modeling power of the ASR models for mask estimation. It generates multiple candidate ASR features at every time frame, with each candidate corresponding to a particular back-end acoustic phonetic unit. The ASR decoder then selects the most appropriate candidate via a maximum likelihood criterion. Ultimately, one could envision an iterative process, where a baseline recognizer will generate first hypotheses for mask estimation. Using the estimated mask, ASR is improved which in turn results in improved mask estimation (Hartmann and Fosler-Lussier, 2012a).

To summarize, the state of the art in missing data techniques has matured in recent years and the method has become a high-performance noise-robust ASR technique also for medium to large vocabulary tasks.

7.5 SUMMARY

We summarize some representative uncertainty processing methods discussed in this chapter in Table 7.1 in a chronological order, with the following highlights:

- Exploiting model-domain uncertainty was popular in 1990s via modifications of the decision rule to work on the model parameter neighborhood with either minimax classification or BPC. However, it is difficult to define the model parameter neighborhood and the methods have involved a very high-computational cost.

Table 7.1 Uncertainty Processing Methods in Chapter 7, Arranged Chronologically

Method	Proposed Around	Characteristics
Minimax classification (Merhav and Lee, 1993)	1993	Modifies the MAP decision rule by considering the extreme case in the neighborhood model space
Missing-feature approaches (Cooke et al., 1994; Lippmann and Carlson, 1997; Raj et al., 2004)	1994	Determines which spectrum cells are unreliable and then either ignores these unreliable cells or fills by estimation of their putative values
Bayesian predictive classification (BPC) (Huo et al., 1997; Huo and Lee, 2000)	1997	Modifies the MAP decision rule by integrating the model parameters in a neighborhood
Observation uncertainty (Arrowood and Clements, 2002; Deng et al., 2002; Stouten et al., 2006)	2002	The uncertainty in feature is passed to the back-end recognizer using a posterior, reflected as a variance bias
Front-end uncertainty decoding (Droppo et al., 2002; Liao and Gales, 2005)	2002	Transforms feature and adds bias covariance according to the selected front-end component which reflects feature uncertainty
Model joint uncertainty decoding (JUD) (Liao and Gales, 2006b)	2006	Uncertainty is reflected by the regression-class-dependent feature transform and bias covariance
Predictive CMLLR (PCMLLR) (Gales and van Dalen, 2007)	2007	Saves model JUD computational cost by applying a CMLLR-like transform on the feature space transformed by model JUD
VTS-JUD (Xu et al., 2009, 2011)	2009	Model space implementation of model JUD, very similar to VTS but with less computational cost
Uncertainty propagation through multilayer perceptrons (Astudillo and da Silva Neto, 2011)	2011	Propagates the front-end uncertainty through the multilayer perceptron with the piecewise exponential approximation to the sigmoid function

- Then, in the beginning of 2000s, observation uncertainty and front-end uncertainty decoding were developed in the feature domain, which enjoyed computational advantage over model-domain uncertainty methods.
- Later, to address the problem of front-end uncertainty decoding in high noise conditions, model-space JUD was developed in 2006. Further techniques were proposed to improve the accuracy and computational cost of model-space JUD.
- Separated from the above developments, the study of missing-feature approaches has been active from 1994 until now. Many challenges in this area become overcome, among which is how to obtain accurate estimates of spectral masks. As a result, it has matured in recent years.

Many techniques for exploiting uncertainty as discussed in this chapter can be regarded as a hybrid of feature-domain and model-domain processing, as argued in Deng (2011). For example, feature enhancement is a process free of ASR models, and it estimates the degree of uncertainty representing the quality of the enhancement method. When such uncertainty is integrated into the ASR model by enlarging the variances of the GMMs, this becomes the operation in the model space. Much of the interpretation and analysis of uncertainty processing techniques discussed in this chapter can be easily framed in the probabilistic generative modeling scheme (Deng and Li, 2013) such as in the GMM-based systems. When the systems move to discriminative and nonprobabilistic DNN-based ones, it becomes more difficult to incorporate the uncertainty concept and methods. In fact, there have been very few studies in the literature that incorporate uncertainty into DNN modeling for this reason, in addition to the nonlinearity in the DNN forward propagation. Uncertainty propagation through multilayer perceptrons discussed in this chapter may provide a possible solution to the problem, but much more work remains to be done in this direction.

REFERENCES

Afify, M., Siohan, O., Lee, C.H., 2002. Upper and lower bounds on the mean of noisy speech: application to minimax classification. IEEE Trans. Speech Audio Process. 10 (2), 79-88.

Arrowood, J.A., Clements, M.A., 2002. Using observation uncertainty in HMM decoding. In: Proc. Interspeech, pp. 1561-1564.

Astudillo, R., Orglmeister, R., 2013. Computing MMSE estimates and residual uncertainty directly in the feature domain of ASR using STFT domain speech distortion models. IEEE Trans. Audio Speech Lang. Process. 21 (5), 1023-1034.

Astudillo, R.F., Abad, A., da Silva Neto, J.P., 2012. Uncertainty driven compensation of multistream MLP acoustic models for robust ASR. In: Proc. Interspeech.

Astudillo, R.F., Abad, A., Trancoso, I., 2014. Accounting for the residual uncertainty of multilayer perceptron based features. In: Proc. International Conference on Acoustics, Speech and Signal Processing (ICASSP), pp. 6859-6863.

Astudillo, R.F., da Silva Neto, J.P., 2011. Propagation of uncertainty through multilayer perceptrons for robust automatic speech recognition. In: INTERSPEECH, pp. 461-464.

Barker, J., Cooke, M., Green, P.D., 2001. Robust ASR based on clean speech models: an evaluation of missing data techniques for connected digit recognition in noise. In: Proc. Interspeech, pp. 213-217.

Barker, J., Josifovski, L., Cooke, M., Green, P.D., 2000. Soft decisions in missing data techniques for robust automatic speech recognition. In: Proc. Interspeech, pp. 373-376.

Barker, J.P., Cooke, M.P., Ellis, D.P.W., 2005. Decoding speech in the presence of other sources. Speech Commun. 45 (1), 5-25.

Cooke, M., Green, P.D., Crawford, M., 1994. Handling missing data in speech recognition. In: Proc. International Conference on Spoken Language Processing (ICSLP), pp. 1555-1558.

Cooke, M., Green, P.D., Josifovski, L., Vizinho, A., 2001. Robust automatic speech recognition with missing and unreliable acoustic data. Speech Commun. 34 (3), 267-285.

Deng, L., 2011. Front-end, back-end, and hybrid techniques for noise-robust speech recognition. In: Robust Speech Recognition of Uncertain or Missing Data: Theory and Application. Springer, New York, pp. 67-99.

Deng, L., Acero, A., Jiang, L., Droppo, J., Huang, X.D., 2001. High-performance robust speech recognition using stereo training data. In: Proc. International Conference on Acoustics, Speech and Signal Processing (ICASSP), pp. 301-304.

Deng, L., Acero, A., Plumpe, M., Huang, X., 2000. Large vocabulary speech recognition under adverse acoustic environment. In: Proc. International Conference on Spoken Language Processing (ICSLP), vol. 3, pp. 806-809.

Deng, L., Droppo, J., Acero, A., 2002. Exploiting variances in robust feature extraction based on a parametric model of speech distortion. In: Proc. Interspeech, pp. 2449-2452.

Deng, L., Droppo, J., Acero, A., 2005. Dynamic compensation of HMM variances using the feature enhancement uncertainty computed from a parametric model of speech distortion. IEEE Trans. Speech Audio Process. 13 (3), 412-421.

Deng, L., Li, X., 2013. Machine learning paradigms in speech recognition: An overview. IEEE Trans. Audio Speech and Lang. Process. 21 (5), 1060-1089.

Droppo, J., Deng, L., Acero, A., 2002. Uncertainty decoding with SPLICE for noise robust speech recognition. In: Proc. International Conference on Acoustics, Speech and Signal Processing (ICASSP), vol. 1, pp. 57-60.

Du, J., Huo, Q., 2012. IVN-based joint training of GMM and HMMs using an improved VTS-based feature compensation for noisy speech recognition. In: Proc. Interspeech, pp. 1227-1230.

El-Maliki, M., Drygajlo, A., 1999. Missing features detection and handling for robust speaker verification. In: Proc. European Conference on Speech Communication and Technology (EUROSPEECH), pp. 975-978.

Gales, M.J.F., University of Cambridge., 1996. The generation and use of regression class trees for MLLR adaptation.

Gales, M.J.F., 1998. Maximum likelihood linear transformations for HMM-based speech recognition. Comput. Speech Lang. 12, 75-98.

Gales, M.J.F., van Dalen, R.C., 2007. Predictive linear transforms for noise robust speech recognition. In: Proc. IEEE Workshop on Automatic Speech Recognition and Understanding (ASRU), pp. 59-64.

Gonzalez, J., Peinado, A., Ma, N., Gomez, A., Barker, J., 2013. Mmse-based missing-feature reconstruction with temporal modeling for robust speech recognition. IEEE Trans. Audio Speech Lang. Process. 21 (3), 624-635.

Hartmann, W., Fosler-Lussier, E., 2011. Investigations into the incorporation of the ideal binary mask in ASR. In: Proc. International Conference on Acoustics, Speech and Signal Processing (ICASSP), pp. 4804-4807.

Hartmann, W., Fosler-Lussier, E., 2012a. ASR-driven top-down binary mask estimation using spectral priors. In: Proc. International Conference on Acoustics, Speech and Signal Processing (ICASSP), pp. 4685-4688.

Hartmann, W., Fosler-Lussier, E., 2012b. Improved model selection for the ASR-driven binary mask. In: Proc. Interspeech, pp. 1203-1206.

Hartmann, W., Narayanan, A., Fosler-Lussier, E., Wang, D., 2013. A direct masking approach to robust ASR. IEEE Trans. Audio Speech Lang. Process. 21 (10), 1993-2005. ISSN 1558-7916. http://dx.doi.org/10.1109/TASL.2013.2263802.

Hermansky, H., Ellis, D.P.W., Sharma, S., 2000. Tandem connectionist feature extraction for conventional HMM systems. In: Proc. International Conference on Acoustics, Speech and Signal Processing (ICASSP), vol. 3, pp. 1635-1638.

Huo, Q., Jiang, H., Lee, C.H., 1997. A Bayesian predictive classification approach to robust speech recognition. In: Proc. International Conference on Acoustics, Speech and Signal Processing (ICASSP), pp. 1547-1550.

Huo, Q., Lee, C.H., 2000. A Bayesian predictive classification approach to robust speech recognition. IEEE Trans. Speech Audio Process. 8 (2), 200-204.

Jiang, H., Deng, L., 2002. A robust compensation strategy against extraneous acoustic variations in spontaneous speech recognition. In: IEEE Trans. Speech and Audio Processing, vol. 10, pp. 9-17.

Jiang, H., Hirose, K., Huo, Q., 1999. Improving Viterbi Bayesian predictive classification via sequential Bayesian learning in robust speech recognition. Speech Commun. 28 (4), 313-326.

Josifovski, L., Cooke, M., Green, P.D., Vizinho, A., 1999. State based imputation of missing data for robust speech recognition and speech enhancement. In: Proc. European Conference on Speech Communication and Technology (EUROSPEECH), pp. 2837-2840.

Kim, D.K., Gales, M.J.F., 2011. Noisy constrained maximum-likelihood linear regression for noise-robust speech recognition. IEEE Trans. Audio Speech Lang. Process. 19 (2), 315-325.

Liao, H., Gales, M.J.F., 2005. Joint uncertainty decoding for noise robust speech recognition. In: Proc. Interspeech, pp. 3129-3132.

Liao, H., Gales, M.J.F., 2006a. Issues with uncertainty decoding for noise robust speech recognition. In: Proc. Interspeech, pp. 1121-1124.

Liao, H., Gales, M.J.F., 2006b. Joint uncertainty decoding for robust large vocabulary speech recognition. University of Cambridge.

Liao, H., Gales, M.J.F., 2007. Adaptive training with joint uncertainty decoding for robust recognition of noisy data. In: Proc. International Conference on Acoustics, Speech and Signal Processing (ICASSP), vol. 4, pp. 389-392.

Liao, H., Gales, M.J.F., 2008. Issues with uncertainty decoding for noise robust automatic speech recognition. Speech Commun. 50 (4), 265-277.

Lippmann, R., Carlson, B., 1997. Using missing feature theory to actively select features for robust speech recognition with interruptions, filtering and noise. In: Proc. European Conference on Speech Communication and Technology (EUROSPEECH), pp. 37-40.

Lu, L., Chin, K., Ghoshal, A., Renals, S., 2013. Joint uncertainty decoding for noise robust subspace Gaussian mixture models. IEEE Trans. Audio Speech Lang. Process. 21 (9), 1791-1804.

Ma, N., Barker, J., 2012. Coupling identification and reconstruction of missing features for noise-robust automatic speech recognition. In: Proc. Interspeech, pp. 2638-2641.

Merhav, N., Lee, C.H., 1993. A minimax classification approach with application to robust speech recognition. IEEE Trans. Speech Audio Process. 1 (1), 90-100.

Narayanan, A., Wang, D., 2013a. Coupling binary masking and robust ASR. In: Proc. International Conference on Acoustics, Speech and Signal Processing (ICASSP), pp. 6817-6821.

Narayanan, A., Wang, D., 2013b. Ideal ratio mask estimation using deep neural networks. In: Proc. International Conference on Acoustics, Speech and Signal Processing (ICASSP), pp. 7092-7096.

Palomäki, K.J., Brown, G.J., Wang, D.L., 2004. A binaural processor for missing data speech recognition in the presence of noise and small-room reverberation. Speech Commun. 43 (4), 361-378.

Povey, D., Burget, L., Agarwal, M., Akyazi, P., Feng, K., Ghoshal, A., et al., 2010. Subspace Gaussian mixture models for speech recognition. In: Proc. International Conference on Acoustics, Speech and Signal Processing (ICASSP), pp. 4330-4333.

Raj, B., Seltzer, M.L., Stern, R.M., 2004. Reconstruction of missing features for robust speech recognition. Speech Commun. 43 (4), 275-296.

Raj, B., Singh, R., 2005. Reconstructing spectral vectors with uncertain spectrographic masks for robust speech recognition. In: Proc. IEEE Workshop on Automatic Speech Recognition and Understanding (ASRU), pp. 65-70.

Raj, B., Stern, R., 2005. Missing-feature approaches in speech recognition. IEEE Signal Process. Mag. 22 (5), 101-116. ISSN 1053-5888. http://dx.doi.org/10.1109/MSP.2005.1511828.

Renevey, P., Drygajlo, A., 1999. Missing feature theory and probabilistic estimation of clean speech components for robust speech recognition. In: Proc. European Conference on Speech Communication and Technology (EUROSPEECH), pp. 2627-2630.

Seltzer, M.L., Raj, B., Stern, R.M., 2004. A Bayesian classifier for spectrographic mask estimation for missing feature speech recognition. Speech Commun. 43 (4), 379-393.

Srinivasan, S., Wang, D., 2007. Transforming binary uncertainties for robust speech recognition. IEEE Trans. Audio Speech Lang. Process. 15 (7), 2130-2140.

Stouten, V., Hamme, H.V., Wambacq, P., 2006. Model-based feature enhancement with uncertainty decoding for noise robust ASR. Speech Commun. 48 (11), 1502-1514.

van Hout, J., Alwan, A., 2012. A novel approach to soft-mask estimation and log-spectral enhancement for robust speech recognition. In: Proc. International Conference on Acoustics, Speech and Signal Processing (ICASSP), pp. 4105-4108.

Van Segbroeck, M., Van Hamme, H., 2011. Advances in missing feature techniques for robust large-vocabulary continuous speech recognition. IEEE Trans. Audio Speech Lang. Process. 19 (1), 123-137. ISSN 1558-7916. http://dx.doi.org/10.1109/TASL.2010.2045235.

Vizinho, A., Green, P.D., Cooke, M., Josifovski, L., 1999. Missing data theory, spectral subtraction and signal-to-noise estimation for robust ASR: an integrated study. In: Proc. European Conference on Speech Communication and Technology (EUROSPEECH), pp. 2407-2410.

Xu, H., Gales, M.J.F., Chin, K.K., 2009. Improving joint uncertainty decoding performance by predictive methods for noise robust speech recognition. In: Proc. IEEE Workshop on Automatic Speech Recognition and Understanding (ASRU), pp. 222-227.

Xu, H., Gales, M.J.F., Chin, K.K., 2011. Joint uncertainty decoding with predictive methods for noise robust speech recognition. IEEE Trans. Audio Speech Lang. Process. 19 (6), 1665-1676.

Joint model training

8

CHAPTER OUTLINE

Most of the noise-robust ASR methods discussed in the preceding chapters assume that the ASR recognizer has been trained from clean speech, and in the test stage noise robustness methods are used to reduce the mismatch between the clean acoustic model and distorted speech with either feature enhancement or model adaptation techniques. However, it is very difficult to collect clean training data in most practical ASR deployment scenarios. Usually, the training set may contain distorted speech data obtained in all kinds of environments. There are several issues related to the acoustic model trained from multi-style training data, which have been elaborated in Deng (2011) and Li et al. (2014). First, the assumption of most robust methods is no longer valid. For example, in the explicit modeling technique discussed in Chapter 6, such as vector Taylor series (VTS), the explicit distortion model assumes that the speech model is only trained from clean data. Another issue is that the trained model is too broad to model the data from all environments. It fails to give a sharp distribution of speech classes because it needs to cover the factors from different environments.

All of these problems can be solved with joint model training which applies the same process at both the training and test stage so that the same sources of variability can be removed consistently. More specifically, the feature compensation or model adaptation technique used in the test stage is also used in the training stage so that a pseudo-clean acoustic model is obtained in training. Using the criterion of whether the ASR models are trained jointly with the process of feature compensation or model adaptation in the test stage, we can categorize most existing noise-robust techniques in the literature into two broad classes: disjoint and joint model training. The disjoint model training methods are straightforward. We will focus on joint model training in this chapter.

Among the joint model training methods, the most prominent set of techniques are based on a paradigm called noise adaptive training (NAT) first published in year 2000 (Deng et al., 2000), which can also be viewed as a hybrid strategy of feature enhancement and model adaptation. One specific example of NAT is the multi-style training of models in the enhanced feature domain, where noisy training data is first cleaned by feature compensation methods, and subsequently the enhanced features are used to retrain the acoustic model for the evaluation of enhanced test features. This feature-space NAT (fNAT) strategy can achieve better performance than the standard noise-matched scheme because it applies consistent process during the training and test phases while eliminating residual mismatch in an otherwise disjoint training paradigm. The feature compensation can be any form of noise reduction or feature enhancement. The model compensation can be any form of MLE or discriminative training, where it is multi-style training operating on the feature-compensated training data. This original fNAT scheme (Deng et al., 2000) was developed and applied with two relatively simple feature compensation techniques, spectral subtraction (SS) (Boll, 1979) and stereo-based piecewise linear compensation for environments (SPLICE) (Deng et al., 2001, 2000, 2003, 2002). Both retrained HMMs (fNAT-SS and fNAT-SPLICE) outperform the noise-matched system under almost all SNR conditions. The only exception is in the clean condition in which the feature compensation methods bring artifacts to the clean speech feature. fNAT has demonstrated its high performance (Deng et al., 2000) on Aurora 2 and has been verified by at least hundreds of additional experiments with all sorts of feature compensation techniques on different databases.

fNAT is popular because it is easy to implement and it has been shown to be very effective. Hence it has been adopted as one of the two major evaluation paradigms, called multi-style training (after denoising), in the popular series of Aurora tasks. However, fNAT decouples the optimization objective of the feature compensation and model training parts, which are not jointly optimized under a common objective function. In contrast, the model-space NAT (mNAT) methods jointly train a canonical acoustic model and a set of transforms or distortion parameters under the same training criteria. All model-space joint model training methods share the same spirit with speaker adaptive training (SAT) (Anastasakos et al., 1996), proposed in 1996 for speaker adaptation. One difference between SAT and NAT methods is whether there is a golden target for canonical model learning. In NAT, the golden target is the truly clean speech features or the model trained from it. However, in SAT, there is no such predefined golden speaker as the target. In this chapter, only model-space joint training methods are presented given its importance and the complexity substantially higher than its counterpart of fNAT.

8.1 SPEAKER ADAPTIVE AND SOURCE NORMALIZATION TRAINING

General adaptation methods, such as MLLR and CMLLR, are initially proposed for speaker adaptation. A speaker-independent GMM acoustic model is obtained from a multi-speaker training set using the standard MLE method. In testing, speaker-dependent transforms are estimated for specific speakers. However, the acoustic model estimated in this way may be a good model for average speakers, but not optimal for any specific speaker. SAT (Anastasakos et al., 1996) is proposed to train a canonical acoustic model with less inter-speaker variability. A compact HMM model Λ_c and the speaker-dependent transforms $\mathcal{W} = (\mathbf{W}^{(1)}, \mathbf{W}^{(2)}, \ldots, \mathbf{W}^{(R)})$ are jointly estimated from an R-speaker training set by maximizing the likelihood of the training data

$$(\hat{\Lambda}_c, \hat{\mathcal{W}}) = \underset{\Lambda_c, \mathcal{W}}{\operatorname{argmax}} \, \Pi_{r=1}^{R} \mathcal{L}(\mathbf{Y}^{(r)}; \mathbf{W}^{(r)}(\Lambda_c)), \tag{8.1}$$

where $\mathbf{Y}^{(r)}$ is the observation sequence of speaker r. A transform $W^{(r)} = [\mathbf{A}^{(r)} \mathbf{b}^{(r)}]$ for speaker r in the training set maps the compact model Λ_c to a speaker-dependent model in the same way as the speaker adaptation methods used in the test stage. With the compact model Λ_c, the speaker-specific variation in the training stage is reduced and the trained compact model represents the phonetic variation more accurately.

Equation 8.1 can be solved with the EM algorithm by maximizing the auxiliary Q function when the MLLR transform is used

$$Q(\hat{\Lambda}; \Lambda) = \sum_{r,t,m} \gamma_t^{(r)}(m) \log \mathcal{N}(\mathbf{y}_t^{(r)}; \mathbf{A}^{(r)} \boldsymbol{\mu}(m) + \mathbf{b}^{(r)}, \boldsymbol{\Sigma}(m)), \tag{8.2}$$

where $\gamma_t^{(r)}(m)$ is the posterior probability of Gaussian component m at time t for utterance r. As shown in Figure 8.1 (after Seltzer 2012), SAT is done with an iterative two-stage scheme. In the first stage, the auxiliary Q function is maximized with respect to the speaker-dependent transforms $\mathcal{W}$ while keeping the Gaussian model parameters of the compact model Λ_c fixed. By setting the derivative of Q with respect to $\mathcal{W}$ to 0, the solution of $\mathcal{W}$ can be obtained as in Section 4.2. In the second stage, the compact model parameters are updated by maximizing the auxiliary Q function while keeping the speaker-dependent transforms $\mathcal{W}$ fixed. By setting the derivative of Q with respect to $\boldsymbol{\mu}(m)$ and $\boldsymbol{\Sigma}(m)$ to 0, the compact model parameters are updated as

$$\hat{\boldsymbol{\mu}}(m) = \left(\sum_{r,t} \gamma_t^{(r)}(m) \mathbf{A}^{(r)T} \boldsymbol{\Sigma}(m)^{-1} \mathbf{A}^{(r)} \right)^{-1} \left(\sum_{r,t} \gamma_t^{(r)}(m) \mathbf{A}^{(r)T} \boldsymbol{\Sigma}(m)^{-1} \left(\mathbf{y}_t^{(r)} - \mathbf{b}^{(r)} \right) \right), \tag{8.3}$$

$$\hat{\boldsymbol{\Sigma}}(m) = \frac{\sum_{r,t} \gamma_t^{(r)}(m) \mathbf{A}^{(r)T} \boldsymbol{\Sigma}(m)^{-1} \left(\mathbf{y}_t^{(r)} - \hat{\boldsymbol{\mu}}(m) \right) \left(\mathbf{y}_t^{(r)} - \hat{\boldsymbol{\mu}}(m) \right)^T}{\sum_{r,t} \gamma_t^{(r)}(m)}. \tag{8.4}$$

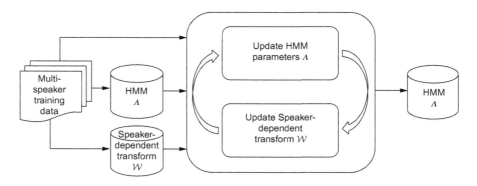

FIGURE 8.1

Speaker adaptive training.

While the SAT formulations are derived with MLLR as the adaptation method, a similar process can also be applied to other adaptation methods, such as CMLLR described in Section 4.2. Although initially proposed to reduce speaker-specific variation in training, SAT can also be used to reduce environment-specific variation when MLLR or CMLLR is used to adapt models in noisy environments.

In some real applications, we need to adapt to a cluster of distortions such as a group of speakers or background noise instead of individual speakers. Source normalization training (SNT) (Gong, 1997) generalizes SAT by introducing another hidden variable to model distortion sources. SNT subsumes SAT by extending the speaker ID to a hidden variable in training and testing. In SNT, the distortion sources (e.g., a speaker or group of speakers) do not have to be tagged; they are discovered by unsupervised training with the EM algorithm. SNT was used to explicitly address environment-specific normalization. In Gong (1997), an environment can refer to speaker, handset, transmission channel, or noise background condition.

MLLR and CMLLR are general adaptation technologies, and cannot work as effectively as the noise-specific explicit modeling methods in noise-robust ASR tasks. Therefore, the SAT-like methods are not as popular as the model space noise adaptive training methods in the next section which are coupled with the explicit distortion model methods in Chapter 6.

8.2 MODEL SPACE NOISE ADAPTIVE TRAINING

As shown in Figure 8.2 (after Seltzer 2012), the model space noise adaptive training (mNAT) scheme is very similar to SAT in Figure 8.1. The speaker-dependent transforms $\mathcal{W}$ are replaced with the distortion model $\Phi = (\Phi^{(1)}, \Phi^{(2)}, \ldots, \Phi^{(R)})$, where R is the total number of training utterances. Every utterance $r = 1 \ldots R$ has its

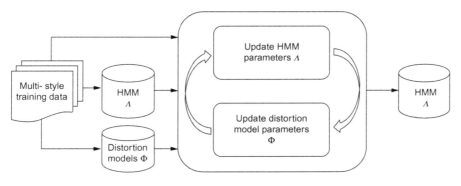

FIGURE 8.2

Noise adaptive training.

own utterance-dependent noise, channel, and adapted HMM parameters. However, all utterances share the same set of canonical HMM parameters. Similar to SAT, the mNAT methods are effective when the same model adaptation methods are used in both training and test stages. The representative mNAT methods are joint adaptive training (JAT) (Liao and Gales, 2007), irrelevant variability normalization (IVN) (Hu and Huo, 2007), and VTS-NAT (Kalinli et al., 2009, 2010). The JAT work uses joint uncertainty decoding (JUD) (Liao and Gales, 2005, 2006) as its model adaptation scheme. The IVN work uses the VTS algorithm presented in Kim et al. (1998) for model adaptation. The VTS-NAT work is coupled with VTS adaptation in Li et al. (2007) and Li et al. (2009), which is described in detail in Section 6.2.1. In the following, we will give details of the VTS-NAT work in Kalinli et al. (2009) and Kalinli et al. (2010).

For example, the adapted static mean and variance parameters for the mth Gaussian in utterance r are obtained with

$$\boldsymbol{\mu}_y^{(r)}(m) \approx \boldsymbol{\mu}_x(m) + \boldsymbol{\mu}_h^{(r)} + \mathbf{g}(\boldsymbol{\mu}_x(m), \boldsymbol{\mu}_h^{(r)}, \boldsymbol{\mu}_n^{(r)}), \tag{8.5}$$

$$\boldsymbol{\Sigma}_y^{(r)}(m) \approx \mathrm{diag}\left(\mathbf{G}^{(r)}(m)\boldsymbol{\Sigma}_x(m)\mathbf{G}^{(r)}(m)^T + \mathbf{F}^{(r)}(m)^T \boldsymbol{\Sigma}_n^{(r)} \mathbf{F}^{(r)}(m)^T\right), \tag{8.6}$$

with

$$\mathbf{G}^{(r)}(m) = \mathbf{C}\,\mathrm{diag}\left(\frac{1}{1 + \exp\left(\mathbf{C}^{-1}\left(\boldsymbol{\mu}_n^{(r)} - \boldsymbol{\mu}_x(m) - \boldsymbol{\mu}_h^{(r)}\right)\right)}\right)\mathbf{C}^{-1}, \tag{8.7}$$

$$\mathbf{F}^{(r)}(m) = \mathbf{I} - \mathbf{G}^{(r)}(m). \tag{8.8}$$

Different from Equations 6.20, 6.21, and 6.18, all the noise, channel, and adapted model parameters, respectively, are now utterance-dependent.

VTS-NAT model estimation is also done with an iterative two-stage scheme. In the first stage, the auxiliary Q function is maximized with respect to the utterance-dependent distortion model parameters $\Phi^{(r)}$ while keeping the canonical Gaussian model parameters fixed. The auxiliary Q function can be written as

$$Q(\hat{\Lambda}; \Lambda) = \sum_{t,r,m} \gamma_t^{(r)}(m) \log p_{\hat{\Lambda}^{(r)}}\left(\mathbf{y}_t^{(r)}|m\right), \qquad (8.9)$$

where $\hat{\Lambda}^{(r)}$ denotes the adapted model for utterance r. Comparing this with the auxiliary function in Equation 6.27, an additional term r is summed in Equation 8.9 to include all the training utterances. The solution is the same as in Section 6.2.1. The noise mean, channel mean, and noise variance parameters of every utterance are updated with Equations 6.29, 6.33, and 6.38, respectively.

In the second stage, the canonical model parameters are updated by maximizing the auxiliary Q function while keeping the utterance-dependent distortion model parameters $\Phi^{(r)}$ fixed. The static mean parameters of the canonical model are obtained by taking the derivative of Q with respect to $\boldsymbol{\mu}_x(m)$ and setting the result to zero.

$$\sum_{t,r} \gamma_t^{(r)}(m)(\mathbf{G}^{(r)}(m))^T \left(\Sigma_{y,0}^{(r)}(m)\right)^{-1}\left(\mathbf{y}_t^{(r)} - \boldsymbol{\mu}_{y,0}^{(r)}(m) - \mathbf{G}^{(r)}(m)(\boldsymbol{\mu}_x(m) - \boldsymbol{\mu}_{x,0}(m))\right) = 0. \qquad (8.10)$$

Then, $\boldsymbol{\mu}_x(m)$ can be obtained as

$$\boldsymbol{\mu}_x(m) = \boldsymbol{\mu}_{x,0}(m) + \mathbf{A}(m)^{-1}\mathbf{b}(m) \qquad (8.11)$$

with

$$\mathbf{A}(m) = \sum_{t,r} \gamma_t^{(r)}(m)(\mathbf{G}^{(r)}(m))^T \left(\Sigma_{y,0}^{(r)}(m)\right)^{-1} \mathbf{G}^{(r)}(m), \qquad (8.12)$$

$$\mathbf{b}(m) = \sum_{t,r} \gamma_t^{(r)}(m)(\mathbf{G}^{(r)}(m))^T \left(\Sigma_{y,0}^{(r)}(m)\right)^{-1} \left(\mathbf{y}_t^{(r)} - \boldsymbol{\mu}_{y,0}^{(r)}(m)\right), \qquad (8.13)$$

where $\boldsymbol{\mu}_{x,0}(m)$ is the mth Gaussian mean parameter of the canonical model in the last update, and $\boldsymbol{\mu}_{y,0}^{(r)}(m)$ and $\Sigma_{y,0}^{(r)}(m)$ are the mth Gaussian mean and variance parameters of adapted model in the last update. The delta and delta-delta mean parameters of the canonical model can be obtained similarly by taking the derivative of Q with respect to $\boldsymbol{\mu}_{\Delta x}(m)$ and $\boldsymbol{\mu}_{\Delta\Delta x}(m)$, and setting the results to zero.

Similar to the solution of noise variance update in VTS in Section 6.2.2 with equations from Equation 6.36 to Equation 6.40, a Newton's method can be used to update the D-dimensional static variance vector $\Sigma_x(m) = \text{diag}(\sigma_x^2(m))$ with $\sigma_x^2(m) = [\sigma_{x,1}^2(m), \sigma_{x,2}^2(m), \dots \sigma_{x,D}^2(m)]^T$. To guarantee the adapted noise variance to be positive, one solution is to transform the static noise variance first with

$$\tilde{\sigma}_x^2(m) = \log(\sigma_x^2(m)), \qquad (8.14)$$

and then update $\tilde{\sigma}_x^2(m)$ with the Newton's method as

$$\tilde{\sigma}_x^2(m) = \tilde{\sigma}_{x,0}^2(m) - \left(\frac{\partial^2 Q}{\partial^2 \tilde{\sigma}_x^2(m)}\right)^{-1} \frac{\partial Q}{\partial \tilde{\sigma}_x^2(m)}, \tag{8.15}$$

The cth element of $\frac{\partial Q}{\partial \tilde{\sigma}_x^2(m)}$ and the (c,e)th element of $\frac{\partial^2 Q}{\partial^2 \tilde{\sigma}_x^2(m)}$ in Equation 8.15 can be obtained with Equations 8.16 and 8.17, respectively.

$$\frac{\partial Q}{\partial \tilde{\sigma}_{x,c}^2(m)} = -\frac{1}{2} \sum_{t,r,m} \gamma_t^{(r)}(m) \sum_{d=1}^{D} \left(\frac{\sigma_{x,c}^2(m) \mathbf{G}_{dc}^{(r)2}(m)}{\tau_d^{(r)}(m)} \left(1 - \frac{\left(\mathbf{y}_{t,d}^{(r)} - \boldsymbol{\mu}_{y,d}^{(r)}(m)\right)^2}{\tau_d^{(r)}(m)}\right)\right), \tag{8.16}$$

$$\frac{\partial^2 Q}{\partial \tilde{\sigma}_{x,c}^2(m) \partial \tilde{\sigma}_{x,e}^2(m)} = -\frac{1}{2} \sum_{t,r,m} \gamma_t^{(r)}(m) \sum_{d=1}^{D} \left(\frac{\sigma_{x,c}^2(m) \mathbf{F}_{dc}^{(r)2}(m)}{\tau_d^{(r)}(m)} \left(1 - \frac{\left(\mathbf{y}_{t,d}^{(r)} - \boldsymbol{\mu}_{y,d}^{(r)}(m)\right)^2}{\tau_d^{(r)}(m)}\right) \delta(c-e) \right.$$
$$\left. + \frac{\sigma_{x,c}^2(m) \mathbf{F}_{dc}^{(r)2}(m) \sigma_{x,e}^2(m) \mathbf{F}_{de}^{(r)2}(m)}{\tau_d^{(r)2}(m)} \left(-1 + 2\frac{\left(\mathbf{y}_{t,d}^{(r)} - \boldsymbol{\mu}_{y,d}^{(r)}(m)\right)^2}{\tau_d^{(r)}(m)}\right)\right), \tag{8.17}$$

where

$$\tau_d^{(r)}(m) = \sum_{i=1}^{D} \left(\sigma_{x,i}^2(m) \mathbf{G}_{di}^{(r)2}(m) + \sigma_{n,i}^{(r)2} \mathbf{F}_{di}^{(r)2}(m) \right). \tag{8.18}$$

$\mathbf{G}_{di}^{(r)}(m)$ and $\mathbf{F}_{di}^{(r)}(m)$ are the dth row and ith column elements of $\mathbf{G}^{(r)}(m)$ and $\mathbf{F}^{(r)}(m)$, respectively. Finally, the static variance in the linear scale can be obtained with

$$\sigma_x^2(m) = \exp(\tilde{\sigma}_x^2(m)). \tag{8.19}$$

$\sigma_{\Delta x}^2(m)$ and $\sigma_{\Delta\Delta x}^2(m)$ can be obtained in a similar way. The detailed formulas are in Kalinli et al. (2010). All of the joint model training techniques learn the canonical model to represent the pseudo-clean speech model, and the transforms are then used to represent the nonlinguistic variability such as environmental variations. It has been well established that joint training methods achieve consistent improvement over the disjoint training methods. The latter are much easier to implement, and easier to train the acoustic model parameters alone without making them compact and without removing the nonlinguistic variability.

Coupled with model JUD instead of VTS, JAT (Liao and Gales, 2007) is another variation of NAT. Similar to VTS-NAT, the adaptive transform in JAT is parameterized, and its parameters are jointly trained with the HMM parameters by the same kind of maximum likelihood criterion. While most noise adaptive training studies are based on the maximum likelihood criterion, discriminative adaptive training can be used to further improve accuracy (Flego and Gales, 2009). To do so, standard MLE-based noise adaptive training (Liao and Gales, 2007) is first performed to get the HMM parameters and distortion model parameters. Then, the distortion model parameters are fixed and the discriminative training criterion is applied to further optimize the HMM parameters. While JAT is initially proposed to handle GMMs, it is extended in Lu et al. (2013) to work with subspace GMMs to get further improvement.

The idea of irrelevant variability normalization (IVN) is a very general concept. The argument is that HMMs trained from a large amount of diversified data, which consists of different speakers, acoustic environments, channels etc., may tend to fit the variability of data irrelevant to phonetic classification. The term IVN is proposed in Huo and Ma (1999) to build a better decision tree that has better modeling capability and generalizability by removing the speaker factors during the decision tree building process. Then from 2002, IVN is widely used as a noise-robustness method for jointly training the front-end and back-end together for stochastic vector mapping (Wu and Huo, 2002), which maps the corrupted speech feature to a clean speech feature by a transform. Six linear feature transform (FT) functions have been studied as in Wu and Huo (2002), Wu and Huo (2006), Huo and Zhu (2006), and Zhu and Huo (2008). Every environment can have a bias vector (Wu and Huo, 2002, 2006), or one environment-dependent transform (Huo and Zhu, 2006), or even multiple environment transforms (Zhu and Huo, 2008). Different from SPLICE discussed in Chapter 5, these transforms are obtained without the need of stereo training data. Instead, they are jointly trained with the HMM parameters. The process is done iteratively with the following two steps until convergence:

- fix HMM parameters, increase the objective function by adjusting the FT function parameters;
- fix FT function parameters, increase the objective function by adjusting the HMM parameters.

The training criterion can be MLE, MAP, and discriminative training criterion. Additional improvements of IVN are the use of sequential estimation of the feature compensation parameters (Shi et al., 2010) by using a moving window-based frame labeling and the use of i-Vector for environment clustering in IVN (Xu et al., 2011).

Although some forms of IVN are similar to SAT, IVN is designed for noise robustness by using environment-dependent transforms and biases to map the corrupted speech feature to the clean speech feature. In Hu and Huo (2007), IVN is further linked with VTS (VTS-IVN) by using explicit distortion modeling to characterize the distortion caused by noise. In principle, both VTS-IVN and VTS-NAT have the same goal of getting a pseudo-clean model from multi-style training data so that the acoustic model can better be coupled with VTS at runtime. There are still some differences between them. First VTS-IVN uses the VTS algorithm presented in Kim et al. (1998) for model adaptation while VTS-NAT uses the VTS adaptation method in Li et al. (2007) and Li et al. (2009), which is described in detail in Section 6.2.1. Second, VTS-IVN only updates static HMM model parameters while VTS-NAT updates all the parameters during training.

8.3 JOINT TRAINING FOR DNN

The standard DNN training procedure takes the acoustic feature extracted from a front-end module, and then input it into the DNN for model training (Dahl et al., 2011, 2012; Deng and Yu, 2014; Hinton et al., 2012; Yu and Deng, 2011, 2014). The front-end extraction is separated from the DNN training procedure. Also, the procedure treats all the input data equally to train the standard DNN. Clearly, this procedure is suboptimal. A better way is to do the joint training of the DNN and the front-end. Depending on whether the DNN is jointly trained with the front-end module or jointly trained with source-dependent data, the joint training method can be classified as (1) joint front-end and DNN training and (2) joint adaptive training.

8.3.1 JOINT FRONT-END AND DNN MODEL TRAINING

The concept of joint front-end and DNN model training is very simple: treat the front-end as a trainable module and combine it with the DNN to form a bigger network as shown in Figure 8.3. Then the parameters in the trainable front-end are combined with the weight and bias of the DNN to form a super set of parameters, which can be optimized with the standard back propagation algorithm. The error signal now back propagates to the trainable front-end to update both DNN and front-end parameters, instead of stopping at the input layer of the DNN. In contrast, although there are also some joint front-end and GMM training works (Li and Lee, 2007; Povey et al., 2005), the optimization is much more complicated.

Figure 8.4 (after (Narayanan and Wang, 2014)) is an example of the joint front-end and DNN model training. The trainable front-end is a DNN described in Section 5.1.3 which estimates the ideal ratio mask (IRM) in Equation 5.19. Then the noisy Mel spectrum is multiplied with the IRM matrix to generate the estimation of clean feature with Equation 5.20. In the next step, the cleaned Mel spectrum is passed into a fixed front-end module which consists a series of standard front-end operation to generate the input feature of the other DNN used for senone classification. The joint training method trains both the DNN for senone classification and the DNN for IRM estimation together by back propagating the classification error signal through both DNNs. This joint training method is reported to have better noise-robustness (Narayanan and Wang, 2014). Similar joint training work has also been done in Gao et al. (2015). In addition to optimizing the noise-robust front-end, the traditional front-end parameters, such as the filter bank weight matrices and delta parameters for dynamic feature, can also be jointly trained with DNN parameters (Sainath et al., 2013, 2014).

8.3.2 JOINT ADAPTIVE TRAINING

In Section 4.2.2, the methods of adapting DNN are described. The adaptation is done in the evaluation stage for an already trained DNN. As its GMM counterpart, this is suboptimal because the source variability is not considered during training. This

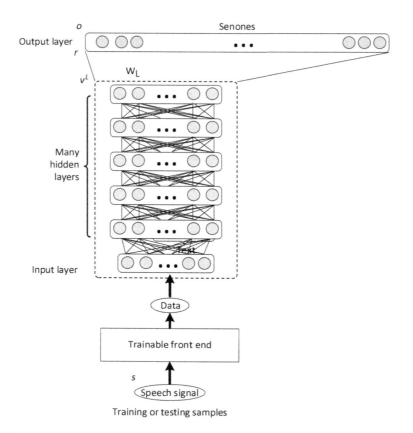

FIGURE 8.3

Joint training of front-end and DNN model.

is solved by joint model training which applies the same adaptation process at both the training and test stage so that the same sources of variability can be removed consistently. In noise-adaptive training of GMM, there is a canonical model and a set of noise-dependent transforms. Similarly, joint adaptive training can also be applied to DNN models as in Figure 8.5, where a set of connection under each layer denotes a weight matrix. There are multiple environment-independent layers and an environment-dependent activation (EDA) layer. Data 1 in Figure 8.5 will only go through EDA 1, and then all above layers. And data n will only go through EDA n and above layers. In this way, only weights and biases connecting to the EDA layer are environment dependent. Then the whole network is trained with the standard back propagation algorithm. It is not necessary that the EDA layer directly connecting the input layer of the DNN can be placed between any hidden layers. In addition to being used for noise adaptive training, the structure in Figure 8.5 can also be applied to speaker adaptive training (Ochiai et al., 2014).

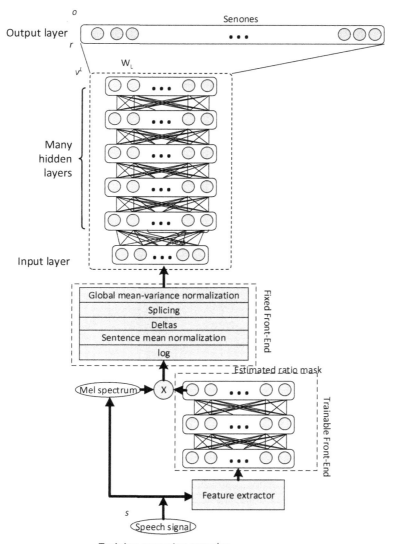

Output layer

o

r

Senones

W_L

v^L

Many
hidden
layers

Input layer

Global mean-variance normalization

Splicing

Deltas

Sentence mean normalization

log

Fixed Front-End

Estimated ratio mask

Mel spectrum   X

Trainable Front-End

Feature extractor

s

Speech signal

Training or testing samples

FIGURE 8.4

An example of joint training of front-end and DNN models.

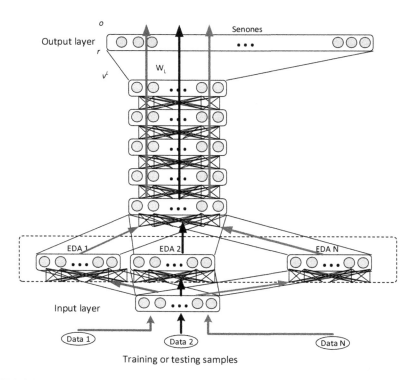

FIGURE 8.5

Adaptive training of DNN.

8.4 SUMMARY

We summarize here the major representative methods with joint model training discussed in this chapter in Table 8.1 in a chronological order, with the following technical highlights:

- In 1997, source normalization training was developed to train a canonical model using SAT-like joint training method to remove the environment-specific variation in the training data.
- The feature-domain NAT, proposed originally in 2000, has been widely used since its invention, and is the simplest method to make the training and test features consistent, both in the form of pseudo-clean speech features.
- IVN was proposed in 2002 for jointly training the GMM-HMM parameters and linear feature transforms which characterize the irrelevant variability of environment. It was further developed around 2007 to exploit explicit distortion modeling with VTS.

Table 8.1 Joint Model Training Methods in Chapter 8, Arranged Chronologically

Method	Proposed Around	Characteristics
Speaker adaptive training (SAT) (Anastasakos et al., 1996)	1996	Trains a canonical acoustic model with less inter-speaker variability by including the speaker-specific transforms in the training stage
Source normalization training (Gong, 1997)	1997	Every environment is modeled with a MLLR transform, and a SAT-like method is used to obtain a canonical model by removing the environment-specific variation
Feature space noise adaptive training (NAT) (Deng et al., 2000)	2000	Noisy training data is cleaned by feature compensation methods, and subsequently the enhanced features are used to retrain the acoustic model for the evaluation of enhanced test features
Irrelevant variability normalization (IVN) (Hu and Huo, 2007; Wu and Huo, 2002; Zhu and Huo, 2008)	2002	Jointly trains the acoustic model parameter and distortion parameter to remove the environment variability while insisting on the mapping from noisy speech to pseudo-clean speech
Joint adaptive training (JAT) for GMM (Liao and Gales, 2007)	2007	Obtains the canonic environment-free model by using model JUD to represent the feature transform during training
VTS-NAT (Kalinli et al., 2009, 2010)	2009	Obtains the canonic environment-free model by using VTS to represent the nonlinear feature transform during training
Joint front-end and DNN model training (Narayanan and Wang, 2014)	2014	Treats the front-end as a trainable module and combines it with the DNN to form a bigger network. Back propagation is used to train both front-end and DNN parameters
Joint adaptive training for DNN	2014	There are multiple environment-independent layers and environment-dependent activation (EDA) layers in the DNN. Back propagation is used to train both environment-independent and environment-dependent parameters

- With the development of explicit distortion modeling, technologies such as JAT and VTS-NAT were used to remove environment variability by coupling explicit distortion modeling in the training process from 2007.
- Although all formulations of joint model training for the GMM are complicated, it is much easier to optimize parameters for joint model training for the DNN. This is because all the components, including the trainable front-end, canonical DNN model, and environment-dependent weight matrix, can be considered as parts of a big ensemble DNN. Therefore, the standard back propagation update can be applied to optimize all these components in a unified manner. This clearly facilitates the joint model training technique and will be a promising future direction.

REFERENCES

Anastasakos, T., McDonough, J., Schwartz, R., Makhoul, J., 1996. A compact model for speaker-adaptive training. In: Proc. International Conference on Spoken Language Processing (ICSLP), vol. 2, pp. 1137-1140.

Boll, S.F., 1979. Suppression of acoustic noise in speech using spectral subtraction. IEEE Trans. Acoust. Speech Signal Process. 27 (2), 113-120.

Dahl, G., Yu, D., Deng, L., Acero, A., 2011. Large vocabulary continuous speech recognition with context-dependent DBN-HMMs. In: Proc. International Conference on Acoustics, Speech and Signal Processing (ICASSP).

Dahl, G., Yu, D., Deng, L., Acero, A., 2012. Context-dependent pre-trained deep neural networks for large-vocabulary speech recognition. IEEE Trans. Audio Speech Lang. Process. 20 (1), 30-42.

Deng, L., 2011. Front-end, back-end, and hybrid techniques for noise-robust speech recognition. In: Robust Speech Recognition of Uncertain or Missing Data: Theory and Application. Springer, New York, pp. 67-99.

Deng, L., Acero, A., Jiang, L., Droppo, J., Huang, X.D., 2001. High-performance robust speech recognition using stereo training data. In: Proc. International Conference on Acoustics, Speech and Signal Processing (ICASSP), pp. 301-304.

Deng, L., Acero, A., Plumpe, M., Huang, X., 2000. Large vocabulary speech recognition under adverse acoustic environment. In: Proc. International Conference on Spoken Language Processing (ICSLP), vol. 3, pp. 806-809.

Deng, L., Droppo, J., A.Acero., 2003. Recursive estimation of nonstationary noise using iterative stochastic approximation for robust speech recognition. IEEE Trans. Speech Audio Process. 11, 568-580.

Deng, L., Wang, K.A., Acero, H.H., Huang, X., 2002. Distributed speech processing in MiPad's multimodal user interface. IEEE Trans. Audio Speech Lang. Process. 10 (8), 605-619.

Deng, L., Yu, D., 2014. Deep Learning: Methods and Applications. Now Publishers, Hanover, MA.

Flego, F., Gales, M.J.F., 2009. Discriminative adaptive training with VTS and JUD. In: Proc. IEEE Workshop on Automatic Speech Recognition and Understanding (ASRU), pp. 170-175.

Gao, T., Du, J., Dai, L.R., Lee, C.H., 2015. Joint training of front-end and back-end deep neural networks for robust speech recognition. In: Proc. International Conference on Acoustics, Speech and Signal Processing (ICASSP).

Gong, Y., 1997. Source normalization training for HMM applied to noisy telephone speech recognition. In: Proc. European Conference on Speech Communication and Technology (EUROSPEECH), pp. 1555-1558.

Hinton, G., Deng, L., Yu, D., Dahl, G.E., Mohamed, A., Jaitly, N., et al., 2012. Deep neural networks for acoustic modeling in speech recognition: The shared views of four research groups. IEEE Signal Process. Mag. 29 (6), 82-97.

Hu, Y., Huo, Q., 2007. Irrelevant variability normalization based HMM training using VTS approximation of an explicit model of environmental distortions. In: Proc. Interspeech, pp. 1042-1045.

Huo, Q., Ma, B., 1999. Irrelevant variability normalization in learning HMM state tying from data based on phonetic decision-tree. In: Proc. International Conference on Acoustics, Speech and Signal Processing (ICASSP), vol. 2, pp. 577-580.

Huo, Q., Zhu, D., 2006. A maximum likelihood training approach to irrelevant variability compensation based on piecewise linear transformations. In: Proc. Interspeech, pp. 1129-1132.

Kalinli, O., Seltzer, M.L., Acero, A., 2009. Noise adaptive training using a vector Taylor series approach for noise robust automatic speech recognition. In: Proc. International Conference on Acoustics, Speech and Signal Processing (ICASSP), pp. 3825-3828.

Kalinli, O., Seltzer, M.L., Droppo, J., Acero, A., 2010. Noise adaptive training for robust automatic speech recognition. IEEE Trans. Audio Speech Lang. Process. 18 (8), 1889-1901.

Kim, D.Y., Un, C.K., Kim, N.S., 1998. Speech recognition in noisy environments using first-order vector Taylor series. Speech Commun. 24 (1), 39-49.

Li, J., Deng, L., Gong, Y., Haeb-Umbach, R., 2014. An overview of noise-robust automatic speech recognition. IEEE/ACM Trans. Audio Speech Lang. Process. 22 (4), 745-777.

Li, J., Deng, L., Yu, D., Gong, Y., Acero, A., 2007. High-performance HMM adaptation with joint compensation of additive and convolutive distortions via vector Taylor series. In: Proc. IEEE Workshop on Automatic Speech Recognition and Understanding (ASRU), pp. 65-70.

Li, J., Deng, L., Yu, D., Gong, Y., Acero, A., 2009. A unified framework of HMM adaptation with joint compensation of additive and convolutive distortions. Comput. Speech Lang. 23 (3), 389-405.

Li, J., Lee, C.H., 2007. Soft margin feature extraction for automatic speech recognition. In: Proc. Interspeech, pp. 30-33.

Liao, H., Gales, M.J.F., 2005. Joint uncertainty decoding for noise robust speech recognition. In: Proc. Interspeech, pp. 3129-3132.

Liao, H., Gales, M.J.F., 2006. Joint uncertainty decoding for robust large vocabulary speech recognition. University of Cambridge.

Liao, H., Gales, M.J.F., 2007. Adaptive training with joint uncertainty decoding for robust recognition of noisy data. In: Proc. International Conference on Acoustics, Speech and Signal Processing (ICASSP), vol. 4, pp. 389-392.

Lu, L., Ghoshal, A., Renals, S., 2013. Noise adaptive training for subspace Gaussian mixture models. In: Proc. Interspeech, pp. 3492-3496.

Narayanan, A., Wang, D., 2014. Joint noise adaptive training for robust automatic speech recognition. In: Proc. International Conference on Acoustics, Speech and Signal Processing (ICASSP).

Ochiai, T., Matsuda, S., Lu, X., Hori, C., Katagiri, S., 2014. Speaker adaptive training using deep neural networks. In: Proc. International Conference on Acoustics, Speech and Signal Processing (ICASSP), pp. 6349-6353.

Povey, D., Kingsbury, B., Mangu, L., Saon, G., Soltau, H., Zweig, G., 2005. fMPE: Discriminatively trained features for speech recognition. In: Proc. International Conference on Acoustics, Speech and Signal Processing (ICASSP), vol. 1, pp. 961-964.

Sainath, T.N., Kingsbury, B., Mohamed, A., Ramabhadran, B., 2013. Learning filter banks within a deep neural network framework. In: Proc. IEEE Workshop on Automatic Speech Recognition and Understanding (ASRU), pp. 297-302.

Sainath, T.N., Kingsbury, B., Mohamed, A., Saon, G., Ramabhadran, B., 2014. Improvements to filterbank and delta learning within a deep neural network framework. In: Proc. International Conference on Acoustics, Speech and Signal Processing (ICASSP), pp. 6839-6843.

Seltzer, M.L., 2012. Acoustic model training for robust speech recognition. In: Techniques for Noise Robustness in Automatic Speech Recognition, Virtanen, T., Singh, R., & Raj, B. (Eds.), John Wiley & Sons, pp. 347-368.

Shi, G., Shi, Y., Huo, Q., 2010. A study of irrelevant variability normalization based training and unsupervised online adaptation for LVCSR. In: Proc. Interspeech, pp. 1357-1360.

Wu, J., Huo, Q., 2002. An environment compensated minimum classification error training approach and its evaluation on Aurora 2 database. In: Proc. Interspeech, pp. 453-456.

Wu, J., Huo, Q., 2006. An environment-compensated minimum classification error training approach based on stochastic vector mapping. IEEE Trans. Audio Speech Lang. Process. 14 (6), 2147-2155.

Xu, J., Zhang, Y., Yan, Z.J., Huo, Q., 2011. An i-vector based approach to acoustic sniffing for irrelevant variability normalization based acoustic model training and speech recognition. In: Proc. Interspeech, pp. 1701-1704.

Yu, D., Deng, L., 2011. Deep learning and its applications to signal and information processing. In: IEEE Signal Processing Magazine, vol. 28, pp. 145-154.

Yu, D., Deng, L., 2014. Automatic Speech Recognition—A Deep Learning Approach. Springer, New York.

Zhu, D., Huo, Q., 2008. Irrelevant variability normalization based HMM training using MAP estimation of feature transforms for robust speech recognition. In: Proc. International Conference on Acoustics, Speech and Signal Processing (ICASSP), pp. 4717-4720.

Reverberant speech recognition

CHAPTER OUTLINE

After detailed discussions and analyses of noise-robust automatic speech recognition (ASR) techniques in the preceding five chapters, all assuming the use of a single microphone to acquire speech signals and assuming no acoustic distortions with reverberation, in this and the following chapters we delve into very different sets of techniques free from the single-microphone and no-reverberation constraints. In this chapter, we provide an overview of techniques for reverberant speech recognition.

9.1 INTRODUCTION

While for some usage scenarios of ASR it is natural that the sound capturing device is close to the speaker's mouth, many other would benefit in terms of user convenience if the microphone need not be held or worn close to the speaker's mouth. Examples include the hands-free control of consumer devices like interactive TVs, automatic meeting note transcription, speech interfaces in smart rooms, or the access of automated services over a telephone, which is operated in hands-free mode. In all these instances, the distance between the speaker and the microphone will no longer be small. Indeed, freeing the user from holding or wearing a microphone will not only increase usability. For some applications, it is even mandatory in order to assure safety of operation. Consider, for example, a voice interface to the car information and entertainment system. Here, hands-free operation is a must, and regulations in many countries prohibit manual dialing and holding a cellphone while driving. Reliable speech recognition with distant microphones is therefore essential for extending the scope of applications and increasing the convenience of existing speech recognition solutions.

However, increasing the distance between speaker and microphone has a major impact on the quality of the captured speech signal. There are three factors that need to be considered.

First, there is the signal attenuation due to the sound propagation from the source to the sensor. In free space, the signal power per unit surface decreases by the square of the distance. In Elko (2001), Elko argued by a simple thought experiment that increasing the distance between speaker and microphone from 2 to 10 cm or 1 m would correspond to an attenuation by 14 and 34 dB, respectively. While this rough calculation may be a bit too pessimistic, since it assumes an omnidirectional sound dissemination, whereas in reality the speaker's mouth is a directional source, it still points to a significant loss of signal power. This will usually come along with a loss in signal-to-noise ratio (SNR) in the same order of magnitude. Noise robustness techniques, the topic of preceding chapters, will have to be employed to compensate for this loss in SNR.

Second, in a distant-talking speech recognition scenario, it is likely that the microphone will capture other interfering sounds, in addition to the desired speech signal. These other acoustic events can be very diverse, hard to predict and very often of nonstationary nature and thus difficult to account for. Consider, for example, an interactive TV scenario with one person controlling the TV by voice and others in the room engaged in a conversation, or children playing in the background. An effective means to mitigate the impact of such interfering sounds on the ASR performance is to exploit the spatial diversity of the desired and undesired signals: If the distortion emanates from different locations than the desired speech, acoustic beamforming with microphone arrays are an effective means to suppress signals from directions other than that of the desired signal. For this second issue, we refer to Chapter 10.

The third effect on the signal caused by increasing the distance between speaker and microphone is the topic of this chapter: In an enclosure, the source signal will travel via multiple paths to the sensor. The wavefront of the speech is repeatedly reflected at the walls and other objects in the room. Thus, the signal at the microphone consists of multiple copies of the source signal, each with a different attenuation and time delay, see Figure 9.1. Reverberation refers to this process of multipath propagation. It alters the acoustic characteristics of the original speech signal in a way that it can mess up the automatic speech recognizer. The major problem

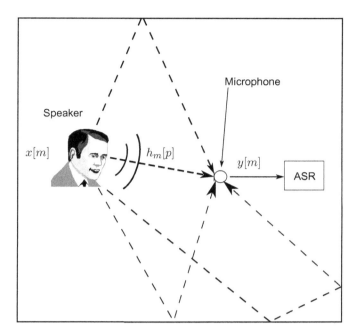

FIGURE 9.1

Hands-free automatic speech recognition in a reverberant enclosure: the source signal travels via a direct path and via single or multiple reflections to the microphone.

with reverberation is that its effect cannot be adequately described by considering a single frame of the short-time signal analysis in isolation. In contrast, the effect of reverberation spans multiple consecutive time frames leading to a temporal dispersion of a speech event over multiple speech feature vectors. It thus requires dedicated approaches which are quite different from what is done to combat additive noise or channel distortions that extend over just a single time frame.

Before discussing the approaches to reverberant speech recognition, we will first present a model of the physical effect of reverberation, both in the time, the frequency, and the feature domain. This model will help to understand the various approaches to handle reverberation. Following the taxonomy proposed in Yoshioka et al. (2012), the techniques to treat reverberation will be categorized according to the place where reverberation is addressed. We will thus discuss signal domain approaches based on linear filtering, followed by magnitude or power spectral domain methods, feature domain methods, and acoustic model domain approaches. While doing so, we will also discuss their properties w.r.t. the other categorizations that are used in this book. We will describe, whether prior knowledge of the distortion (here: reverberation) is used, whether an explicit, that is, physically motivated, or implicit, that is, data-driven modeling of reverberation is done, and whether disjoint or joint model training is executed.

9.2 ACOUSTIC IMPULSE RESPONSE

In Equation 3.1, we have introduced a model of the microphone signal, which we repeat here for convenience:

$$y[m] = x[m] * h[m] + n[m]. \tag{9.1}$$

According to this equation, the discrete-time microphone signal $y[m]$ at time instant m is a superposition of the convolution of the clean speech signal $x[m]$ with what we called in earlier chapters *convolutive channel distortion* $h[m]$ and of additive noise $n[m]$. This model is fairly general and serves also as the starting point of our discussion here. However, if reverberation is considered, the term $h[m]$ will have quite different properties than assumed so far. We will call $h[m]$ the acoustic impulse response (AIR) in the following. It represents the acoustic reaction of the room at the location of the microphone in response to a impulsive sound at the location of the speaker. Knowing $h[m]$, the changes the speech signal exhibits when radiated from the speaker's mouth and traveling through the room to the microphone can be computed according to Equation 9.1.

The AIR is in general time-variant. The time variance is caused by, for example, movements of the speaker, movements in the environment and even by temperature changes. We, therefore, generalize Equation 9.1 by assuming a time-variant AIR:

$$y[m] = \sum_{p=0}^{\infty} h_m[p]x[m - p] + n[m], \tag{9.2}$$

with the time-variant AIR $h_m[p]$. Here, the subscript m indicates the time variability, whereas $p \in \mathbb{N}_0$ is the ordinary lag index of an impulse response, that is, the time difference between the occurrence of exciting impulse and the reaction of the system.

To illustrate the time variance, consider a signal recorded at a sampling rate of $1/T_s = 8$ kHz. The distance traveled during a sampling period is $d = c \cdot T_s = 4.3$ cm, where c denotes the velocity of sound, that is, 343 m/s. Therefore, if the length of an echo path changes by 4.3 cm, for example, because the speaker slightly moves his head, the related contribution of this echo path to the AIR moves by one sampling interval. Thus, the AIR is highly sensitive to even smallest movements. Based on a statistical model of room acoustics, Radlovic et al. (1999) demonstrated that even a small movement of the speaker of the order of a tenth of the acoustic wavelength can cause significant changes in the AIR.

Figures 9.2 and 9.3 are typical examples of acoustic impulse responses. The AIR can be divided into three portions, as shown in the figure: The first impulse is the direct sound, which travels along the line-of-sight propagation path to the microphone. Next follow several strong reflections, called early reflections, which occur within 50 ms after the direct path. After that comes a series of numerous multiple reflections which are called late reverberation and which appear as a noise-like signal with an exponentially decaying envelope. The distinction between the different portions of the AIR is important both from a perceptual point of view and from a signal processing point of view.

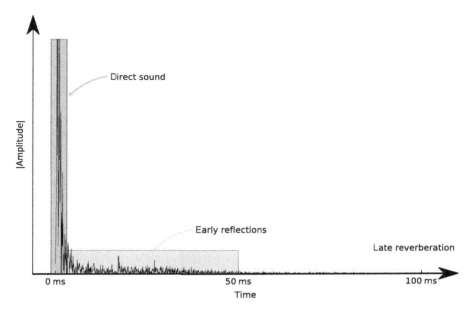

FIGURE 9.2

A typical acoustic impulse response for a small room with short distance between source and sensor (0.5 m). This impulse response has the parameters $T_{60}=250$ ms and $C_{50}=31$ dB. The impulse response is taken from the REVERB challenge data.

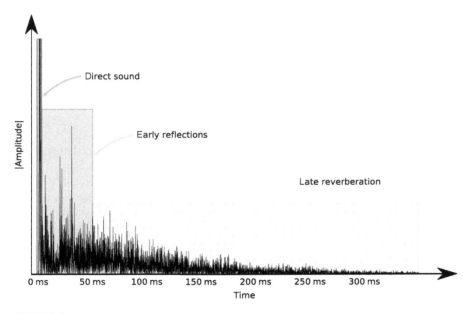

FIGURE 9.3

A typical acoustic impulse response for a large room with large distance between source and sensor (2 m). This impulse response has the parameters $T_{60}=700$ ms and $C_{50}=6.6$ dB. The impulse response is taken from the REVERB challenge data.

A signal consisting of just the direct sound is perceived as being "dry" and is, perhaps surprisingly, perceptually less pleasing than a signal which consists of a direct sound and early reflections. The early reflections that reach the listener within 50 ms are integrated with the direct sound by the auditory system and perceived as a single sound event, thus having a positive effect on perceived speech clarity. The late reverberation, on the contrary, is perceived as disturbing. It degrades the intelligibility of the signal.

Technically speaking, the effect of reverberation is thus twofold. The early reflections cause a coloration of the signal, that is, a slight shift of formants, similar to the channel distortion mentioned in earlier chapters, while the late reverberation leads to a temporal dispersion of the signal. It is the latter effect that bothers automatic (and human) speech recognition most!

It is also important to note that the characteristics of the early reflections depend strongly on the speaker and microphone positions, while the energy decay in the late reverberation is independent of the positions. The latter is determined by the room properties only. The time it takes until the energy in the late reverberation decays to 60 dB of the total energy of the AIR is called *room reverberation time* and is usually given the symbol T_{60}. For typical office and home rooms, the reverberation time ranges from 200 to 700 ms. The insensitivity of the late reverberation to the speaker position is exploited in various algorithms, as will be shown later.

There are several related measures that characterize the direct sound and early reflections. One of the most commonly used one is the *clarity index C_{50}*. It measures the ratio of the energy of the AIR in the first 50 ms versus the energy in the remaining part:

$$C_{50} = 10 \log_{10} \left(\frac{\sum\limits_{m=0}^{m_{50}} h^2[m]}{\sum\limits_{m=m_{50}+1}^{\infty} h^2[m]} \right), \tag{9.3}$$

where m_{50} is the sample index corresponding to 50 ms. The clarity index is, as mentioned, highly dependent on the speaker and microphone positions.

Because the effect of reverberation is characterized by the AIR $h_m[p]$, it is not surprising that many approaches to the recognition of reverberated speech require some knowledge of the AIR. A direct measurement of the AIR by exciting the room with an impulse or a noise like signal, however, is not an option for most applications of ASR. It would require a careful measurement setup before the speech recognition can be used, which is, of course, impractical. Furthermore, since the AIR is highly sensitive to even the slightest changes in the speaker-to-microphone enclosure, the true AIR present in the reverberated speech may be quite different from the measured one.

There are also blind channel identification techniques that estimate the AIR from an observed reverberated speech signal. These techniques usually require multi-channel recordings. However, the channel is only identifiable under two conditions:

the acoustic transfer functions (ATFs) of the source to the sensors must not share any common zeros in the z-plane, and the autocorrelation matrix of the source signal must be full rank. The ATF is the Fourier transform of the AIR. If these conditions are met, Huang and Benesty developed efficient frequency domain multi-channel adaptive filtering approaches for blind channel identification (Huang and Benesty, 2002). They are based on the construction of an error signal from the cross correlations between different channels, whose power is minimized using *least mean square* (LMS) and Newton algorithms. More on this and other methods for blind multi-channel identification can be found in Huang et al. (2008).

Due to the difficulties in estimating the AIR, researchers have developed models of it that can be used to approximate the impact of the AIR on the microphone signal. In almost all of these models, the AIR is assumed to be time invariant.

In a simplistic model, the effect of multiple reflections of sound in an enclosure can be described by an exponential decay of the acoustic energy (Kuttruff, 2009). The AIR can thus be modeled as a realization of a zero-mean white Gaussian process of unit variance $\zeta[m]$ with an exponentially decaying envelope

$$h[m] \approx \sigma_h \cdot \zeta[m] e^{-\frac{m}{M_0}}, \tag{9.4}$$

a model which was originally proposed by Polack (1988). It was used in the context of reverberant ASR, for example, in Krueger and Haeb-Umbach (2010) and Leutnant et al. (2014). Here, σ_h is a normalizing constant which determines the AIR's overall energy. The time constant M_0 is related to the room reverberation time T_{60} via

$$M_0 = \frac{T_{60}}{3 \log(10) \cdot T_s}. \tag{9.5}$$

In Hirsch and Finster (2008), an even coarser model of the AIR is used. The square of the AIR is approximated by its envelope only:

$$h^2[m] \approx \sigma_h^2 \cdot e^{-2m/M_0}. \tag{9.6}$$

With these simplified models, a measurement of the AIR in the target room is no longer required. The models are parameterized by T_{60} and σ_h. These parameters are the only properties that needs to be known about the acoustic environment. This not only makes algorithms employ these models independent of measurements but also increases their flexibility. A change of the room where the ASR system is used can be accounted for by only re-estimating these two parameters. Indeed, there exist methods to estimate these parameters blindly without the need for any calibration sentences, measurements, or offline processing.

The first parameter, σ_h^2, controls the energy E_h of the AIR:

$$E_h := E\left[\sum_{m=0}^{\infty} h^2[m] \right] = \sigma_h^2 \frac{1}{1 - e^{-2/M_0}}. \tag{9.7}$$

On databases containing artificially reverberated data, the estimation of E_h is often superfluous since the AIR that had been used to reverberate the data was normalized to unit energy ($E_h = 1$). However, this does not hold for true recordings of reverberant speech.

Let us assume that the training data has the property that $E_h = 1$, while this is not the case for the test data. To avoid a mismatch, the test data is rescaled to $E_h = 1$ as follows: The average power of the noise-free reverberant speech, σ^2, is related to that of nonreverberant speech, σ_x^2, through $\sigma^2 = E_h \sigma_x^2$. Let us assume that the average power of the test data prior to reverberation, $\sigma_{x,\text{TEST}}^2$, is equal to that of the training data, $\sigma_{x,\text{TRAIN}}^2$. To ensure $E_h = 1$ on the reverberant test data, the noisy reverberant test data is multiplied by $\sqrt{\sigma_{x,\text{TRAIN}}^2/\sigma_{\text{TEST}}^2}$ before feature enhancement, where the average power of reverberant speech of the test data, σ_{TEST}^2, can be computed from the average power of the noisy reverberant speech in the test data, $\sigma_{y,\text{TEST}}^2$, and the average power of the noise in the test data, $\sigma_{n,\text{TEST}}^2$, by $\sigma_{\text{TEST}}^2 \approx \sigma_{y,\text{TEST}}^2 - \sigma_{n,\text{TEST}}^2$ and by using a voice activity detector (Krueger et al., 2012).

The second parameter is the room reverberation time T_{60}. Depending on how this parameter can be estimated, there are a number of techniques where the blind approaches that estimate it from the reverberant speech signal, rather than from the AIR, are more relevant in practice. A maximum likelihood technique has been proposed by Ratnam et al. (2003). Wen et al. (2009) estimated the parameters of a two-slope model, while Löllmann and Vary (2008) treated the estimation of T_{60} in the presence of additive noise. In Gaubitch et al. (2012), a performance comparison of algorithms is presented for blind room reverberation time estimation from speech.

The authors of Hirsch and Finster (2008) proposed to select the T_{60} parameter based on the likelihood computed by the ASR decoder: T_{60} is estimated after multiple recognitions of an utterance with HMMs adapted under different hypotheses for the value of T_{60}. They then decided on that value of T_{60}, whose corresponding set of HMMs achieved the highest likelihood in a forced alignment of the feature vector sequence with the recognized sentence.

However, it must be conjectured that coarse models of the AIR will lead to some performance loss in the speech recognizer. Indeed, it has been observed that the model of Equation 9.4, when used for enhancing reverberated speech features prior to recognition, performs significantly better if the distance between speaker and microphone is large than if the speaker is close to the microphone (Heymann et al., 2015). The reason is clearly the poor modeling of the direct path and the early reflections. As described earlier, the decaying envelope model is, strictly speaking, only appropriate for the late reverberation. A model that better accounts for the direct path has been proposed in Habets et al. (2009), where the energy contribution of the direct path was explicitly estimated.

Furthermore, the simplified models of Equations 9.4 and 9.6 do not account for the frequency dependence of the reverberation time. The frequency dependence is caused by the fact that materials often have frequency-dependent acoustic properties, leading to different reflection and attenuation patterns at different frequencies. However, it has been observed in Erkelens and Heusdens (2011) that a frequency-dependent model was hardly superior to a frequency-independent model in terms of reverberation suppression.

9.3 A MODEL OF REVERBERATED SPEECH IN DIFFERENT DOMAINS

Equations 9.1 and 9.2 describe the microphone signal in the time domain. Here, we are seeking the corresponding relation in the feature domain, where we assume the computation of MFCC features as described in Section 3.2. In the following, we assume a time-invariant AIR of finite length L_h for simplicity

$$y[m] = \sum_{p=0}^{L_h-1} h[p]x[m-p] + n[m]. \tag{9.8}$$

In Equation 3.2, it was assumed that the *multiplicative transfer function (MTF)* approximation holds. The MTF approximation states that a convolution in the time domain is well approximated by a multiplication in the short-time Discrete Fourier transform (STDFT) domain, see Avargel and Cohen (2007). This is true if the length of the impulse response is much smaller than the window length of the STDFT.

This assumption is most likely invalid in the case of reverberation. For normal rooms, the AIR extends over several hundred milliseconds, while a typical window length used in ASR is $L_w = 25$ ms. If the MTF approximation is not used, the following expression has been derived for the STDFT of Equation 9.8 (Krueger et al., 2011; Nakatani et al., 2010b):

$$\dot{y}[t][k] \approx \sum_{\tau=0}^{L_H} \dot{h}[\tau][k]\dot{x}[t-\tau][k] + \dot{n}[t][k]. \tag{9.9}$$

Here, t denotes the frame index, while k is the frequency bin index. $\dot{x}[t][k]$ is the STDFT of $x[m]$, $\dot{n}[t][k]$ is the STDFT of the noise $n[m]$, and $\dot{h}[\tau][k]$ is an STDFT domain representation of the time-domain AIR (see Krueger and Haeb-Umbach (2010) for details).

This is the so-called convolutive transfer function (CTF) approximation, which states that the STDFT of the reverberant and noisy speech signal $\dot{y}[t][k]$ depends not only on the current STDFT of the clean speech STDFT $\dot{x}[t][k]$ but also on the L_H past STDFTs, where

$$L_H = \left\lfloor \frac{L_h + L_w - 2}{B} \right\rfloor, \tag{9.10}$$

where L_w is the length of the STDFT analysis window and B is the frame advance.

Note that Equation 9.9 is still an approximation because the contributions from neighboring frequencies to $\dot{y}[t][k]$ have been neglected (Avargel and Cohen, 2007).

In a further approximation disregarding the cross terms between different time frames and disregarding the "phase term," see Equation 3.3 for an explanation, the power spectral density (PSD) domain model of noisy reverberated speech is given by

$$|\dot{y}[t][k]|^2 \approx \sum_{\tau=0}^{L_H} |\dot{h}[\tau][k]|^2 |\dot{x}[t-\tau][k]|^2 + |\dot{n}[t][k]|^2. \tag{9.11}$$

After applying the Mel-filter-bank and the logarithmic compression, the following expression for the log-Mel power spectral coefficients (LMPSCs) of speech and noise can be derived

$$\tilde{\mathbf{y}}[t] = \log\left(\sum_{\tau=0}^{L_H} e^{\tilde{\mathbf{h}}[\tau]+\tilde{\mathbf{x}}[t-\tau]} + e^{\tilde{\mathbf{n}}[t]}\right) + \tilde{\mathbf{v}}[t]$$

$$= f\left(\tilde{\mathbf{x}}[t-L_H:t], \tilde{\mathbf{h}}[0:L_H], \tilde{\mathbf{n}}[t]\right) + \tilde{\mathbf{v}}[t], \tag{9.12}$$

where

$$f\left(\tilde{\mathbf{x}}[t-L_H:t], \tilde{\mathbf{h}}[0:L_H], \tilde{\mathbf{n}}[t]\right) = \log\left(\sum_{\tau=0}^{L_H} e^{\tilde{\mathbf{h}}[\tau]+\tilde{\mathbf{x}}[t-\tau]} + e^{\tilde{\mathbf{n}}[t]}\right) \tag{9.13}$$

and where

$$\tilde{\mathbf{x}}[t-L_H:t] = \tilde{\mathbf{x}}[t-L_H], \ldots, \tilde{\mathbf{x}}[t], \tag{9.14}$$

$$\tilde{\mathbf{h}}[0:L_H] = \tilde{\mathbf{h}}[0], \ldots, \tilde{\mathbf{h}}[L_H]. \tag{9.15}$$

Here, $\tilde{\mathbf{h}}[\tau]$ is an approximate LMPSC domain representation of the AIR. The error term $\tilde{\mathbf{v}}[t]$ captures all approximations made in the course of the derivation of Equation 9.12 from Equation 9.8.

In the above equations, we have used a vector notation, where the vector comprises the Mel-filter-bank outputs, for example, $\tilde{\mathbf{y}}[t] = (\tilde{y}[t][1], \ldots, \tilde{y}[t][K_{FB}])^T$, where K_{FB} is the number of filters in the filter bank. Further, the operations $\log(\cdot)$ and $\exp(\cdot)$ are independently applied to each vector element.

The last step in the feature extraction is the application of the DCT to obtain the MFCC feature vector $\mathbf{y}[t]$.

Disregarding the "phase term," that is, the cross term between the (reverberated) speech and the noise in the PSD domain, is common practice as mentioned in Section 3.2. However, if the noise power is in the same order of magnitude as the power of reverberated speech, the phase term plays a significant role. In Leutnant et al. (2014), the phase factor has been considered. Further, it has been shown that the statistics of the error term $\tilde{\mathbf{v}}[t]$ in Equation 9.12, which is the difference between the feature domain representation of $y[m]$ and that predicted from $\tilde{x}[t - L_H : t]$, $\tilde{h}[0 : L_H]$ and $\tilde{n}[t]$, depends on the instantaneous ratio of the power of the reverberated speech and the noise, which is time variant.

9.4 THE EFFECT OF REVERBERATION ON ASR PERFORMANCE

Figure 9.4 shows the spectrogram of a nonreverberant speech signal, and the same speech signal under mild and strong reverberation, respectively. From the figure, the temporal dispersion caused by reverberation can be clearly observed. Consider, for example, the glottal stop at the beginning of the word "company," which is clearly visible in the clean speech spectrogram, while it has disappeared due to the temporal smearing of the speech energy in the reverberant case.

The effect of reverberation on ASR performance is dramatic. In Krueger et al. (2012), the following numbers have been reported for the Aurora 5 connected digit recognition task (Hirsch, 2007) : If training and test have been performed on clean, nonreverberant digits, a digit error rate of 0.6% has been obtained. When the test data are artificially reverberated with acoustic impulse responses of reverberation times between 300 and 400 ms, the word error rate increased to 6.4% and when the

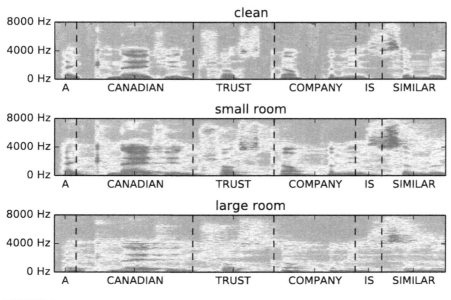

FIGURE 9.4

Spectrogram of a clean speech signal (top), a mildly reverberated signal (T_{60}=250 ms, middle) and a severely reverberated signal (T_{60}=700 ms, bottom). The dashed lines indicated the word boundaries.

reverberation times of the AIRs ranged from 400 to 500 ms, the word error rate even rose to 15.0 %.

On the RealData set of the REVERB challenge, the word error rate rose from about 10% for speech recorded by a headset to over 80% for the same speech, if it was recorded by a microphone 2.5 m away from the speaker (Delcroix et al., 2014; Kinoshita et al., 2013), if no means to combat reverberation were employed.

These numbers indicate an obvious need to improve the reverberation robustness of ASR. In the following, we will give an overview of algorithms to improve the performance of ASR to speech degraded by reverberation. We categorize the methods according to the place where the reverberation is treated:

- Linear filtering approaches: these are signal domain approaches to remove the effect of reverberation, which are either implemented in the time or in the STDFT domain.
- Magnitude or power spectrum domain techniques: methods in this category attempt to remove the effect of reverberation on the magnitude or power spectrum of the signal, while the phase is left unprocessed.
- Feature domain approaches attack reverberation either by devising reverberation robust features, conducting an appropriate normalization, or by dereverberating the LMPSC or MFCC feature vectors
- Acoustic model-based approaches modify the acoustic models to account for reverberation.

9.5 LINEAR FILTERING APPROACHES

Linear filtering techniques aim at dereverberating the speech signal. There exist a large variety of algorithms (Huang et al., 2008; Naylor and Gaubitch, 2010). Here, we concentrate on methods that have been successfully applied as a front end processing technique for ASR.

Linear filtering is a signal domain technique that is either realized in the time domain or in the STDFT domain. Unlike magnitude or power spectrum domain techniques, which also aim at dereverberating the speech signal, they account for the phase of the reverberated signal. Accounting for the phase can improve the accuracy of dereverberation because both phase and amplitude of a signal are affected by reverberation.

A characteristic of many linear filtering techniques is that they can make effective use of multiple microphones. The signals arriving at the sensors of a spatially compact microphone array differ for the most part in the phase, not in the amplitude. These additional observations can be exploited for dereverberation and/or for beamforming (where the latter has also some dereverberating effect). On the other hand, spectrum, feature, or acoustic model enhancement techniques are not likely to profit much from observing additional channels, unless the distance between the microphones is large enough.

While the exploitation of the signal phase can be seen as an advantage, it adds to the complexity and vulnerability of the algorithms, though. Small speaker movements will hardly affect the power spectral density of the received signal. However, they can have a major impact on the signal phase, as discussed earlier. We have already mentioned the difficulties in blindly estimating the AIR. Note that for typical sampling rates of 8 or 16 kHz and typical room reverberation times of 300 ms to 700 ms the AIR can easily have a length of several thousand taps! The *multiple input/output inverse theorem* (MINT) states, that if the AIR is known, a perfect channel inversion is in principle possible, if the number of microphones is larger than the number of signal sources and if the transfer functions from the source to microphones do not share common zeros (Miyoshi and Kaneda, 1988). However, it has been observed that even small AIR estimation errors can lead to significant signal distortions, making it a less attractive approach in a real-world system.

An explicit estimation of the AIR can be avoided by directly estimating the linear filter that inverts the channel, that is, by attempting to whiten the microphone signal. Here, the crucial issue is, however, that a complete whitening would destroy the correlation structure of the speech. Ideally, one would like to keep the spectral properties of speech, while eliminating the effect of the channel.

We start our discussion on channel deconvolution techniques with Equation 9.9, the STDFT representation of the noisy reverberated speech signal. Equation 9.9 states that the effect of reverberation can be considered in each frequency bin independently, where in each frequency bin a convolution of the unreverberated signal with the sequence of channel coefficients $\dot{h}[\tau][k]$, $\tau = 0, \dots, L_H$ is performed. L_H is the number of frames over which the reverberation disperses the signal and which corresponds to roughly L_h/B, see Equation 9.10. Attempting a dereverberation in the STDFT domain rather than the time domain thus has the advantage that the number L_H of coefficients per frequency bin k is much smaller than the number L_h of time domain samples of the AIR, considerably simplifying the estimation problem.

Neglecting the additive noise term, an obvious approach to dereverberation would be to determine a linear filter $\dot{b}[\tau][k]$ such that

$$\sum_{\tau=T_l}^{T_u} \dot{b}[\tau][k]\dot{h}[t-\tau][k] = \delta[\tau], \tag{9.16}$$

where $\delta[\tau]$ is the discrete-time unit impulse, which is one for $\tau = 0$ and zero else. This is the task of blind deconvolution, also called channel equalization, which has been investigated extensively in digital communications for equalizing, for example, a multipath channel. It has also been investigated in the context of speech signals, for example, in Hopgood and Rayner (2003).

It is well known that a single-input multiple-output filter can be equalized blindly by applying multi-channel linear prediction (LP) to its output when the input is white. However, when applied to a non-white input signal such as speech, channel equalization as in Equation 9.16 will destroy the correlation structure of speech because the equalizer cannot distinguish between the correlation introduced by the

AIR and the correlation of the speech signal itself. The deconvolution filter $\dot{b}[\tau][k]$ will thus both remove the reverberation and whiten the source.

This unwanted effect can be avoided by introducing knowledge about the source signal. In Triki and Slock (2005, 2006), the spatial diversity provided by multi-channel input and the speech signal's non-stationarity is exploited to estimate the source signal correlation structure. It is used to determine a source whitening filter. Multi-channel linear prediction is then applied to the microphone signals after filtering by the source whitening filter. This delivers the coefficients of a deconvolution filter that removes the correlations introduced by reverberation. Thus if this deconvolution filter is applied to the reverberated speech signal, the reverberation tends to be removed, while the source correlation structure is maintained.

Long-term linear prediction of speech is used in Nakatani et al. (2010b), Kinoshita et al. (2009), Yoshioka et al. (2011), and Yoshioka and Nakatani (2012) for dereverberation while maintaining the correlation structure of speech. Here, the reverberated signal is predicted by long-term (multistep) linear prediction, where the current reverberated observation $\dot{y}[\tau][k]$ is predicted as the sum of the clean speech signal $\dot{x}[\tau][k]$ and a signal predicted by the last T_u but T_l frames, see Equation 9.16. The most recent frames $\dot{y}[\tau][k], \ldots, \dot{y}[\tau - T_l][k]$ are excluded to avoid whitening the speech signal. A speech model is employed to facilitate the estimation of the deconvolution filter $\dot{b}[\tau][k]$. In the formulation for a single-channel input, dereverberation is achieved by

$$\dot{x}[t][k] = \dot{y}[t][k] - \sum_{\tau=T_l}^{T_u} \dot{b}[\tau][k]\dot{y}[t - \tau][k], \tag{9.17}$$

where a typical value of T_l is 3, while T_u is chosen between 7 and 40 (for a window length of $L_w = 32$ ms and frame shift of $B = 8$ ms) (Delcroix et al., 2014).

Using the concept of weighted prediction error (WPE) minimization (Yoshioka and Nakatani, 2012; Yoshioka et al., 2011), the linear predictor $\dot{b}[\tau][k]$ can be found by optimizing an objective function, which is obtained by assuming the prediction error $\dot{y}[t][k] - \sum_{\tau=T_l}^{T_u} \dot{b}[\tau][k]\dot{y}[t - \tau][k]$ is Gaussian with a time-variant variance. In Yoshioka and Nakatani (2012), the single-channel dereverberation algorithm of Equation 9.17 was generalized to multi-channel input and even extended to produce the same number of output signals as input signals. Thus a beamformer can follow this multiple-input multiple-output (MIMO) dereverberation filter for further noise reduction.

This dereverberation algorithm has been successfully applied to different re-verberant speech tasks, such as an automatic meeting transcription task and the REVERB challenge tasks. It has been shown to improve the word error rate performance of deep neural network (DNN) acoustic models (Dahl et al., 2011, 2012; Hinton et al., 2012; Yu and Deng, 2011, 2014) trained on nearly matched reverberant conditions (Delcroix et al., 2014; Yoshioka et al., 2014). On the REVERB challenge data, the multi-channel WPE algorithm was able to reduce the WER on the RealData by 25% (see Section 9.9 for a description of the data set).

It is long known that dereverberation can be achieved by processing the linear prediction residual (Naylor and Gaubitch, 2010). Reverberation causes extraneous peaks in the LP residual. Consequently, dereverberation is achieved by processing the LP residual to attenuate these peaks. In one form, known as correlation shaping, a multi-channel input is assumed, and the microphone signals are processed by an adaptive linear filter whose coefficients are determined via gradient descent. The objective function to be minimized is the weighted mean square error between the actual output autocorrelation and the desired output autocorrelation of the filter, where the filter is fed with the LP residual. The desired output autocorrelation is zero correlation, that is, a dirac delta impulse (Gillespie and Atlas, 2003). This approach improves both the audible quality and ASR performance of reverberant speech. On the REVERB challenge data, correlation shaping reduced the ASR word error rate by up to 25%, both for a GMM-HMM and a recurrent neural network back-end (Geiger et al., 2014b).

9.6 MAGNITUDE OR POWER SPECTRUM ENHANCEMENT

Magnitude or power spectrum dereverberation techniques usually operate on single channel input.

In a popular approach, the AIR $h[m]$ is subdivided in two parts, the early part $h^e[m]$ and the late part $h^\ell[m]$. Neglecting additive noise Equation 9.1 then reads

$$y[m] = h^e[m] * x[m] + h^\ell[m] * x[m - \Delta m] = h^e[m] * x[m] + r[m]. \qquad (9.18)$$

The late reverberation component $r[m]$ is assumed to be uncorrelated with the direct sound and early reflections component. This approximation is justified by the fact that the correlation structure of clean speech hardly extends beyond 50 ms. Thus the multiple reflections arriving at large time lags can indeed be considered uncorrelated with the "desired" signal. The key idea is to estimate the magnitude or power spectrum of the late reverberation component and subsequently to subtract it from that of the reverberant signal in much the same way as is done in spectral subtraction approaches for noise suppression, see Section 4.1.3. This approach has originally been proposed by Lebart et al. (2001) and was later refined in various ways (Erkelens and Heusdens, 2010a,b; Habets, 2007; Habets et al., 2009).

A major issue is the estimation of the power spectrum of the highly nonstationary late reverberation component $r[m]$, called late reverberant spectral variance (LRSV). One method to do so is by correlation analysis, where the observed correlation introduced by late reverberation is used to estimate the LRSV without having to assume a specific model for the AIR (Erkelens and Heusdens, 2010b).

Other methods start from the model of Equation 9.11, which describes the relationship between the observed PSD of reverberated speech and the PSD of current and past clean speech segments. From this, a recursive estimator of the LRSV can be derived, which requires an estimate of the PSD representation of the AIR. For this, several options exist. One option is to employ models of the AIR, such as the model given in Equation 9.4. Then only the room reverberation time (and the AIR energy) need to be known or estimated. We have mentioned earlier, however, that this model does not account for the direct sound and early reflections well. An improved model, which accounts for this, would require an estimate of C_{50} or a related parameter. With this improved AIR model, better performance has been reported in particular for small speaker-to-microphone distances (Habets et al., 2009).

In an experimental study, it was shown that ASR performance was maximized if the time lag beyond which the AIR is declared to represent late reverberation is set to $\Delta m \cdot T_s = 50$ ms. In that study, the remaining reverberation and artifacts were accounted for by training on moderately reverberant speech (Maas et al., 2012).

PSD enhancement can also be accomplished without an explicit model of the AIR. In Kumar et al. (2011a), a blind factorization is derived to decompose the PSD of reverberated speech into the PSD of clean speech and the PSD of the AIR. The approach of Kumar et al. (2011a) operates in the PSD domain, however, after applying a Gammatone filter bank to the magnitude spectrum. The reverberant Gammatone magnitude spectrum $|\ddot{y}[t][k]|$ is decomposed into its convolutive constituents $|\dot{h}[\tau][k]|$ and $\dot{x}[t][k]|$ by a variant of non-negative matrix factorization (NMF), assuming absence of additive noise. With this approach, however, *a priori* knowledge of the AIR (e.g., in the form of Equation 9.4 or Equation 9.6) cannot be easily incorporated. As an alternative to NMF, in Kumar et al. (2011b), an iterative least squares approach is employed for the decomposition of the magnitude spectrum of reverberated speech.

9.7 FEATURE DOMAIN APPROACHES

In this section, we distinguish between reverberation robust features, feature normalization, and feature enhancement.

9.7.1 REVERBERATION ROBUST FEATURES

It is commonly understood that the speech perception of humans is much more robust to reverberated speech than an ASR system is. In an attempt to mimic this robustness, researchers have tried to develop representations of the speech signal that share some of the properties of human hearing. One approach is to enhance the modulation spectrum of speech in the range of 2-16 Hz with a maximum at 4 Hz, the syllable rate of speech (Kingsbury et al., 1998). Others suggest to employ a Gammatone instead of a Mel-filter-bank which is supposedly closer to the processing in the human auditory system (Kumar et al., 2011a; Maganti and Matassoni, 2010). The authors of Kim and Stern (2010) suggest a nonlinear filtering of the modulation spectrum in a way which mimics the precedence effect. This psychoacoustic effect states that similar sounds arriving from different directions are solely localized in the direction of the first sound wave arriving at our ears (Litovsky et al., 1999). Motivated by this, they introduced an algorithm to suppress the slowly varying components and the falling edge of the power envelope in each frequency band to enhance speech recognition accuracy under reverberant environments. This nonlinear highpass filtering applied to the modulation spectrum to detect the first-arriving wavefront has been successful in suppressing reverberation, however at the cost of introducing distortions leading to increased ASR error rate on clean, nonreverberant speech.

In follow-up work, temporal masks were constructed to suppress reflected wavefronts. The perceived peak sound level was estimated in each frequency band, after applying a power-law nonlinearity, and a temporal mask was applied based on this. Experimental results showed that this algorithm was slightly superior to the earlier nonlinear highpass filtering approach and much more robust to reverberation than standard MFCC features (Kim et al., 2014).

9.7.2 FEATURE NORMALIZATION

Probably the most well-known approach to feature normalization is the cepstral mean normalization (CMN) that has already been discussed in Section 4.1.2 in the context of improving noise robustness.

It has also been applied to normalize a feature vector w.r.t. to reverberation. Indeed, if the MTF approximation (Avargel and Cohen, 2007) holds, the convolution of the source signal with the AIR turns into a multiplication with the ATF in the STDFT domain, while in the LMPSC or cepstral domain, the contribution by the ATF is additive and can thus be removed by CMN. With typical STDFT analysis window sizes of $L_w = 25$ ms and reverberation times well beyond 100 ms, the MTF approximation does not hold, and CMN or related approaches like RASTA are not very effective (Kingsbury and Morgan, 1997). As a consequence, it has been suggested to use very long analysis window sizes of up to 2 s (Avendano et al., 1997; Gelbart and Morgan, 2001), which, however, leads to an undesired latency in the signal processing chain.

While CMN treats each feature vector component separately, fMLLR (*feature-space maximum likelihood linear regression*) considers all components of the feature vector jointly and applies a transformation matrix to them, see Section 4.2. Again, as fMLLR does not consider the temporal smearing caused by reverberation and transforms each feature vector separately, independent of the preceding vectors, the gains achieved by fMLLR in terms of reverberation robustness are rather moderate (Krueger and Haeb-Umbach, 2010).

9.7.3 MODEL-BASED FEATURE ENHANCEMENT

Feature enhancement aims at removing the effect of reverberation from the computed speech features. We first discuss an approach that is based on an explicit, physically motivated model of reverberation, while in the subsequent section we will consider data-driven approaches, which do not require such a physical model.

The model-based approach exploits knowledge of the physical process of reverberation as described in Section 9.3. In Krueger and Haeb-Umbach (2010) and Leutnant et al. (2013, 2014), a Bayesian feature enhancement (BFE) approach is proposed. The method computes the posterior PDF of the clean LMPSC feature vector $\tilde{\mathbf{x}}[t]$ given the observed reverberated feature vectors $\tilde{\mathbf{y}}[1], \ldots, \tilde{\mathbf{y}}[t], \ldots, \tilde{\mathbf{y}}[t + L_c]$. Note that a latency of L_c frames (a typical value would be $L_c = 6$) is introduced which greatly improves the estimate's quality, compared to an estimation without looking into the future. The observation, that a latency of a few frames greatly improves the estimate of clean speech, has also been made in Schwartz et al. (2014) in the context of linear filtering-based signal dereverberation.

A state vector is introduced

$$\tilde{\mathbf{z}}[t] := \left((\tilde{\mathbf{x}}[t])^T, \ldots, (\tilde{\mathbf{x}}[t - L_c + 1])^T, (\tilde{\mathbf{n}}[t])^T \right)^T \tag{9.19}$$

containing the most recent L_c LMPSC vectors of the clean speech and the current noise LMPSC vector. The *a posteriori* probability density function $p(\tilde{\mathbf{z}}[t]|\tilde{\mathbf{y}}[1 : t])$ is computed recursively in a Kalman filter-like fashion by alternating between a prediction step

$$p(\tilde{\mathbf{z}}[t]|\mathbf{y}[1 : t - 1]) \tag{9.20}$$

$$= \int p(\tilde{\mathbf{z}}[t]|\tilde{\mathbf{z}}[t - 1], \tilde{\mathbf{y}}[1 : t - 1]) p(\tilde{\mathbf{z}}[t - 1]|\tilde{\mathbf{y}}[1 : t - 1]) \mathrm{d}\tilde{\mathbf{z}}[t - 1] \tag{9.21}$$

and an update step

$$p(\tilde{\mathbf{z}}[t]|\tilde{\mathbf{y}}[1 : t]) \propto p(\tilde{\mathbf{y}}[t]|\tilde{\mathbf{z}}[t], \tilde{\mathbf{y}}[1 : t - 1]) p(\tilde{\mathbf{z}}[t]|\tilde{\mathbf{y}}[1 : t - 1]). \tag{9.22}$$

The prediction step requires an *a priori* model $p(\tilde{\mathbf{z}}[t]|\tilde{\mathbf{z}}[t - 1], \tilde{\mathbf{y}}[1 : t - 1])$ for the clean speech and noise LMPSCs. For speech, a switching linear dynamic model (SLDM) was employed, which is able to capture the characteristic dynamics of clean speech feature trajectories. Switching between multiple dynamic models leads to a much better representation of the speech feature dynamics than using a single dynamic model. A GMM as *a priori* model, on the contrary, is unable to model the correlation between successive speech frames, and indeed leads to much worse ASR performance. The GMM *a priori* model is no longer able to help in distinguishing feature trajectories typical of speech from deviations from it introduced by reverberation (Krueger et al., 2011).

The update step calls for an observation model $p(\tilde{\mathbf{y}}[t]|\tilde{\mathbf{z}}[t], \tilde{\mathbf{y}}[1 : t - 1])$. According to Equation 9.12, the relation between the observed noisy reverberant feature vectors and those of clean speech and of noise is highly nonlinear. To arrive at a tractable solution, iterative vector Taylor series approximation is applied and $p(\tilde{\mathbf{y}}[t]|\tilde{\mathbf{z}}[t], \tilde{\mathbf{y}}[1 : t - 1])$ is modeled to be a normal PDF, however, with time-variant mean and covariance matrix. The parameters of the PDF depend on the AIR. To avoid estimating the AIR, the simplified model of Equation 9.4 was used. Several variants of the observation model were derived, a recursive and a nonrecursive one,

one that neglects the contribution of the additive noise $\tilde{\mathbf{n}}[t]$ to the observation error $\tilde{\mathbf{v}}[t]$, see Equation 9.12, and one that includes its contribution (Leutnant et al., 2014).

Exact inference in the presence of a switching dynamic model is computationally intractable, as the number of hypotheses to consider grows exponentially in time. Thus only an approximate computation of the clean speech posterior can be realized, using algorithms like the Generalized Pseudo Bayesian approach of first (GPB1) or second order (GPB2) or the Interacting Multiple Model algorithm (Bar-Shalom et al., 2001). They consist of a bank of Kalman filters and an algorithm to compute the contribution of each Kalman filter to the final posterior estimate. The estimate forwarded to the ASR back-end is the component $\tilde{\mathbf{x}}[t - L_c + 1]$ of the state vector of Equation 9.19.

The charm of this approach to feature enhancement is that it is a parametric approach. One only needs to know the reverberation time T_{60} and the energy parameter σ_h of the target environment, both of which can be estimated rather easily. If the testing environment changes, the feature enhancement can be adjusted just by setting these two parameters.

Furthermore, it is interesting to note that the BFE algorithm estimates the clean speech posterior, the mean of which is the MMSE estimate of the clean, noise-free and unreverberated speech. The variance of the posterior is related to the estimation error of the MMSE estimate and can thus be employed for uncertainty decoding, as shown in Krueger and Haeb-Umbach (2010), thus placing this approach in the category of uncertainty modeling techniques, which are considered in the context of noise-robust ASR in Chapter 7.

BFE has shown very good performance in ASR tasks with a large mismatch between training and test conditions. For example, the word error rate of a GMM-HMM recognizer on the multi-channel Wallstreet Journal Audiovisual database (Lincoln et al., 2005), which had been trained on clean, nonreverberant speech, was reduced by 50% by employing BFE (Krueger et al., 2012).

A Bayesian approach was also taken in Wölfel and McDonough (2009). In the observation model used there, reverberation is treated as an additive distortion in the Mel power spectral domain, as is suggested by the late reverberation model of Equation 9.18. Its parameters were estimated by multistep linear prediction (Kinoshita et al., 2009). The *a priori* model of speech was a higher-order autoregressive process, and the inference was carried out with a particle filter.

9.7.4 DATA-DRIVEN ENHANCEMENT

As the prominent example of a data-driven method to dereverberating speech features, we are going to discuss denoising autoencoders. They have been proposed to map noisy or reverberated speech features to clean features for robust ASR. In the context of reverberant speech they have been applied either to the power spectrum coefficients, the LMPSC coefficients, or the MFCC coefficients of clean speech.

An autoencoder (AE) is a neural network that is trained to encode the input in some representation so that the input can be reconstructed from that representation.

The mapping from the input to the hidden representation is called encoder, and the reconstruction of the input from that representation the decoder. The encoder is a mapping $f(\mathbf{x})$ that transforms a D dimensional input vector $\mathbf{x}$ to a representation $\boldsymbol{\xi}$. The mapping consists of an affine transformation with weight matrix $\mathbf{W}$ and bias vector $\mathbf{b}$, followed by a nonlinearity (Bengio et al., 2007; Deng et al., 2010)

$$\boldsymbol{\xi} = s(\mathbf{x}) = s\,(\mathbf{Wx} + \mathbf{b})\,. \tag{9.23}$$

The nonlinearity $s(\cdot)$ can be a sigmoid nonlinearity, but other nonlinearities have also been used, such as a rectified linear unit. The parameters of this mapping are the weight matrix and the bias vector: $\theta = \{\mathbf{W}, \mathbf{b}\}$. The resulting hidden representation $\boldsymbol{\xi}$ is then mapped to a reconstructed vector $\hat{\mathbf{x}}$ in the original input space by the decoder operation $g(\boldsymbol{\xi})$, whose typical form is an affine transformation, optionally followed again by a nonlinearity, that is,

$$\hat{\mathbf{x}} = g(\boldsymbol{\xi}) = \mathbf{W}'\boldsymbol{\xi} + \mathbf{b}' \tag{9.24}$$

or

$$\hat{\mathbf{x}} = g(\boldsymbol{\xi}) = s\,(\mathbf{W}'\boldsymbol{\xi} + \mathbf{b}') \tag{9.25}$$

with parameters $\theta' = \{\mathbf{W}', \mathbf{b}'\}$. The parameters of the encoder and decoder are learnt as to minimize a loss function, such as the mean squared reconstruction error: $J(\theta, \theta') = \|\hat{\mathbf{x}} - \mathbf{x}\|^2$.

Note that if there were no nonlinearity, neither in the encoder nor in the decoder, then the hidden units would learn to project the input to the first principal components, that is, conduct a Principal Component Analysis (PCA). However, due to the nonlinearity, the autoencoder behaves quite differently, with the ability to capture multi-modal aspects in the input distribution (Japkowicz et al., 2000).

The denoising autoencoder (DAE) differs from the autoencoder discussed so far by the fact that it is trained to reconstruct the clean input $\mathbf{x}$ from an observed noisy version $\mathbf{y}$ of it. The corrupted input $\mathbf{y}$ is mapped as described above to a hidden representation, from which $\hat{\mathbf{x}}$ is reconstructed by the decoder, see Figure 9.5. The parameters θ and θ' are trained to minimize the average quadratic deviation of the reconstructed feature vector from the clean feature vector: $J(\theta, \theta') = \|\hat{\mathbf{x}} - \mathbf{x}\|^2$. Note that one requires stereo data for the training of the DAE: the clean and the distorted version of the speech data. This, however, may not be a big disadvantage in case of reverberation, because reverberated data can be generated from clean data by artificially convolving the clean data with an AIR, which may be either measured or generated artificially, for example, by the image method (Allen and Berkley, 1979). There may, however, still remain a mismatch between the artificially reverberated training data and the data of the testing environment, even if the T_{60} and C_{50} parameters match, because the test data will be true reverberant speech, while the training data has been artificially reverberated.

Denoising autoencoders have been analyzed by Vincent et al. (2008) and Bengio (2009) and applied to different pattern classification tasks, such as optical

Decode layer

Encoding layers

Input layer

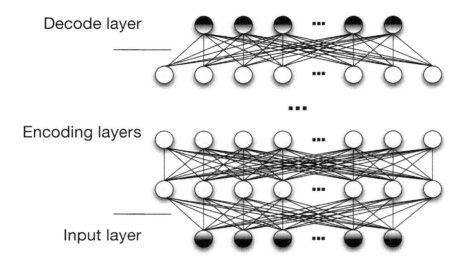

FIGURE 9.5

Principle structure of a denoising autoencoder.

character recognition. Multiple encoding and decoding layers can be stacked to form stacked denoising autoencoders (Vincent et al., 2010), also called deep denoising autoencoders.

DAEs have been used in the context of reverberation robust speech recognition and applied to different representations of the input signal. In Ishii et al. (2013), a stacked DAE with five layers was applied to reconstruct the clean speech spectrum from reverberant speech spectra. In order to capture context effects, multiple short-windowed spectral frames were concatenated to form a single input vector. Additionally, a combination of short- and long-term spectra was used to account for long acoustic impulse responses. In Mimura et al. (2014), a stacked DAE with three encoding and decoding layers was applied to a window of LMPSC coefficients, while the authors (Feng et al., 2014) operated the DAE on MFCC features.

To capture the context and the temporal dispersion caused by reverberation well, multiple frames of noisy features are applied to the input of the DAE. Typical values range from 7 to 15 frames, where each feature vector includes static, delta, and delta-delta features. For the output, several options have been proposed. A natural choice is to output as many frames as there are at the input and then forward only one of them, for example, the center frame, to the decoder. Rather than selecting a single frame, another option is to compute the average of all frames and to forward this as the final clean feature vector estimate to the decoder. Finally, one can have an asymmetric network with a single frame at the output, as proposed in Feng et al. (2014).

Several training strategies are possible. In one strategy, which proved successful, the network is pre-trained layer-by-layer, while fixing the weights of the already trained layers. During pre-training, the clean corpus is used and the input for the trainable layer is corrupted by randomly setting 50% of it to zero. Afterward fine-tuning is performed using the *Newbob* learning strategy, where the learning rate is adapted according to the performance of the net on a held-out development set. During fine-tuning, the input is the multicondition and the output is the clean training data (Heymann et al., 2015).

Denoising autoencoders have been applied to various reverberant ASR tasks. The DAE can be followed by either a conventional GMM-HMM back-end (Feng et al., 2014) or a DNN back-end (Mimura et al., 2014). Feng et al. employed a GMM-HMM back-end and used a DAE with two encoder and one decoder layer on the data of the 2nd CHiME challenge, a database with two-channel recordings of reverberant and noisy speech with an SNR ranging from −6 to 9 dB in steps of 3 dB. Acoustic models were trained for each SNR condition separately, while a single denoising DAE was trained and applied to all SNR conditions. The acoustic models were retrained with the features processed by the DAE. They reported word error rate improvements by the DAE over unprocessed features of 15-20%.

Mimura et al. (2014) used a denoising autoencoder with three encoding and three decoding layers. They obtained improvements in recognition accuracy of more than 50% if the acoustic models had been trained on clean data and recognition performed on reverberated data. This was true both for a GMM-HMM and a DNN back-end. With a DNN trained on multicondition data, the DAE applied to the test data prior to decoding did not always lead to improved word accuracy. Improvements were only observed in the more adverse conditions with large reverberation times and large distances between speaker and microphone.

In Weninger et al. (2014a,b), a deep recurrent neural network (RNN) was used as denoising autoencoder. To be specific, the long short-term memory (LSTM) RNN was employed which features neural network nodes with an internal feedback loop (Hochreiter and Schmidhuber, 1997). LSTM RNNs are able to exploit a self-learnt amount of temporal context and may therefore be well suited to treat reverberated speech. In the utterance batch mode, where the ASR result is output only after the whole utterance has been processed, future context of a current frame t can be considered by adding a second set of layers which processes the input features backward in time, that is, from $t = T$ to $t = 1$. This leads to the so-called bidirectional LSTM (BLSTM). In a deep BLSTM, activations from both processing directions are collected in a single activation vector before passing them on as inputs to the next layer. The LSTM autoencoder was applied to LMPSC feature vectors to obtain an estimate $\hat{\mathbf{x}}[t]$ of the clean feature vector from the sequence of observed feature vectors $\mathbf{y}[1], \ldots, \mathbf{y}[\mathbf{T}]$. On single channel input data, the LSTM autoencoder was able to reduce the WER by 50% on the RealData of the REVERB challenge, using a GMM-HMM back-end trained on clean data. If the back-end is trained on multicondition data, the improvement was much smaller (about 10%).

In general, it can be stated that the improvement in word error rate obtained by feature enhancement is much smaller if the acoustic model has been trained on multicondition data than if the acoustic model has been trained on clean data.

Improvements with multicondition acoustic models hardly exceed 10%, and in some cases, even lead to a degradation (Geiger et al., 2014a). An interesting question is whether, in the case of multicondition training, feature enhancement should also be applied to the training data. The enhanced speech feature may have different characteristics than the original reverberant speech. Therefore, a retraining of the acoustic model using the processed features is promising. On the other hand, it has been argued that the speech enhancement front-end should not be used in multicondition training to preserve the acoustic variability of the training data. Experiments using BFE and a GMM-HMM back-end showed that the WER could not be improved if the training was carried out on the BFE enhanced feature vectors (Krueger et al., 2011). A similar observation was made in Mimura et al. (2014) using a denoising autoencoder: retraining on the processed feature vectors has led to only negligible performance improvement, if at all. Also, Seltzer et al. (2013) reported that using speech enhancement on a noisy speech recognition task may even degrade the performance of a DNN back-end. On the other hand, Geiger et al. (2014b) observed improvements in recognition accuracy due to retraining the multicondition recognizer on enhanced features, using an enhancement scheme based on non-negative matrix factorization and considering noise rather than reverberation. These mixed results indicate that a definite statement cannot yet been made whether retraining is beneficial.

In Heymann et al. (2015), a comparison is conducted between the model-based Bayesian feature enhancement and the data-driven DAE-based feature enhancement using the data of the REVERB challenge. It turned out that the DAE outperformed the parametric feature enhancement. However, it also turned out that in the case of a large mismatch between the data on which the DAE was trained and the testing data, the WER performance of the DAE enhanced features dropped significantly. Note that the BFE can be easily adjusted to a new environment by proper parameter selection, and such mismatch situations can therefore be avoided. Interestingly, the BFE can generate the clean target features for the training of the DAE. Thus a mismatch between the data on which the DAE is trained and the testing conditions can be reduced by an unsupervised adaptation of the DAE using the features obtained by BFE on the test data as clean speech targets: First, the features to be decoded are enhanced by BFE, resulting in a cleaned-up version of the LMPSC or MFCC features. These are then used as a targets to retrain the DAE. After this process, the DAE can be used to enhance the test data. This approach delivered a 10% WER improvement on the RealData of the REVERB challenge over the unadapted DAE-DNN system (Heymann et al., 2015).

9.8 ACOUSTIC MODEL DOMAIN APPROACHES

A simple way to obtain an acoustic model appropriate for the recognition of reverberant speech is to train the recognizer with reverberant data recorded in the target environment. Often, the properties of the target room are not known at training time, or the recognizer may be used in different rooms. Then, a pool of models for different reverberation conditions can be trained in advance, and the one of the trained models that is considered appropriate is selected during recognition (Couvreur and Couvreur, 2004). Alternatively, a single acoustic model is trained using all available training data representing a variety of acoustic conditions. This latter approach, that is, the multicondition training, is preferred for DNN back-ends, at least if the mismatch between training and test data is not too large, while for GMM-HMM back-ends, the selection of an environment-specific acoustic model out of a variety of models has often been advocated. The reason is that DNNs tend to be more robust to a (moderate) mismatch between training and test conditions than GMM-HMM models are. Furthermore, DNN acoustic models tend to be very large, and the amount of storage required for multiple models may not be available.

Compared to training on clean, nonreverberant data, multicondition training leads to a significant increase in recognition accuracy of reverberated speech. For example, on the REVERB challenge data, an error rate reduction on the order of 50% has been observed, both for a GMM-HMM and a DNN back-end.

To avoid the effort of recording a database in one or more target environments, alternatively only the AIR can be measured and the reverberant training data is generated by convolving the nonreverberant database with the measured AIR (Stahl et al., 2001). To mimic the time variance of the AIR, the AIR used to reverberate the training data may change from utterance to utterance (Hirsch, 2007; Kinoshita et al., 2013). This, however, does not account completely for the time variance of the AIR when a speaker speaks in a reverberant environment because the AIR is still time invariant within an utterance.

To even save the effort of an AIR measurement, artificially reverberated speech can be generated by the image method (Allen and Berkley, 1979).

There are several studies that demonstrate the high correlation of the reverberation time and a measure of the direct-to-reverberation ratio, such as C_{50}, with the word error rate. In Brutti and Matassoni (2014), it has been shown empirically that a combination of the two measures, called early-to-late reverberation ratio (ELR), which is the ratio of the energy in the first 110 ms of the AIR to the energy in the remaining part of the AIR can serve as a single good predictor of the word error rate. ELR can be used to select an acoustic model for a target environment: There is no need to match both T_{60} and C_{50} between training and test data. As long as the ELR is the same, the best recognition performance is achieved (Matassoni et al., 2014), using a GMM-HMM back-end.

Alternatively, an acoustic model that is trained on nonreverberant data can be adapted to the reverberant test conditions. This can be done either statically, that is, prior to recognition, or dynamically during the actual recognition.

For the adaptation of GMM acoustic models, several methods can be found in the literature. *Maximum likelihood linear regression* (MLLR) (Leggetter and Woodland, 1995) is an example of static adaptation. Originally meant to adapt an acoustic model to a new speaker, it was employed for the recognition of reverberant speech in Toh et al. (2006). However, the same restrictions apply that were mentioned earlier in the context of feature domain MLLR: MLLR assumes that the distortion of the acoustic features can be described by affine transforms applied to individual feature vectors, which, as we explained earlier, is an invalid assumption for reverberated speech.

A more exact model must account for the temporal convolution in the spectral domain, as is seen from Equations 9.9 and 9.11. Thus, the modifications of the emission probabilities of the acoustic model necessary to account for reverberation greatly depend on the preceding context. One way to do so is to split a HMM state into substates to take into account the energy dispersion due to reverberation (Raut et al., 2005a,b). The number of substates is chosen according to the average state occupancy duration within an HMM state.

Hirsch and Finster (2008) proposed to adapt the mean vectors of the emission PDFs of a GMM-HMM recognizer, that had been trained on nonreverberant data, to the reverberant test data prior to recognition in a way that is reminiscent of the well-known *parallel model combination (PMC)* method (Gales, 1995): the mean of the emission probability of a certain HMM state is adapted to the target environment by transforming it back into the power spectral domain, adding the contribution of the preceding states caused by reverberation and transforming the result to the feature, that is, cepstral, domain. However, this static adaptation does not work well for phoneme-based HMMs since the preceding states for a particular search hypothesis are only known during decoding, but not prior to recognition.

A better treatment of the energy dispersion is possible with dynamic adaptation, where the adaptation of the models is carried out in parallel to the decoding of the word sequence. During decoding, different hypotheses about the left context of a current HMM state are maintained, and for each context, its specific impact on the emission probability of the current state due to reverberation can be estimated (Sehr et al., 2011; Takiguchi and Nishimura, 2004). This more precise modeling comes, however, at the cost of greatly increased computational complexity. Therefore, dynamic acoustic model adaptation techniques have been mostly used in the context of small vocabulary recognition tasks only. Furthermore, this technique is so far only applicable to GMM-HMM back-ends.

However, an observation from the REVERB challenge was that DNN recognizers clearly outperformed GMM-HMM systems, both on simulated and on real reverberant speech data (Kinoshita et al., 2014). While DNNs seem to be more robust toward some mismatch between training and testing conditions than GMM-HMM systems, they nevertheless perform poorly in case of a strong mismatch, just as GMM-HMM systems do. This indicates that DNNs should also benefit from adapting their acoustic model to the target environment.

It has been empirically shown that an effect similar to model adaptation in GMM-HMM systems can be obtained simply by additional backpropagation training using the test data (Xiao et al., 2012). In Mimura et al. (2014), unsupervised adaptation of the DNN is carried out by 10 epochs of additional backpropagation training using the test utterances within a common test condition. The target labels of backpropagation training were generated from the recognition results using a nonadapted denoising autoencoder followed by a DNN back-end. This led to improvements in word accuracy under all test conditions of the REVERB challenge. This technique has also been investigated for speaker adaptation, where it was shown that the retraining of the whole network provided better performance compared to retraining only the input or the output layer (Liao, 2013). On the REVERB challenge data, the authors of Delcroix et al. (2014) obtained better results if only the input layer was adapted.

An alternative to an unsupervised retraining of the DNN on the test data is the retraining of the data-driven feature enhancement, as discussed in Heymann et al. (2015) and as mentioned in the last section. This has the potential advantage of requiring less adaptation data, since the DAE network is usually smaller than the DNN.

A particularly interesting acoustic model architecture for reverberant speech processing is a recurrent neural network. Due to their feedback connections, RNNs can encompass a (theoretically infinitely) long temporal window of input features to compute the output. In this way, the temporal dispersion caused by reverberation can be accounted for. A variant is the LSTM RNN, which can also be used as a DAE architecture as was mentioned in the last section. The LSTM uses so-called memory blocks instead of the conventional activation functions in the hidden layers. Each memory block consists of a memory cell and three gate units, the input gate, the output gate, and the forget gate, which control the behavior of the memory block. With this architecture, the network is capable of storing input over a longer period of time and thus exploits a self-learned amount of long-range temporal context (Geiger et al., 2014b; Hochreiter and Schmidhuber, 1997). LSTMs have been originally used for phoneme prediction in a multistream GMM-HMM framework. More recently, they have been used to predict HMM states and used in a hybrid setup (Geiger et al., 2014b). LSTMs have been successfully used as a back-end in several reverberant recognition tasks (Geiger et al., 2013, 2014b). They have shown comparable performance to feed-forward DNN back-ends. A definite statement of the superiority of one over the other is difficult to make.

9.9 THE REVERB CHALLENGE

In 2013, a group of researchers interested in reverberant speech processing defined a common evaluation framework for both speech enhancement (SE) and ASR techniques in reverberant environments (Kinoshita et al., 2013). This framework was used for the REVERB (reverberant voice enhancement and recognition benchmark) challenge. The challenge aimed at providing an opportunity to researchers in the field to carry out a comprehensive evaluation of their methods based on a common system and database. This allowed to compare existing techniques, and highlight their respective advantages. By performing both SE and ASR evaluation at a time, one can draw potential successful research directions for each field and create synergy between the two areas.

Another unique feature of the challenge was that it was based not only on the evaluations in simulated conditions but also on real recordings of speakers included in a reverberant environment. This allowed the participants to challenge their algorithms in terms of generality and practicability.

The scenarios assumed in the challenge were a single stationary distant talking speaker, whose signal is captured by a single microphone, and by a 2-channel and 8-channel microphone array in a reverberant meeting room. The provided databases consisted of three subsets: a simulation data set (SimData), a real-recording data set (RealData), and the corresponding training data set (TrainData). All the data sets were based on the test/training data set of the Wall Street Journal Cambridge (WSJCAM0) corpus (Robinson et al., 1995), which has been used for evaluating large vocabulary continuous speech recognition systems.

SimData contained a set of reverberant speech signals that were created by the convolution of clean speech and measured acoustic impulse responses, and the addition of noise. This data set included several reverberant conditions. There were three different rooms/reverberation times of approximately 250, 500, and 700 ms denoted small, medium, and large room, and for each room there were two microphone-speaker distances: near, which corresponded to a distance of about 0.5 m, and far with a distance of about 2 m. Some of the reverberant conditions simulated in SimData (large room, far speaker) were similar to those of RealData, which makes the results obtained based on SimData and RealData comparable to some extent.

RealData contained the recordings of speakers speaking in a reverberant environment. The data were taken from the Multi-channel Wall Street Journal Audio-Visual (MC-WSJ-AV) corpus (Lincoln et al., 2005). Note that the text prompts of the utterances used in RealData and part of the SimData are the same, but the utterances are spoken by different British English speakers. The prompts are based on the Wall Street Journal 5k corpus (Paul and Baker, 1992). Therefore, the participants could use the same language and acoustic models for both SimData and RealData.

There were two types of TrainData, one clean nonreverberant training set (clean training) and one data set which was artificially reverberated with similar, however, different AIRs than the SimData (multicondition training).

In the following, we are going to interpret the results of the challenge w.r.t. to a number of questions often raised in the community. These questions are discussed assuming a DNN back-end. Note that the comments made here may be subjective although an attempt was made to draw only conclusions that are corroborated by the data. We are going to discuss the following:

- on the benefit of front-end enhancement on ASR performance;
- on the benefit of multi-channel vs. single-channel input data;
- on the virtue of SimData vs. RealData.

Concerning the importance of enhancement techniques on ASR performance, it can be clearly stated that enhancement is very important if there is a strong mismatch between training data and test data, in particular if the training has been conducted on clean, nonreverberant data, while test is carried out on reverberated speech. In case of multicondition training, the situation is less clear. However, it is safe to say that good linear filtering-based signal dereveberation techniques improve performance even in the presence of multicondition training and a strong DNN back-end (Delcroix et al., 2014; Geiger et al., 2014b; Yoshioka and Gales, 2015). With spectrum or feature enhancement techniques, the situation is less clear: They do not always lead to performance improvement if an acoustic model trained on multi condition data is used. Concerning the question whether the acoustic model should be (re)trained with enhanced multicondition training data, it is difficult to make a definite statement. Some researchers argue that it is beneficial to expose the acoustic model training to the artifacts introduced by speech enhancement, others argue that the speech enhancement front-end should not be used during training to preserve the acoustic variability of the data. Consequently, some researchers reported improvements by enhancing the multicondition training data while others did not observe a gain in word accuracy.

With respect to multi-channel versus single-channel input, one should first mention that many signal dereverberation methods based on linear filtering require multi-channel input. Thus, multi-channel data are beneficial if one wants to reap the benefit of linear dereverberation filtering. Furthermore, it has been shown that acoustic beamforming is beneficial to improve the noise robustness of ASR, irrespective of whether the back-end is GMM-HMM or DNN. Note that the foremost target of beamforming is improving the robustness toward additive noise. The effect of beamforming on dereverberation is less pronounced: there is some dereverberating effect if reflections from other directions than that of the line-of-sight signal are suppressed by beamforming. Concerning the benefit of using two vs. many (eight) channels, it was observed that the largest improvement in WER performance is when going from one to two channels, but there is still a lot to gain when going from two to eight channels. According to Delcroix et al. (2014), most of the performance gain obtained by the weighted prediction error (WPE) dereverberation algorithm is observed when going from one to two microphones, while the MVDR beamformer still significantly improved the WER when moving from two to eight channels.

Note that the data of the REVERB challenge employed a compact microphone array with a small spatial extent—in contrast to microphones distributed over a larger space. In a compact microphone, array differences between the channels are mostly visible in the phase of the signals. For this situation, one may conjecture that only signal enhancement techniques, which process both phase and amplitude of the signals, pay off in terms of improved WER. Spectrum of feature enhancement, on the contrary, disregards the phase. This could be one reason why they have a hard time to improve the performance of a multicondition/matched condition trained DNN.

Finally, a note on the virtue of simulated reverberant data versus recordings of a speaker in a real environment. The challenge revealed that techniques that improve performance on SimData do in general so also on RealData. In particular, a strong correlation was observed between the scores obtained on SimData, room 3, far, and RealData. The SimData of room 3 with a microphone far away from the speech source is supposedly closest to RealData. However, there was still a considerable mismatch between SimData of that condition and RealData. A recognizer trained on simulated reverberant training data still achieved much lower WER on the SimData than on the RealData. Thus, adapting the denoising autoencoder used for feature enhancement or retraining the DNN models in an unsupervised way to the RealData led to noticeable WER improvements. One should, however, mention that the mismatch between SimData and RealData was not only due to simulated vs. real reverberant speech but also due to the fact that MC-WJ-AV corpus was recorded at a different site and with different speakers than the WSJCAM0 corpus.

9.10 TO PROBE FURTHER

Those who would like to learn more about the fascinating subject of "making machines understand us in reverberant room," we wish to refer to the publication with exactly this title in the IEEE Signal Processing Magazine (Yoshioka et al., 2012). There is also a book on distant speech recognition (Wölfel and McDonough, 2009), and there are book chapters devoted to ASR in reverberant environments in recent books (Haeb-Umbach and Krueger, 2012; Krueger and Haeb-Umbach, 2011). Those more interested in speech enhancement are referred to a book by Naylor and Gaubitch (2010) and by book chapters in Nakatani et al. (2005) and Huang et al. (2008). Further, a special issue of the IEEE Transactions on Audio, Speech, and Language processing has been devoted to processing reverberant speech. The special issue features articles related to ASR but also related to the enhancement of reverberated speech for human-to-human communication (Nakatani et al., 2010a).

We further recommend to study the webpages and publications related to two recent challenges related to reverberant speech processing. The CHiME-2 challenge (http://spandh.dcs.shef.ac.uk/chime_challenge/) used two-channel simulated data employing a set of binaural impulse responses. The reverberated utterances have then been mixed with binaural recordings of genuine room noise made over a period of days in a family living room. The focus of the challenge was more on the robustness to nonstationary noise than on reverberation, although reverberation played also a significant role (Vincent et al., 2013).

The other challenge is the REVERB challenge which we discussed in the last section (Kinoshita et al., 2014). Publications and information on data and software can be found on the challenge website at http://reverb2014.dereverberation.com/.

9.11 SUMMARY

Table 9.1 summarizes approaches to the recognition of reverberated speech, arranged chronologically. Most of the techniques operate in the signal and feature domain. They are thus applicable to both GMM and DNN-based back-ends, although most of them have originally been developed with a GMM back-end in mind, except for the denoising autoencoder, which was first developed for a DNN back-end. However, even the DAE can be employed with either back-end. The model-based techniques, static or dynamic model adaptation, have been developed for GMM back-ends and cannot be easily transferred to a DNN back-end.

Table 9.1 Approaches to the Recognition of Reverberated Speech, Arranged Chronologically

Method	Proposed Around	Characteristics
Long-term cepstral mean normalization (CMN) (Gelbart and Morgan, 2001) or RASTA (Kingsbury and Morgan, 1997)	1997	Removes cepstral mean by normalization or bandpass filtering
Spectral subtraction (Erkelens and Heusdens, 2010b; Habets, 2007; Lebart et al., 2001)	2001	Treats late reverberation as additive distortion
Training model selection (Couvreur and Couvreur, 2004)	2004	Trains models under different reverberation conditions and selects at test time
Dynamic model adaptation (Takiguchi and Nishimura, 2004)	2004	Modifies emission probabilities of a GMM acoustic model dynamically during recognition to account for energy contribution of preceding states caused by reverberation
Signal deconvolution (Triki and Slock, 2005, 2006)	2005	Whitens reverberated signal, however keeping the speech correlation structure
Static model adaptation (Hirsch and Finster, 2008)	2008	Modifies emission probabilities of a GMM acoustic model to account for energy contribution of preceding states caused by reverberation
Long-term multiple-step linear prediction (Kinoshita et al., 2009; Nakatani et al., 2010b; Yoshioka et al., 2011)	2009	Estimates late reverberation by multistep linear prediction and removes reverberation via spectral subtraction or a linear prediction filter
Bayesian feature adaptation (Krueger and Haeb-Umbach, 2010)	2010	Removes the effect of reverberation on speech features using a state-space model for clean speech feature trajectories in a Kalman-filter like inference machine
Reverberation robust features (Kim et al., 2014; Kim and Stern, 2010)	2010	Nonlinear filtering of the power spectral coefficients of each frequency band to mimic the precedence effect
Denoising autoencoder (Feng et al., 2014; Mimura et al., 2014)	2014	Learns the mapping from reverberant features to clean features with a deep denoising autoencoder

REFERENCES

Allen, J.B., Berkley, D., 1979. Image method for efficiently simulating small-room acoustics. J. Acoust. Soc. Amer. 65 (4), 943-950.

Avargel, Y., Cohen, I., 2007. System identification in the short-time Fourier transform domain with crossband filtering. IEEE Trans. Audio Speech Lang. Process. 15 (4), 1305-1319.

Avendano, C., Tibrewala, S., Hermansky, H., 1997. Multiresolution channel normalization for ASR in reverberant environments. In: Proc. of European Conference on Speech Communication and Technology (EUROSPEECH), Rhodes, Greece, pp. 1107-1110.

Bar-Shalom, Y., Li, X.R., Kirubarajan, T., 2001. Estimation with Applications to Tracking and Navigation: Theory, Algorithms, and Software. Wiley, New York.

Bengio, Y., Hanover, M.A., 2009. Learning deep architectures for AI. Found. Trends Mach. Learn. 2 (1), 1-127. ISSN 1935-8237. http://dx.doi.org/10.1561/2200000006. URL http://dx.doi.org/10.1561/2200000006.

Bengio, Y., Lamblin, P., Popovici, D., Larochelle, H., 2007. Greedy layer-wise training of deep networks. In: Advances in neural information processing systems, pp. 153-160, URL http://www.iro.umontreal.ca/lisa/pointeurs/BengioNips2006All.pdf.

Brutti, A., Matassoni, M., 2014. On the use of early-to-late reverberation ratio for ASR in reverberant environments. In: Proc. of IEEE International Conference on Acoustics, Speech and Signal Processing (ICASSP), pp. 4638-4642.

Couvreur, L., Couvreur, C., Hingham, M.A., 2004. Blind model selection for automatic speech recognition in reverberant environments. J. VLSI Signal Process. 36 (2/3), 189-203.

Dahl, G., Yu, D., Deng, L., Acero, A., 2011. Large vocabulary continuous speech recognition with context-dependent DBN-HMMs. In: Proc. International Conference on Acoustics, Speech and Signal Processing (ICASSP).

Dahl, G., Yu, D., Deng, L., Acero, A., 2012. Context-dependent pre-trained deep neural networks for large-vocabulary speech recognition. IEEE Trans. Audio Speech Lang. Process. 20 (1), 30-42.

Delcroix, M., Yoshioka, T., Ogawa, A., Kubo, Y., Fujimoto, M., Ito, N., 2014. Linear prediction-based dereverberation with advanced speech enhancement and recognition technologies for the reverb challenge. In: REVERB Challenge Workshop, Florence, Italy.

Deng, L., Seltzer, M., Yu, D., Acero, A., Mohamed, A., Hinton, G., 2010. Binary coding of speech spectrograms using a deep auto-encoder. In: Proc. Annual Conference of International Speech Communication Association (INTERSPEECH).

Elko, G.W., 2001. Microphone arrays. In: Proc. International Workshop on Hands-free Speech Communication, Kyoto, Japan.

Erkelens, J., Heusdens, R., 2010a. Correlation-based and model-based blind single-channel late-reverberation suppression in noisy time-varying acoustical environments. IEEE Trans. Audio Speech Lang. Process. 18 (7), 1746-1765.

Erkelens, J., Heusdens, R., 2010b. Noise and late-reverberation suppression in time-varying acoustical environments. In: Proc. of IEEE International Conference on Acoustics, Speech and Signal Processing (ICASSP), pp. 4706 -4709.

Erkelens, J., Heusdens, R., 2011. A statistical room impulse response model with frequency dependent reverberation time for single-microphone late reverberation suppression. In: Proc. Interspeech, Florence, Italy.

Feng, X., Kumantani, K., McDonough, J., 2014. The CMU-MIT REVERB channel 2014 system: description and results. In: REVERB Challenge Workshop, Florence, Italy.

Gales, M., 1995. Model-Based Techniques for Noise Robust Speech Recognition. Ph.D. thesis, Cambridge University.

Gaubitch, N., Löllmann, H.W., Jeub, M., Falk, T., Naylor, P.A., Vary, P., et al., 2012. Performance comparison of algorithms for blind reverberation time estimation from speech. In: Proc. of International Workshop on Acoustic Signal Enhancement (IWAENC), Aachen, Germany.

Geiger, J., Gemmeke, J., Schuller, B., Rigoll, G., 2014a. Investigating NMF speech enhancement for neural network based acoustic models. In: Proc. INTERSPEECH.

Geiger, J., Weninger, F., Hurmalainen, A., Gemmeke, J., Wöllmer, M., Schuller, B., et al., 2013. The TUM + TUT + KUL approach to the 2nd CHiME challenge: multi-stream ASR exploiting BLSTM networks and sparse NMF. In: The 2nd International Workshop on Machine Listening in Multisource Environments, Vancouver, Canada.

Geiger, J., Zhang, Z., Weninger, F., Schuller, B., Rigoll, G., 2014b. Robust speech recognition using long short-term memory recurrent neural networks for hybrid acoustic modelling. In: Fifteenth Annual Conference of the International Speech Communication Association.

Gelbart, D., Morgan, N., 2001. Evaluating long-term spectral subtraction for reverberant ASR. In: Proc. of IEEE Workshop on Automatic Speech Recognition and Understanding (ASRU), Madonna Di Campiglio, Italy, pp. 103-106.

Gillespie, B., Atlas, L., 2003. Strategies for improving audible quality and speech recognition accuracy of reverberant speech. In: Proc. of IEEE International Conference on Acoustics, Speech and Signal Processing (ICASSP), vol. 1, pp. I-676-I-679.

Habets, E., 2007. Single- and Multi-Microphone Speech Dereverberation Using Spectral Enhancement. Ph.D. thesis, Technische Universiteit Eindhoven.

Habets, E., Gannot, S., Cohen, I., 2009. Late reverberant spectral variance estimation based on a statistical model. IEEE Signal Process. Lett. 16 (9), 770-773. ISSN 1070-9908. http://dx.doi.org/10.1109/LSP.2009.2024791.

Haeb-Umbach, R., Krueger, A., 2012. Reverberant speech recognition. In: Virtanen, T, Singh, R, Raj, B. (Eds.), Noise Robustness in Automatic Speech Recognition. Wiley, West Sussex, UK.

Heymann, J., Haeb-Umbach, R., Golik, P., Schlüter, R., 2015. Unsupervised adaptation of a denoising autoencoder by Bayesian feature enhancement for reverberant ASR under mismatch conditions. In: Proc. of IEEE International Conference on Acoustics, Speech and Signal Processing (ICASSP).

Hinton, G., Deng, L., Yu, D., Dahl, G.E., Mohamed, A., Jaitly, N., et al., 2012. Deep neural networks for acoustic modeling in speech recognition: the shared views of four research groups. IEEE Signal Process. Mag. 29 (6), 82-97.

Hirsch, H., 2007. Aurora-5 experimental framework for the performance evaluation of speech recognition in case of a hands-free speech input in noisy environments. Niederrhein University of Applied Sciences.

Hirsch, H., Finster, H., Amsterdam, The Netherlands, 2008. A new approach for the adaptation of HMMs to reverberation and background noise. Speech Commun. 50 (3), 244-263.

Hochreiter, S., Schmidhuber, J., 1997. Long short-term memory. Neural Comput. 1735-1780.

Hopgood, J., Rayner, P., 2003. Blind single channel deconvolution using nonstationary signal processing. IEEE Trans. Speech Audio Process. 11 (5), 476-488. ISSN 1063-6676. http://dx.doi.org/10.1109/TSA.2003.815522.

Huang, Y., Benesty, J., Amsterdam, 2002. Adaptive multi-channel least mean square and Newton algorithms for blind channel identification. Signal Processing 82, 1127-1138.

ISSN 0165-1684. http://dx.doi.org/10.1016/S0165-1684(02)00247-5. URL http://portal. acm.org/citation.cfm?id=607012.607020.

Huang, Y., Benesty, J., Chen, J., 2008. Dereverberation. In: Benesty, J, Sondhi, M, Huang, Y. (Eds.), Springer Handbook of Speech Processing. Springer, New York.

Ishii, T., Komiyama, H., Shinozaki, T., Horiuchi, Y., Kuroiwa, S., 2013. Reverberant speech recognition based on denoising autoencoder. In: INTERSPEECH, pp. 3512-3516.

Japkowicz, N., Jose Hanson, S., Gluck, M., 2000. Nonlinear autoassociation is not equivalent to PCA. Neural Comput. 12 (3), 531-545. ISSN 0899-7667. http://dx.doi.org/10.1162/089976600300015691. URL http://dx.doi.org/10.1162/089976600300015691.

Kim, C., Chin, K.K., Bacchiani, M., Stern, R., 2014. Robust speech recognition using temporal masking and thresholding algorithm. In: Proc. Interspeech, Singapore.

Kim, C., Stern, R.M., 2010. Nonlinear enhancement of onset for robust speech recognition. In: Proc. of Annual Conference of the International Speech Communication Association (Interspeech), Makuhari, Japan.

Kingsbury, B., Morgan, N., Greenberg, S., 1998. Robust speech recognition using the modulation spectrogram. Speech Commun. 25 (1-3), 117-132.

Kingsbury, B.E.D., Morgan, N., 1997. Recognizing reverberant speech with RASTA-PLP. In: Proc. of IEEE International Conference on Acoustics, Speech and Signal Processing (ICASSP), vol. 2, pp. 1259-1262.

Kinoshita, K., Delcroix, M., Nakatani, T., Miyoshi, M., 2009. Suppression of late reverberation effect on speech signal using long-term multiple-step linear prediction. IEEE Trans. Audio Speech Lang. Process. 17 (4), 534-545.

Kinoshita, K., Delcroix, M., Yoshioka, T., Nakatani, T., Habets, E., Haeb-Umbach, R., et al., 2013. The REVERB challenge: a common evaluation framework for dereverberation and recognition of reverberant speech. In: Proc. of IEEE ASSP Workshop on Applications of Signal Processing to Audio and Acoustics, pp. 1-4.

Kinoshita, K., Delcroix, M., Yoshioka, T., Nakatani, T., Habets, E., Haeb-Umbach, R., et al., 2014. Summary of the REVERB challenge. In: REVERB Workshop, Florence, Italy.

Krueger, A., Haeb-Umbach, R., 2010. Model-based feature enhancement for reverberant speech recognition. IEEE Trans. Audio Speech Lang. Process. 18 (7), 1692-1707.

Krueger, A., Haeb-Umbach, R., 2011. A model-based approach to joint compensation of noise and reverberation for speech recognition. In: Haeb-Umbach, R, Kolossa, D. (Eds.), Robust Speech Recognition of Uncertain or Missing Data. Springer, New York.

Krueger, A., Walter, O., Leutnant, V., Haeb-Umbach, R., 2012. Bayesian feature enhancement for ASR of noisy reverberant real-world data. In: Proc. Interspeech, Portland, USA, URL http://nt.uni-paderborn.de/public/pubs/2012/KrWaLeHa2012.pdf.

Krueger, A., Warsitz, E., Haeb-Umbach, R., 2011. Speech enhancement with a gsc-like structure employing eigenvector-based transfer function ratios estimation. IEEE Trans. Audio Speech Lang. Process. 19 (1), 206-219. ISSN 1558-7916. http://dx.doi.org/10.1109/TASL.2010.2047324.

Kumar, K., Raj, B., Singh, R., Stern, R., 2011a. An iterative least-squares technique for dereverberation. In: Proc. of IEEE International Conference on Acoustics, Speech and Signal Processing (ICASSP), pp. 5488-5491.

Kumar, K., Singh, R., Raj, B., Stern, R., 2011b. Gammatone sub-band magnitude-domain dereverberation for ASR. In: Proc. of IEEE International Conference on Acoustics, Speech and Signal Processing (ICASSP), pp. 4604-4607.

Kuttruff, H., 2009. Room Acoustics. Taylor & Francis Group, Boca Raton, FL.

Lebart, K., Boucher, J, Denbigh, P., 2001. A new method based on spectral subtraction for speech dereverberation. Acta Acustica united with Acustica 87, 359-366(8).

Leggetter, C., Woodland, P., 1995. Maximum likelihood linear regression for speaker adaptation of continuous density hidden Markov models. Comput. Speech Lang. 9 (2), 171-185.

Leutnant, V., Krueger, A., Haeb-Umbach, R., 2013. Bayesian feature enhancement for reverberation and noise robust speech recognition. IEEE Trans. Audio Speech Lang. Process. 21 (8), 1640-1652.

Leutnant, V., Krueger, A., Haeb-Umbach, R., 2014. A New Observation Model in the Logarithmic Mel Power Spectral Domain for the Automatic Recognition of Noisy Reverberant Speech. IEEE Trans. Audio Speech Lang. Process. 22 (1), 95-109.

Liao, H., 2013. Speaker adaptation of context dependent deep neural networks. In: Proc. of IEEE International Conference on Acoustics, Speech and Signal Processing (ICASSP), pp. 7947-7951.

Lincoln, M., McCowan, I., Vepa, J., Maganti, H., 2005. The multi-channel Wall Street Journal audio visual corpus (MC-WSJ-AV): specification and initial experiments. In: IEEE Workshop on Automatic Speech Recognition and Understanding, pp. 357-362.

Litovsky, R., Colburn, H., Yost, W., Guzman, S., 1999. The precedence effect. J. Acoust. Soc. Amer. 106 (4), 1633-1654.

Löllmann, H., Vary, P., 2008. Estimation of the reverberation time in noisy environments. In: Proc. of International Workshop on Acoustic Echo and Noise Control (IWAENC), Seattle, WA.

Maas, R., Habets, E., Sehr, A., Kellermann, W., 2012. On the application of reverberation suppression to robust speech recognition. In: Proc. of IEEE International Conference on Acoustics, Speech and Signal Processing (ICASSP), pp. 297-300.

Maganti, H., Matassoni, M., 2010. An auditory based modulation spectral feature for reverberant speech recognition. In: Proc. of Annual Conference of the International Speech Communication Association (Interspeech), Makuhari, Japan.

Matassoni, M., Brutt, A., Svaizer, P., 2014. Acoustic modeling based on early-to-late reverberation ratio for robust ASR. In: Proc. International Workshop on Acoustic Signal Enhancement (IWAENC), Juan Les Pins, France.

Mimura, M., Sakai, S., Kawahara, T., 2014. Exploring deep neural networks and deep autoencoders in reverberant speech recognition. In: 4th Joint Workshop on Hands-free Speech Communication and Microphone Arrays (HSCMA), pp. 197-201.

Miyoshi, M., Kaneda, Y., 1988. Inverse filtering of room acoustics. IEEE Trans. Acoust. Speech Signal Process. 36 (2), 145-152.

Nakatani, T., Kellermann, W., Naylor, P., Miyoshi, M., Juang, B. (Eds.), 2010a. IEEE Transactions on Audio, Speech, and language Processing; Special Issue on Processing Reverberant Speech: Methodologies and Applications. IEEE, Piscataway, NJ.

Nakatani, T., Miyoshi, M., Kinoshita, K., 2005. Single-microphone blind dereverberation. In: Benesty, J., Makino, S., Chen, D. (Eds.), Speech Enhancement. Springer, New York.

Nakatani, T., Yoshioka, T., Kinoshita, K., Miyoshi, M., Juang, B.H., 2010b. Speech dereverberation based on variance-normalized delayed linear prediction. IEEE Transa. Audio Speech Lang. Process. 18 (7), 1717-1731. ISSN 1558-7916. http://dx.doi.org/10.1109/TASL.2010.2052251.

Naylor, P., Gaubitch, N., 2010. Speech Dereverberation. Springer, New York.

Paul, D.B., Baker, J.M., 1992. The design for the Wall Street Journal-based CSR corpus. In: International Conference on Spoken Language Processing (ICSLP), pp. 899-902.

Polack, J., 1988. La Transmission de l'énergie Sonore dans les Salles. Dissertation, Université du Maine. Dissertation.

Radlovic, B., Williamson, R., Kennedy, R., 1999. On the poor robustness of sound equalization in reverberant environments. In: Proc. of IEEE International Conference on Acoustics, Speech and Signal Processing (ICASSP), vol. 2, pp. 881-884.

Ratnam, R., Jones, D., Wheeler, B., O'Brien Jr., W., Lansing, C.R., Feng, A., 2003. Blind estimation of reverberation time. J. Acoust. Soc. Amer. 114 (5), 2877-2892. http://dx.doi.org/10.1121/1.1616578. URL http://link.aip.org/link/?JAS/114/2877/1.

Raut, C., Nishimoto, T., Sagayama, S., 2005a. Acoustic model adaptation for reverberant speech by state splitting of HMM and convolution of distributions. In: Techn. Report of Institute of Electronics, Information and Communication Engineers (IEIC), pp. 37-42.

Raut, C., Nishimoto, T., Sagayama, S., 2005b. Maximum likelihood based HMM state filtering approach to model adaptation for long reverberation. In: Proc. of IEEE Workshop on Automatic Speech Recognition and Understanding (ASRU), pp. 353-356.

Robinson, T., Fransen, J., Pye, D., Foote, J., Renals, S., 1995. WSJCAM0: a British English speech corpus for large vocabulary continuous speech recognition. In: Proc. of IEEE International Conference on Acoustics, Speech and Signal Processing (ICASSP), vol. 1, pp. 81-84.

Schwartz, B., Gannot, S., Habets, E., 2014. LPC-based speech dereverberation using Kalman-EM algorithm. In: Proc. International Workshop on Signal Enhancement (IWAENC), Juan Les Pins, France.

Sehr, A., Maas, R., Kellermann, W., 2011. Frame-wise HMM adaptation using state-dependent reverberation estimates. In: Proc. of IEEE International Conference on Acoustics, Speech and Signal Processing (ICASSP), Prague, Czech Republic.

Seltzer, M.L., Yu, D., Wang, Y., 2013. An investigation of deep neural networks for noise robust speech recognition. In: Proc. International Conference on Acoustics, Speech and Signal Processing (ICASSP), pp. 7398-7402.

Stahl, V., Fischer, A., Bippus, R., 2001. Acoustic synthesis of training data for speech recognition in living room environments. In: Proc. of IEEE International Conference on Acoustics, Speech and Signal Processing (ICASSP), vol. 1, pp. 21-24.

Takiguchi, T., Nishimura, M., 2004. Acoustic model adaptation using first order prediction for reverberant speech. In: Proc. of IEEE International Conference on Acoustics, Speech and Signal Processing (ICASSP), Montreal, Quebec, Canada, pp. 869-872.

Toh, A., Togneri, R., Nordholm, S., 2006. Combining MLLR adaptation and feature extraction for robust speech recognition in reverberant environments. In: Proc. of International Conference on Speech Science and Technology (SST), Auckland, New Zealand.

Triki, M., Slock, D., 2005. Blind dereverberation of a single source based on multichannel linear prediction. In: Proc. of International Workshop on Acoustic Echo and Noise Control (IWAENC), Eindhoven, The Netherlands, pp. 173-176.

Triki, M., Slock, D., 2006. Delay and predict equalization for blind speech dereverberation. In: Proc. of IEEE International Conference on Acoustics, Speech and Signal Processing (ICASSP), vol. 5.

Vincent, E., Barker, J., Watanabe, S., Le Roux, J., Nesta, F., Matassoni, M. (Eds.), 2013. The 2nd International Workshop on Machine Listening in Multisource Environments.

Vincent, P., Larochelle, H., Bengio, Y., Manzagol, P., 2008. Extracting and composing robust features with denoising autoencoders. In: Proc.of the 25th international conference on Machine learning, pp. 1096-1103.

Vincent, P., Larochelle, H., Lajoie, I., Bengio, Y., Manzagol, P., 2010. Stacked denoising autoencoders: learning useful representations in a deep network with a local denoising criterion. J. Mach. Learn. Res. 11, 3371-3408. ISSN 15324435. http://dx.doi.org/10.1111/1467-8535.00290.

Wen, J., Sehr, A., Naylor, P., Kellermann, W., 2009. Blind estimation of a feature-domain reverberation model in non-diffuse environments with variance adjustment. In: Proc. of European Signal Processing Conference (EUSIPCO), pp. 175-179.

Weninger, F., Watanabe, S., Le Roux, J., Hershey, J., Tachioka, Y., Rigoll, G., 2014a. The MERL/MELCO/TUM system for the REVERB challenge using deep recurrent neural network feature enhancement. In: REVERB Challenge Workshop, Florence, Italy.

Weninger, F., Watanabe, S., Tachioka, Y., Schuller, B., 2014b. Deep recurrent de-noising auto-encoder and blind de-reverberation for reverberated speech recognition. In: Proc. of IEEE International Conference on Acoustics, Speech and Signal Processing (ICASSP), pp. 4623-4627.

Wölfel, M., McDonough, J., 2009. Distant Speech Recognition. Wiley, West Sussex, UK.

Xiao, Y., Zhang, Z., Cai, S., Pan, J., Yan, Y., 2012. A initial attempt on task-specific adaptation of deep neural network-based large vocabulary continuous speech recognition. In: Proc. Interspeech, Portland, OR.

Yoshioka, T., Chen, X., Gales, M., 2014. Impact of single-microphone dereverberation on dnn-based meeting transcription systems. In: Proc. of IEEE International Conference on Acoustics, Speech and Signal Processing (ICASSP), pp. 5527-5531.

Yoshioka, T., Gales, M., 2015. Environmentally robust {ASR} front-end for deep neural network acoustic models. Comput. Speech Lang. 31 (1), 65-86. ISSN 0885-2308. http://dx.doi.org/http://dx.doi.org/10.1016/j.csl.2014.11.008. URL http://www.sciencedirect.com/science/article/pii/S0885230814001259.

Yoshioka, T., Nakatani, T., 2012. Generalization of multi-channel linear prediction methods for blind MIMO impulse response shortening. IEEE Trans. Audio Speech Lang. Process. 20 (10), 2707-2720. ISSN 1558-7916. http://dx.doi.org/10.1109/TASL.2012.2210879.

Yoshioka, T., Nakatani, T., Miyoshi, M., Okuno, H., 2011. Blind separation and dereverberation of speech mixtures by joint optimization. IEEE Transa. Audio Speech Lang. Process. 19 (1), 69-84. ISSN 1558-7916. http://dx.doi.org/10.1109/TASL.2010.2045183.

Yoshioka, T., Sehr, A., Delcroix, M., Kinoshita, K., Maas, R., Nakatani, T., et al., 2012. Making machines understand us in reverberant rooms: robustness against reverberation for automatic speech recognition. IEEE Signal Process. Mag. 29 (6), 114-126. ISSN 1053-5888. http://dx.doi.org/10.1109/MSP.2012.2205029.

Yu, D., Deng, L., 2011. Deep learning and its applications to signal and information processing. In: IEEE Signal Processing Magazine, vol. 28, pp. 145-154.

Yu, D., Deng, L., 2014. Automatic Speech Recognition—A Deep Learning Approach. Springer, New York.

Multi-channel processing

10

CHAPTER OUTLINE

In the previous chapters of this book, we have assumed that the sound capturing device for an ASR system is a single microphone that receives noisy and distorted speech signals; that is, the robust algorithms deal with a single-channel input. However, the recent availability of cheap hardware makes it practical nowadays to exploit multiple sound capturing channels. While some earlier studies showed that using different types of microphones improves robustness of ASR systems (Zhang et al., 2004; Zheng et al., 2003), the majority of studies in this area involves the use of multiple microphones of the same type; that is, microphone arrays. The processing scheme of the latter is more standardized and more widely used. Hence, we devote this chapter to only the latter class of multi-microphone techniques for speech enhancement and noise-robust ASR.

10.1 INTRODUCTION

At the beginning of the last chapter, we mentioned the different effects that increasing the distance between the sound source and the sensor have on the received signal. While we discussed in detail the phenomenon of multipath propagation resulting in

Robust Automatic Speech Recognition. http://dx.doi.org/10.1016/B978-0-12-802398-3.00010-6

a reverberated microphone signal and its mitigation in that chapter, we are concerned with another effect of distant-talking speech acquisition here: If the distance between the target speaker and the sound capturing device increases, it is likely that the microphone will capture other acoustic events, in addition to the desired speech signal. Further, due to the propagation loss, the signal power, and consequently also the signal-to-noise ratio (SNR), will be decreased. Interfering acoustic events and decreased SNR will lead to a significant increase in the word error rate of an ASR system, if no appropriate measures are taken. Thus, while moving from the use of close-talking microphones to distant speech recognition is advantageous in terms of usability, the increased error rate caused by reduced SNR can easily diminish, if not override the advantages of an otherwise more convenient hands-free human-machine interface.

Multi-channel speech processing algorithms aim at exploiting the spatial diversity of sound sources to extract one or more signals of interest. To this end, a number of microphones are placed in an acoustic enclosure. One popular approach is to arrange the microphones in a fixed compact form, for example, equidistantly along a line forming a so-called uniform linear array, or equidistantly on a circle forming a uniform circular array, and to carry out multi-channel signal processing to obtain an enhanced single output signal, a concept commonly referred to as acoustic beamforming.

Just like digital filters process temporal samples of a continuous-time signal to obtain desired spectral, that is, frequency selective characteristics, the processing of the spatially sampled signal by the beamformer aims at forming a desired spatial selectivity, a so-called beampattern. By pointing a beam of increased sensitivity to a desired source and/or pointing a null toward an interferer, the signal-to-interference ratio can be significantly improved. By exploiting spatial diversity, even highly nonstationary interferers can be effectively suppressed, which is known to be a rather difficult task for single-channel speech enhancement. While the main goal is interference/noise suppression, beamforming has also some dereverberating effect by suppressing echoes from directions other than the target direction.

An alternative to a compact microphone array is the distribution of the microphones in a certain region of interest. Here, the microphones should be placed such that at least one microphone is close to each signal of interest. It is also advantageous to place microphones close to interfering sound sources to provide noise reference signals. The goal of the signal processing is to form a beam toward the desired signal or to select the most appropriate microphone and to enhance the target signal. Microphones close to an interfering source provide reference signals which help to identify the interfering signal components present in the desired source signal. An example of particular interest is the placement of multiple microphones on a mobile phone (Jeub et al., 2012; Nelke et al., 2013), where secondary microphones are employed to provide noise references. If many sensors are distributed in an area, possibly at unknown locations, the concept is known as acoustic sensor network (Bertrand, 2011).

If multiple microphones are available for a speech recognition task, the question arises as where to condense the multi-channel data to arrive at a single word recognition result. Sofar, and this is the majority of works, we have been discussing linear or nonlinear filtering of the time-domain signals to arrive at an enhanced signal, which is then forwarded to the ASR unit. An alternative is to extract features for ASR in each input channel separately, commonly referred to as multi-stream ASR, and merge the streams either in the feature domain, during the score computation, or to perform recognition on each stream separately and merge the recognition outputs to come up with a single hypothesized word sequence. A significant advantage of the processing of the time domain signals is that phase information can be exploited. If the array size is small, the microphone signals will mostly differ in the phase, not in the amplitude. Since typical ASR features, like MFCCs, are based on magnitude or power spectral densities, they are unable to reflect these phase differences and will thus lead to almost identical feature streams in the different channels. If on the other hand the distance between the microphones is large, features derived from the different channels may have quite different characteristics.

10.2 **THE ACOUSTIC BEAMFORMING PROBLEM**

We consider the uniform linear array (ULA) of microphones depicted in Figure 10.1. Let us assume a point source emitting the monofrequent signal $x(t) = e^{j\omega_0 t}$ in the far field of the array. The far field assumption implies that the ratio of the distance of the source to the microphones versus the size of the array is so large that the wavefront curvature observed at the array position is very small and the sound wave arriving at the sensors can be modeled as a planar wave. We further assume wave propagation in free field, that is, the signals received at the sensors differ only in the delays, which are caused by the different distances the sensors have from the source.

The beamformer consists of the beamforming weights (or, more generally, filters) $w_0, \ldots, w_{M-1}$ followed by the summation of the filter results. The beamformer output signal is given by

$$
\begin{aligned}
z(t) &= \sum_{m=0}^{M-1} w_m^* e^{j\omega_0 (t-\tau_m)} \\
&= e^{j\omega_0 t} \sum_{m=0}^{M-1} w_m^* e^{-j\omega_0 \frac{dm\cos\theta}{c}} \\
&= e^{j\omega_0 t} \sum_{m=0}^{M-1} w_m^* e^{-j2\pi \frac{d}{\lambda_0} m \cos\theta},
\end{aligned} \tag{10.1}
$$

where $\lambda_0 = \frac{2\pi c}{\omega_0}$ is the wavelength of the source signal, and where $\tau_m = dm\cos\theta/c$ is the time difference between the arrival of the signal at the mth microphone and the arrival of the signal at the first microphone. Further, M denotes the number of microphones, and θ is the angle of the impinging wavefront relative to the endfire position (i.e., a position in the direction of the line connecting the microphones).

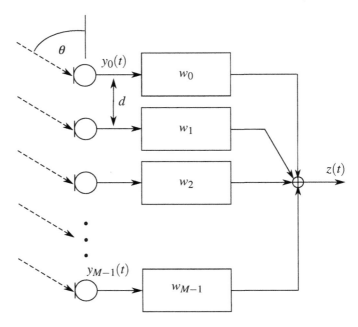

FIGURE 10.1

Uniform linear array with a source in the far field.

It is common practice to define the beamforming weights as complex conjugate coefficients w_m^*, as is done here. Equation 10.1 can be compactly written in vector notation

$$z(t) = e^{j\omega_0 t} \mathbf{w}^H \mathbf{v}, \tag{10.2}$$

where $(\cdot)^H$ denotes complex conjugate transpose. Here, $\mathbf{w}^H \mathbf{v}$ is the scalar product between the beamforming weight vector

$$\mathbf{w} = \begin{pmatrix} w_0 & \dots & w_{M-1} \end{pmatrix}^T \tag{10.3}$$

and the steering vector

$$\mathbf{v} = \begin{pmatrix} 1 & e^{-j\phi} & \dots & e^{-j(M-1)\phi} \end{pmatrix}^T, \tag{10.4}$$

where $\phi = 2\pi \frac{d}{\lambda_0} \cos\theta$. The steering vector describes the propagation from the source to the sensors. Note that we have set the time of flight from the source to the first sensor to zero, a simplification, which does not change any of the following considerations.

Computing the weighted summation of Equation 10.2 yields spatial selectivity. Let us assume a *delay-sum beamformer* (DSB) as an example. In a DSB, the weights are pure delays

$$\mathbf{w} = \frac{1}{M} \begin{pmatrix} 1 & e^{-j\phi_0} & \cdots & e^{-j(M-1)\phi_0} \end{pmatrix}^T,$$ (10.5)

where $\phi_0 = 2\pi \frac{d}{\lambda_0} \cos\theta_0/c$. With these weights, the signals at the beamforming filter outputs add constructively for signals arriving at an angle $\theta = \theta_0$, while they do so to a lesser extent for other angles of arrival. The spatial selectivity of the array can be expressed by the *beam pattern*, which is the magnitude response of a beamformer steered to some geometric angle θ_0, as a function of the angle of arrival θ. For the DSB, the beam pattern is given by

$$H(\theta) = \frac{1}{M} \cdot \frac{\sin\left(\frac{M}{2} 2\pi \frac{d}{\lambda_0} (\cos\theta - \cos\theta_0)\right)}{\sin\left(\frac{1}{2} 2\pi \frac{d}{\lambda_0} (\cos\theta - \cos\theta_0)\right)}.$$ (10.6)

Figure 10.2 displays sample DSB beampatterns. They show that the beamwidth depends on the ratio d/λ_0 (the larger the ratio of the inter-element distance to the

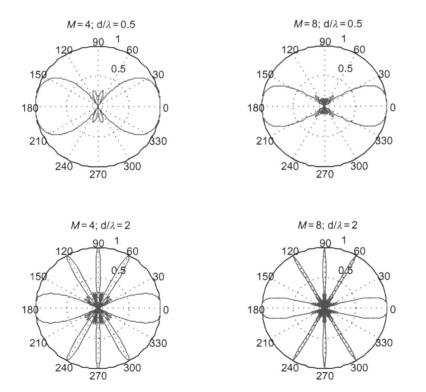

FIGURE 10.2

Sample beam patterns of a Delay-Sum Beamformer steered toward $\theta_0 = 0$.

wavelength the smaller is the beamwidth), the number of microphones (the more microphones the smaller is the beamwidth), and on the angle θ_0, with the endfire position $\theta_0 = 0$ delivering a broader beam than, for example, $\theta_0 = \pi/2$, the broadside direction.

The DSB is the simplest of all beamformers. While providing a "distortionless" response in the desired direction, that is, $\mathbf{w}^H \mathbf{v} = 1$ for $\theta = \theta_0$, it does not provide a means to place a spatial null in the direction of an interferer.

Before discussing alternative beamformer designs, we wish to generalize the propagation model. The steering vector Equation 10.4 assumes a free field or anechoic propagation: The received signal is solely characterized by the direction of arrival θ. Acoustic sound propagation in rooms, however, is much more involved. As discussed in Chapter 9, the transmitted signal arrives at the sensor via different propagation paths with individual delays and attenuations. The delay elements of the steering vector thus need to be generalized to acoustic transfer functions (ATFs), which comprise the effect of all propagation paths from the source to the sensor.

Compared to antenna array beamforming, acoustic beamforming is more involved due to a number of reasons:

- Acoustic sound propagation in rooms suffers from multipath propagation. The transmitted signal arrives at the sensors via different propagation paths with individual delays and attenuation. It is only inadequately described by the delay model of Equation 10.4.
- Antenna array design can be done under a narrowband assumption: the bandwidth of the signal is small compared to the carrier frequency. Thus, the beamformer has essentially the same gain for all frequencies. This is not the case in acoustic beamforming, which is a wideband beamforming problem: Here, the beamwidth "seen" by an input signal depends on the frequency of the input signal; see the dependency of $\mathbf{w}$ on the wavelength λ_0 in Equation 10.5.
- The interfering signals to be suppressed are often also speech, thus having the same spectral characteristics as the desired signal. The interferers and the acoustic noise usually cover the same frequency range as the desired signal.
- We are often confronted with time-varying scenarios: the target speaker or the interferers may move, causing changes in the ATF and requiring an adaptive system to let the beam of sensitivity follow the speaker.
- Finally, the human ear has a very high dynamic range and is very sensitive to artifacts introduced by the processing.

These all render acoustic beamforming in an uncontrolled real-world scenario a challenging problem. Many beamforming algorithms have been proposed, which may be categorized into fixed beamforming, data-dependent beamforming, and parametric beamforming. Fixed beamforming, where the beamformer coefficients are computed *a priori* according to some design criterion, is appropriate in static scenarios, where the position of the source and the acoustic scene is known and fixed. For dynamic scenarios with unknown and moving speech or noise sources data-dependent methods are preferred, where the beamforming coefficients are computed from the statistics of the incoming signal. They are able to adapt to changing source positions or changing interference patterns. They are also able to at least partly compensate for unknown sensor characteristics. However, they may be too slow to track fast changes. Fast adaptation to changing sound fields is possible with parametric beamforming approaches, where a perceptually or physically motivated model of the sound field is assumed, and only the parameters of the model are learnt from the input data. This requires less data and results in faster adaptation compared to purely data-driven methods, where the whole statistical characterization of the scene, e.g., power spectral densities of desired and interfering signals, has to be estimated.

In the following, we present the basic beamforming design criteria of data-driven beamforming. We will concentrate on criteria based on second-order statistics. They form the basis of most beamforming solutions found in the literature, although beamforming based on higher-order statistics has also been proposed (Kumatani et al., 2009).

10.3 FUNDAMENTALS OF DATA-DEPENDENT BEAMFORMING
10.3.1 SIGNAL MODEL AND OBJECTIVE FUNCTIONS

Before discussing different beamformer designs, we wish to generalize the wave propagation model. The steering vector in Equation 10.4 assumes a free-field anechoic sound propagation. However, sound propagation in enclosures is more involved, as discussed in Chapter 9. To account for multipath propagation, the delay elements of the steering vector have to be generalized to acoustic transfer functions, which comprise the effect of all propagation paths from the source to the sensors.

We consider an array of M microphones. Assuming that the multiplicative transfer function approximation holds, see Equation 9.8, the signal received at the mth microphone can be expressed in the STDFT domain as follows:

$$\dot{y}_m[t][k] = \dot{x}[t][k]\dot{h}_m[k] + \dot{n}_m[t][k]; \qquad m = 0,\ldots,M-1, \qquad (10.7)$$

where t denotes the STDFT frame index, while k is the frequency bin index. According to this model, the received signal consists of a component that is dependent on the target source signal $\dot{x}[t][k]$ and a distortion $\dot{n}_m[t][k]$. The impact of the signal transmission from the source to the sensor is captured by the acoustic transfer function (ATF) $\dot{h}_m[k]$, which is assumed to be time-invariant. While this may not be always true, see remarks in Section 9.2, it can often be assumed that the time variance of the ATF is low compared to the rate of change of the statistics of the speech source or of speech interferers.

Gathering all microphone signals in a vector $\dot{\mathbf{y}}[t][k] = \left(\dot{y}_0[t][k] \ \ldots \ \dot{y}_{M-1}[t][k]\right)^T$, and similar so for $\dot{\mathbf{n}}[t][k]$ and $\dot{\mathbf{h}}[k]$, we can write the system of equations (Equation 10.7) compactly as

$$\dot{\mathbf{y}}[t][k] = \dot{x}[t][k]\dot{\mathbf{h}}[k] + \dot{\mathbf{n}}[t][k]. \tag{10.8}$$

Beamforming is applied independently to each frequency bin. In the following, we therefore omit the frequency bin index k. We may also omit the time frame index to simplify the notation.

Assuming that speech and noise are mutually uncorrelated, the power spectral density (PSD) matrix of the microphone signals of Equation 10.8 is given by

$$\mathbf{\Phi}_{\dot{\mathbf{y}}\dot{\mathbf{y}}} = \phi_{\dot{x}\dot{x}}\dot{\mathbf{h}}\dot{\mathbf{h}}^{H} + \mathbf{\Phi}_{\dot{\mathbf{n}}\dot{\mathbf{n}}}. \tag{10.9}$$

where the PSD matrix of the microphone signals is defined by $\mathbf{\Phi}_{\dot{\mathbf{y}}\dot{\mathbf{y}}} = E[\dot{\mathbf{y}}\dot{\mathbf{y}}^{H}]$, the PSD of the source signal by $\phi_{\dot{x}\dot{x}} = E[|\dot{x}|^2]$, and the PSD matrix of the noise by $\mathbf{\Phi}_{\dot{\mathbf{n}}\dot{\mathbf{n}}} = E[\dot{\mathbf{n}}\dot{\mathbf{n}}^{H}]$.

In the *linearly constrained minimum variance* (LCMV) beamformer (Van Veen and Buckley, 1988), the filter coefficients $\mathbf{w} = (w_0, \ldots, w_{M-1})^T$, from which the beamformer output is computed via

$$\dot{z} = \mathbf{w}^{H}\dot{\mathbf{y}}, \tag{10.10}$$

are obtained by optimizing the following criterion: Minimize the noise power spectral density at the beamformer output subject to a set of linear constraints

$$J^{(\text{LCMV})}(\mathbf{w}) = \mathbf{w}^{H}\mathbf{\Phi}_{\mathbf{nn}}\mathbf{w} \quad \text{subject to} \quad \mathbf{C}^{H}\mathbf{w} = \mathbf{g} \tag{10.11}$$

where the constraints matrix $\mathbf{C}$ is of size $M \times P$, where P is the number of linear constraints. A constraint could be to have a distortionless response toward a certain source (see the MVDR beamforming criterion further below) or to form a spatial null toward the direction of an interfering source.

The optimization of Equation 10.11 is straightforward and yields (Haykin, 2001; Van Trees, 2002)

$$\mathbf{w}^{(\text{LCMV})} = \mathbf{\Phi}_{\mathbf{nn}}^{-1}\mathbf{C}\left(\mathbf{C}^{H}\mathbf{\Phi}_{\mathbf{nn}}^{-1}\mathbf{C}\right)^{-1}\mathbf{g}. \tag{10.12}$$

It is worthwhile to note that the minimization of the output power $\mathbf{\Phi}_{\dot{\mathbf{y}}\dot{\mathbf{y}}}$ instead of the minimization of the output noise power will lead to the same optimal solution (Van Trees, 2002).

The *minimum variance distortionless response (MVDR)* beamformer (Capon, 1969; Hoshuyama et al., 1999; Van Veen and Buckley, 1988) is a special case of the LCMV beamformer. The output power is minimized under a single linear constraint: a distortionless response is enforced toward the desired signal $\dot{x}$:

$$J^{(\text{MVDR})}(\mathbf{w}) = \mathbf{w}^{H}\mathbf{\Phi}_{\dot{\mathbf{y}}\dot{\mathbf{y}}}\mathbf{w} \quad \text{subject to} \quad \dot{\mathbf{h}}^{H}\mathbf{w} = 1. \tag{10.13}$$

This leads to the beamformer coefficients

$$\mathbf{w}^{(\mathrm{MVDR})} = \frac{\boldsymbol{\Phi}_{\dot{\mathbf{y}}\dot{\mathbf{y}}}^{-1}\dot{\mathbf{h}}}{\dot{\mathbf{h}}^{H}\boldsymbol{\Phi}_{\dot{\mathbf{y}}\dot{\mathbf{y}}}^{-1}\dot{\mathbf{h}}} = \frac{\boldsymbol{\Phi}_{\dot{\mathbf{n}}\dot{\mathbf{n}}}^{-1}\dot{\mathbf{h}}}{\dot{\mathbf{h}}^{H}\boldsymbol{\Phi}_{\dot{\mathbf{n}}\dot{\mathbf{n}}}^{-1}\dot{\mathbf{h}}}. \tag{10.14}$$

Another popular design criterion is the *multi-channel wiener filter (MWF)* (Doclo and Moonen, 2002). The MWF aims at minimizing the mean squared error between a desired signal d and the actual beamformer output $\dot{z}$. If the desired signal is the speech signal, $d = \dot{x}$, the objective function to be minimized is

$$\begin{aligned} J^{(\mathrm{MWF})}(\mathbf{w}) &= E\left[|\dot{x} - \mathbf{w}^{H}\dot{\mathbf{y}}|^{2}\right] \\ &= E\left[|\dot{x} - \mathbf{w}^{H}(\dot{\mathbf{h}}\dot{x} + \dot{\mathbf{n}})|^{2}\right] \\ &= \phi_{\dot{x}\dot{x}}|1 - \mathbf{w}^{H}\dot{\mathbf{h}}|^{2} + \mathbf{w}^{H}\boldsymbol{\Phi}_{\dot{\mathbf{n}}\dot{\mathbf{n}}}\mathbf{w}. \end{aligned} \tag{10.15}$$

First note that the Wiener filter does not enforce a distortionless response: $\mathbf{w}^{H}\dot{\mathbf{h}}$ is allowed to deviate from unity. It rather seeks an optimal compromise between equalization and noise reduction. Further, note that, in contrast to the single-channel Wiener Filter, the MWF estimates coefficients for a single time tap and multiple sensor taps.

In the *speech distortion weighted multi-channel Wiener filter (SDW-MWF)*, this trade-off is controlled by a design parameter μ (Doclo et al., 2005):

$$J^{(\mathrm{SDW\text{-}MWF})}(\mathbf{w}, \mu) = \phi_{\dot{x}\dot{x}}|1 - \mathbf{w}^{H}\dot{\mathbf{h}}|^{2} + \mu\mathbf{w}^{H}\boldsymbol{\Phi}_{\dot{\mathbf{n}}\dot{\mathbf{n}}}\mathbf{w}. \tag{10.16}$$

This leads to the following beamformer coefficients:

$$\mathbf{w}^{(\mathrm{SDW\text{-}MWF})} = \frac{\phi_{\dot{x}\dot{x}}\boldsymbol{\Phi}_{\dot{\mathbf{n}}\dot{\mathbf{n}}}^{-1}\dot{\mathbf{h}}}{\mu + \phi_{\dot{x}\dot{x}}\dot{\mathbf{h}}^{H}\boldsymbol{\Phi}_{\dot{\mathbf{n}}\dot{\mathbf{n}}}^{-1}\dot{\mathbf{h}}}. \tag{10.17}$$

Here, $\mu = 1$ corresponds to the MWF solution, while $\mu = 0$ gives the MVDR solution.

If the weighting parameter μ becomes very large, the speech distortion term in the objective function (Equation 10.16) can be neglected and the goal is to minimize the noise or, put differently, maximize the signal-to-noise ratio at the beamformer output, irrespective of the induced speech distortion. Noting that the SNR at the beamformer output

$$\bar{J}^{(\mathrm{maxSNR})} = \frac{\mathbf{w}^{H}\boldsymbol{\Phi}_{\dot{\mathbf{y}}\dot{\mathbf{y}}}\mathbf{w}}{\mathbf{w}^{H}\boldsymbol{\Phi}_{\dot{\mathbf{n}}\dot{\mathbf{n}}}\mathbf{w}}, \tag{10.18}$$

is the largest eigenvalue of the generalized eigenvalue problem

$$\boldsymbol{\Phi}_{\dot{\mathbf{y}}\dot{\mathbf{y}}}\mathbf{w} = \lambda\boldsymbol{\Phi}_{\dot{\mathbf{n}}\dot{\mathbf{n}}}\mathbf{w}, \tag{10.19}$$

we observe an elegant way of determining the beamformer coefficients: They are obtained by solving the generalized eigenvalue problem (Equation 10.19). Since the criterion (Equation 10.18) does not control the speech distortion, an appropriate postfilter is required to limit or eliminate it (Warsitz and Haeb-Umbach, 2007).

It is important to note that the different objective functions, except for the LCMC criterion with multiple linear constraints, all lead to the same multi-channel filter, up a complex constant (Van Veen and Buckley, 1988). For example, the SDW-MWF filter can be transformed to a MVDR beamformer by postfiltering the beamformer output signal with a filter with transfer function

$$\dot{h}_{\text{SDW-MWF\_to\_MVDR}} = \frac{\mu + \phi_{\dot{x}\dot{x}} \dot{\mathbf{h}}^H \boldsymbol{\Phi}_{\dot{n}\dot{n}}^{-1} \dot{\mathbf{h}}}{\phi_{\dot{x}\dot{x}} \dot{\mathbf{h}}^H \boldsymbol{\Phi}_{\dot{n}\dot{n}}^{-1} \mathbf{h}}. \tag{10.20}$$

10.3.2 GENERALIZED SIDELOBE CANCELLER

The LCMV beamformer, of which the MVDR beamformer is a special case, has been obtained as the solution of an optimization problem under P linear constraints: $\mathbf{C}^H \mathbf{w} = \mathbf{g}$. This constrained optimization can be reformulated as an unconstrained problem, which is computationally more efficient, leading to the so-called generalized sidelobe canceller (GSC) (Gannot et al., 2001; Griffiths and Jim, 1982). To this end, the beamforming vector $\mathbf{w}$ is split into two orthogonal components, one in the constraints subspace, that is, the subspace spanned by $\mathbf{C}$: $\mathbf{w}_0 \in \text{Span}\{\mathbf{C}\}$, and the other lying in the orthogonal subspace, the null subspace of the constraints subspace: $\mathbf{w}_n \in \text{Null}\{\mathbf{C}\}$:

$$\mathbf{w}^{(\text{LCMV})} = \mathbf{w}_0 - \mathbf{w}_n. \tag{10.21}$$

The component $\mathbf{w}_0$ is obtained by projecting $\mathbf{w}^{(\text{LCMV})}$ onto $\mathbf{C}$:

$$\mathbf{w}_0 = \mathbf{C}(\mathbf{C}^H \mathbf{C})^{-1} \mathbf{C}^H \mathbf{w}^{(\text{LCMV})} = \mathbf{C}(\mathbf{C}^H \mathbf{C})^{-1} \mathbf{g}. \tag{10.22}$$

This component of the beamformer is called the *fixed* beamformer.

The weights $\mathbf{w}_n$ are obtained by projecting $\mathbf{w}^{(\text{LCMV})}$ onto the null space:

$$- \mathbf{w}_n = (\mathbf{I} - \mathbf{C}(\mathbf{C}^H \mathbf{C})^{-1} \mathbf{C}^H) \mathbf{w}^{(\text{LCMV})}. \tag{10.23}$$

Let $\mathbf{B}$ be a matrix, whose $M - P$ columns span the null space of $\mathbf{C}$. The columns can, for example, be found by the Gram-Schmidt orthonormalization method. Then, $\mathbf{w}_n$ can be written as

$$\mathbf{w}_n = \mathbf{B}\mathbf{q}, \tag{10.24}$$

where the vector $\mathbf{q}$ contains the $M - P$ coefficients of $\mathbf{w}_n$ along the coordinate axes given by the columns of $\mathbf{B}$. The matrix $\mathbf{B}$ is termed *blocking matrix*. It blocks the components of the input signal, which are in the span of $\mathbf{C}$. If, for example, the single linear constraint of the MVDR beamformer is considered, then $\mathbf{C} = \dot{\mathbf{h}}$, and the blocking matrix blocks the desired signal $\dot{x}$, and the output of the blocking matrix consists of noise signal components only.

Consequently, the $M - P$ entries of $\mathbf{q}$ are called *noise cancellation filters*. Their purpose is to determine an estimate of the noise in the input signal filtered by $\mathbf{w}_0$. The coefficients can be determined by the orthogonality principle, by which the noise canceller estimation error has to be uncorrelated (orthogonal) with the input signals of the noise cancellation filters. Let $\mathbf{u} = \mathbf{B}^H \dot{\mathbf{y}}$ be the output of the blocking matrix, which is the input to the noise cancellation filters. According to the orthogonality principle, the optimal filter coefficients $\mathbf{q}$ must satisfy

$$E[\mathbf{u}(\mathbf{w}_0^H \dot{\mathbf{y}} - \mathbf{u}^H \mathbf{q})] = \mathbf{0}. \tag{10.25}$$

This leads to

$$\begin{aligned}
\mathbf{q} &= E[\mathbf{u}\mathbf{u}^H]^{-1} E[\mathbf{u}\dot{\mathbf{y}}^H]\mathbf{w}_0 \\
&= (\mathbf{B}^H \mathbf{\Phi}_{\dot{\mathbf{y}}\dot{\mathbf{y}}} \mathbf{B})^{-1} \mathbf{B}^H \mathbf{\Phi}_{\dot{\mathbf{y}}\dot{\mathbf{y}}} \mathbf{w}_0.
\end{aligned} \tag{10.26}$$

Figure 10.3 shows the block diagram of the generalized sidelobe canceller.

The purpose of the fixed beamformer (FBF) is to satisfy the constraints (e.g., distortionless response in the direction of the desired signal in the case of the MVDR beamformer), while the blocking matrix (BM) blocks the desired signal and delivers $N - P$ noise reference signals at its output. The noise cancellation (NC) filters, which are usually implemented as adaptive filters, generate an estimate of the residual noise in the FBF output, which is then subtracted from the FBF output to generate the overall output signal.

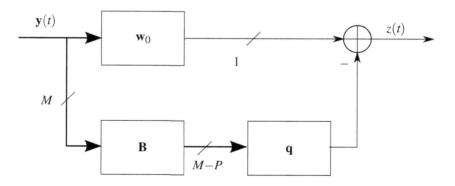

FIGURE 10.3

Block diagram of a generalized sidelobe canceller with fixed beamformer (FBF) $\mathbf{w}_0$, blocking matrix **B** and noise cancellation filters **q**.

10.3.3 **RELATIVE TRANSFER FUNCTIONS**

The beamformers developed in the last section all hinge on the availability of the acoustic transfer functions $\dot{\mathbf{h}}[k]$. However, the acoustic transfer functions are notoriously hard to estimate, as we discussed in Chapter 9. Furthermore, although we neglected time variance in our development, they are usually time-variant, asking for an adaptive tracking procedure to follow changes in the ATF.

In light of these problems, researchers have formulated alternative constraints which avoid the need to estimate the ATFs. With this reformulation, so-called relative transfer functions (RTF) are required, which are somewhat easier to estimate.

To understand this, remember the constraint imposed in the MVDR beamformer design, which ensured a distortionless response in a target direction, was $\dot{\mathbf{h}}^H \mathbf{w} = 1$. This constraint actually postulates that the channel should be completely equalized, resulting not only in a spatial focusing but also in a dereverberation of the signal: The beamformer output should be identical to the clean speech signal, as it is generated at the source position, up to residual noise.

We now give up the dereverberation requirement and postulate instead

$$\dot{\mathbf{h}}^H \mathbf{w} = \dot{h}_0^*. \tag{10.27}$$

Instead of a transfer characteristic of one from the source to the beamformer output, we would like to see the source signal as captured by a reference microphone, here microphone #0, at the beamformer output.

Similarly, for the MWF objective function, we identify the desired signal with the speech component of the reference microphone, $d = \dot{h}_0 \dot{x}$, instead of the speech source signal.

Rearranging Equation 10.27, we obtain

$$\underbrace{\left(1 \quad \frac{\dot{h}_1^*}{\dot{h}_0^*} \quad \cdots \quad \frac{\dot{h}_{M-1}^*}{\dot{h}_0^*} \right)}_{\tilde{\mathbf{h}}^H} \mathbf{w} = 1. \tag{10.28}$$

The transfer function ratios in the vector $\tilde{\mathbf{h}}$ are the aforementioned RTFs. The RTFs can be thought of as generalizing the concept of time difference of arrival to a ratio of acoustic transfer functions.

A comparison of Equation 10.28 with $\mathbf{C}^H \mathbf{w} = \mathbf{g}$ shows that $\mathbf{g} = 1$ and

$$\mathbf{C} = \left(1 \quad \frac{\dot{h}_1}{\dot{h}_0} \quad \cdots \quad \frac{\dot{h}_{M-1}}{\dot{h}_0} \right)^H. \tag{10.29}$$

From this observation, a GSC realization of a MVDR beamformer using the modified constraint Equation 10.27 can be readily obtained. The fixed beamformer is given by

$$\mathbf{w}_0 = \frac{\tilde{\mathbf{h}}}{\|\tilde{\mathbf{h}}\|}, \tag{10.30}$$

while the columns of the blocking matrix have to span the orthogonal complement to Equation 10.30. A possible choice is

$$\mathbf{B} = \begin{pmatrix} -\tilde{h}_1^* & -\tilde{h}_2^* & \cdots & -\tilde{h}_{M-1}^* \\ 1 & 0 & \cdots & 0 \\ 0 & 1 & \cdots & 0 \\ & \cdots & \ddots & \\ 0 & 0 & \cdots & 1 \end{pmatrix}. \tag{10.31}$$

Obviously, if $\dot{\mathbf{y}} = \dot{\mathbf{h}}x$, then $\mathbf{u} = \mathbf{B}^H \dot{\mathbf{y}} = \mathbf{0}$, that is, the desired signal is blocked.

Why are RTFs easier to estimate than ATFs? If the microphone array is compact, the components of $\dot{\mathbf{y}}$, that is, the ATFs from the source to the M microphones are rather similar, and, more importantly, change similarly over time. Consequently, the RTFs can be considered less time variant than the ATFs. It has further been observed that the RTFs are shorter than the ATFs (Gannot et al., 2001; Reuven et al., 2008).

In Gannot et al. (2001), a method is proposed to estimate RTFs which is based on the nonstationarity of speech, while assuming the noise to be stationary. Using Equation 10.31, we find

$$u_m = \dot{y}_m - \tilde{h}_m \dot{y}_0; \quad m = 1, \ldots, M - 1. \tag{10.32}$$

Rearranging terms we obtain

$$\dot{y}_m = u_m + \tilde{h}_m \dot{y}_0; \quad m = 1, \ldots, M - 1. \tag{10.33}$$

We can now set up an overdetermined system of equations by considering different time frames $t_1, \ldots, t_N$:

$$\dot{y}_m[t_1][k] = u_m[t_1][k] + \tilde{h}_m[k]\dot{y}_0[t_1][k]$$
$$\dot{y}_m[t_2][k] = u_m[t_2][k] + \tilde{h}_m[k]\dot{y}_0[t_2][k]$$
$$\vdots$$
$$\dot{y}_m[t_N][k] = u_m[t_N][k] + \tilde{h}_m[k]\dot{y}_0[t_N][k]$$

and compute the cross power spectral density estimates between $\dot{y}_m$ and $\dot{y}_0$ for each observed frame:

$$\begin{pmatrix} \Phi_{\dot{y}_m \dot{y}_0}[t_1][k] \\ \Phi_{\dot{y}_m \dot{y}_0}([t_2][k] \\ \vdots \\ \Phi_{\dot{y}_m \dot{y}_0}([t_N][k] \end{pmatrix} = \begin{pmatrix} \Phi_{\dot{y}_0 \dot{y}_0}[t_1][k] & 1 \\ \Phi_{\dot{y}_0 \dot{y}_0}([t_2][k] & 1 \\ \vdots & \\ \Phi_{\dot{y}_0 \dot{y}_0}([t_N][k] & 1 \end{pmatrix} \begin{pmatrix} \tilde{h}_m \\ \Phi_{u_m \dot{y}_0} \end{pmatrix}. \tag{10.34}$$

Now we assume that the noise is stationary and uncorrelated with the speech. From this, it follows that $\phi_{u_m \dot{y}_0}$ is determined by the noise statistics only and can thus be considered to be constant. This is true since the noise reference u_m ideally does not contain the desired speech component. Thus, $\phi_{u_m \dot{y}_0}$ can be considered independent of time. If this holds, Equation 10.34 can be solved by standard Least Squares techniques, delivering the vector of unknowns $\left(\tilde{h}_m \quad \Phi_{u_m \dot{y}_0} \right)^T$ and therefore $\tilde{h}_m$.

Another approach to estimating RTFs is based on an eigenvector decomposition of the power spectral density matrix $\mathbf{\Phi}_{\dot{y}\dot{y}}$ of the microphone signals (Krueger et al., 2011; Warsitz and Haeb-Umbach, 2007). To see how this works, let us rewrite Equation 10.8 by using the vector of RTFs introduced in Equation 10.28

$$\dot{\mathbf{y}}[t][k] = \dot{x}'[t][k]\tilde{\mathbf{h}}[k] + \dot{\mathbf{n}}[t][k], \tag{10.35}$$

where $\dot{x}'[t][k] = \dot{x}[t][k]\dot{h}_0[k]$. The power spectral density matrix of the microphone signals is then

$$\mathbf{\Phi}_{\dot{y}\dot{y}}[t][k] = \phi_{\dot{x}'\dot{x}'}[t][k]\tilde{\mathbf{h}}[k]\tilde{\mathbf{h}}^H[k] + \mathbf{\Phi}_{\dot{n}\dot{n}}[k]. \tag{10.36}$$

It follows that the principal eigenvector of $\mathbf{\Phi}_{\dot{n}\dot{n}}^{-1}\mathbf{\Phi}_{\dot{y}\dot{y}}$ is proportional to $\mathbf{\Phi}_{nn}^{-1}\tilde{\mathbf{h}}$. If the noise is assumed to be stationary, the noise PSD $\mathbf{\Phi}_{\dot{n}\dot{n}}$ can be estimated in speech pauses. Thus, $\tilde{\mathbf{h}}$ can be readily obtained by eigenvector decomposition, up to an arbitrary complex constant, though. The presence of this constant stems from the fact that an arbitrarily scaled eigenvector remains to be an eigenvector. The scaling can be fixed by ensuring a distortionless response to the desired signal (Krueger et al., 2011; Warsitz and Haeb-Umbach, 2007).

The beamforming criteria discussed so far depend on second-order statistics. The use of beamforming criteria based on higher-order statistics has been investigated in Kumatani et al. (2009). The motivation for using higher-order statistics is that speech is non-Gaussian, while noise usually is Gaussian. It may, therefore, help in retrieving speech from noisy input if the beamformer is designed to maximize the non-Gaussianity of its output signal. A disadvantage, though, is that more data are required to estimate higher-order statistics than to estimate second-order statistics, making the approach less attractive in dynamic scenarios.

When beamforming is used as a front-end to ASR, it seems to be natural to employ beamforming design criteria that are related to the objective of ASR. While an iterative optimization of the beamformer coefficients toward minimizing the word error rate is yet out of reach, in Seltzer et al. (2004) the beamformer coefficients have been adjusted to maximize the likelihood of the correct word sequence.

Since the power spectral densities are not known in advance and can be time variant, adaptive algorithms are used to estimate them. Basing the beamformer coefficients on statistics estimated from the data has the beneficial side effect that sensor calibration becomes superfluous. Any deviation, for example, from the nominal position, will be accounted for automatically and absorbed in the filter coefficients when they are computed from estimated power spectral densities and RTFs. However, in acoustic beamforming, the number of filter coefficients tends to be large, which makes the adaptive algorithms unable to follow fast changes in the signal statistics.

Parametric methods, on the other hand, assume adherence to a certain array geometry or sound propagation model. An example is the assumption of anechoic sound propagation. Then only the time differences of arrival need to be estimated, see Equation 10.4, which can be done, for example, with the GCC-PHAT method (Knapp and Carter, 1976). Another example is to assume the noise field to be of a certain type, such as spatially white noise or diffuse noise. This determines the parametric form of the noise PSD matrix $\mathbf{\Phi_{\ddot{n}\ddot{n}}}$ and only its parameters need to be estimated from the data. The effectiveness of these parametric models crucially depends on their underlying assumptions being correct. Further, parametric methods are susceptible to microphone calibration errors. For example, if the distance between the microphones of an array varies, beamforming coefficients computed on the assumption of constant inter-element distance will not perform optimally.

A compromise between purely parametric and purely data-driven beamforming is the concept of informed spatial filtering. It is a data-driven beamforming approach, where side information is employed to enable the tracking in fast changing acoustic environments (Taseska and Habets, 2014). Such side information can be the position of desired and interfering sound sources or information about the sound field given by the direct-to-diffuse ratio, room reverberation time, or the coherence properties of the noise field.

10.4 MULTI-CHANNEL SPEECH RECOGNITION

Although being the most popular, acoustic beamforming is not the only option to leverage multiple channels in an ASR context. An alternative to condensing the microphone signals to a single beamformed output signal is to forward all channels to the ASR system and to do a fusion or selection within the recognizer, either at the feature, the score, or at the decoder stage. In the following, we will first discuss ASR on beamformed signals before looking at multi-stream ASR.

10.4.1 ASR ON BEAMFORMED SIGNALS

Acoustic beamforming has been considered as front-end processing of ASR in many works. During the 1990s, a number of pioneering studies have been conducted on the use of microphone arrays for ASR (Adcock et al., 1996; Compernolle et al., 1990; Omologo et al., 1997) showing significant word error rate improvements due to acoustic beamforming. This has been recently reconfirmed in the course of the REVERB challenge (Kinoshita et al., 2013): Several groups have employed acoustic beamforming on the two- and eight-channel data provided and all have reported improvements in recognition accuracy due to acoustic beamforming. The employed beamforming concepts ranged from DSB to MVDR (Delcroix et al., 2014; Leng et al., 2014), and even beamforming based on optimizing higher-order objective functions (Feng et al., 2014). It was observed in Delcroix et al. (2014) that beamforming on the 8-channel data delivered significantly better recognition results than beamforming on 2-channel data (Delcroix et al., 2014).

The amount of ASR word error rate improvement due to acoustic beamforming depends on many factors: the objective function the beamformer is based on, the array geometry, the properties of the noise field, the presence of interfering sounds, etc. It is therefore impossible to give universal numbers on how much the WER will drop due to beamforming. In general, one may state, however, the following:

- In a purely diffuse noise field, beamforming is less effective than in the presence of directional noise. Diffuse noise refers to the fact that the noise impinges on the array equally strong from all directions. For this scenario, but also for others, effective single-channel postfilters applied to the beamformed output signal, have been developed (McCowan and Bourlard, 2003; Simmer et al., 2001; Zelinski, 1988).
- More elaborated criteria, such as the MVDR, are more effective than a simple DSB, in particular if the noise is directional. These more elaborated criteria are able to place a spatial null toward the direction of the interfering sound source.
- In a reverberant enclosure, the free field sound propagation model is not appropriate. Therefore, beamformer coefficients based on (relative) acoustic transfer functions instead of time differences of arrival will deliver better noise and reverberation suppression and consequently better ASR performance.
- The last two statements, however, hinge on the quality of the estimated statistics. In highly dynamic scenarios with moving and changing sound sources, it may be advantageous to adopt a parametric beamforming approach, since changes in parameters, such as TDOA, can be faster followed than changes in second-order statistics.
- The size of the array has also a significant impact. If the microphones are close to each other, there is no danger of spatial aliasing. On the other hand, the larger the aperture the more the array is able to suppress low-frequency noise. If the microphones are distributed over a large space, it can be more effective to select a microphone signal than to combine the signals. However, how to determine the microphone that delivers the best signal is an issue by itself (Katsamanis et al., 2014). A selection based on the SNR as in Wölfel et al. (2006) is probably the most popular method. Note that also the choice of the reference microphone for the MWF can have a significant impact on the performance of the beamformer (Lawin-Ore and Doclo, 2012).

10.4.2 MULTI-STREAM ASR

Here we consider techniques, where the fusion or selection of the multi-channel data is carried out inside the recognizer. Considering multiple parallel data streams in a recognizer is a rather general concept (Bourlard et al., 1996; Janin et al., 1999), not restricted to streams being identified with individual microphone channels. The data streams could also come from different front-end processing methods applied to the same signal, different feature extraction methods, etc.

Eventually, the streams have to be merged to arrive at an hypothesis on the spoken word sequence. This can be done at the feature, the acoustic score, or at the decoded word level. A study of channel selection techniques for multi-microphone input has been conducted in Wolf and Nadeu (2014). They developed measures extracted from a given distorted signal that either estimates its quality or measures how well it fits the acoustic models of the recognition system. In Wolf and Nadeu (2013), microphone channel selection is based on N-Best hypotheses of the decoded word sequence.

There have also been attempts to combine multi-channel processing with acoustic modeling in the context of deep neural networks or DNNs (Dahl et al., 2011, 2012; Deng and Yu, 2014; Hinton et al., 2012; Yu and Deng, 2011, 2014). In Swietojanski et al. (2013), a DNN is used to take the audio from multiple distant microphones, and the resulting recognition accuracy is better than the DNN trained with the audio from a single distant channel, however worse than a DNN trained on the output of an acoustic beamformer. In Swietojanski et al. (2014), it has been found that convolutional neural networks achieve a lower WER than conventional DNNs. Convolutional neural networks (CNNs) employ weight sharing. It was found that channel-wise convolution, with weight sharing along the frequency dimension, followed by a cross-channel max-pooling performed better than multi-channel convolution. In multi-channel convolution, each convolutional band is composed of filter activations spanning all input (microphone) channels.

In Renals and Swietojanski (2014), the authors compare the WER performance of an ASR system operating on a single beamformed output with multiple distant microphone signals combined by the neural network as described in the last paragraph. They employed two quite different databases, which allow to draw some interesting conclusions. The first database is the AMI meeting corpus (Carletta, 2007), which was recorded using an 8-element circular microphone array with 10 cm radius. The second corpus is the ICSI meeting corpus (Janin et al., 2003), which contains recordings of four boundary microphones placed about 1 m apart along the tabletop. Compared to the recording conditions of the AMI corpus, the microphones are spread over a larger area and they are at unknown positions. For both corpora, the authors observed that using multiple channels led to improved WER compared to using a single microphone signal, which comes at no surprise. What, however, is interesting is, that in the case of the AMI corpus, ASR on the beamformed output performed clearly better than combining the multiple channels in the CNN as described above, while with the ICSI corpus the two approaches performed comparably. A possible explanation is that in the case of the compact array used for the AMI corpus, the individual microphone signals differ mostly in the phase but not in the magnitude. Unlike acoustic beamforming, phase differences are not captured by ASR feature extraction. Therefore there is not that much diversity in the input channels, unlike the ICSI corpus, where the microphones were further apart. Here, the features of different channels are probably more diverse, and the network is able to learn which channel provides most information.

10.5 **TO PROBE FURTHER**

There are several textbooks on multi-channel processing and beamforming, such as the ones by Johnson and Dudgeon (1993) and Van Trees (2002), and a good tutorial article is the one by Van Veen and Buckley (1988). They are, however, mostly concerned with antenna array beamforming. A collection of articles about microphone arrays and acoustic beamforming is the book edited by Brandstein and Ward (2001). An excellent tutorial on linear and parametric microphone array processing, authored by Gannot and Habets, can be found in Gannot and Habets (2013).

Table 10.1 Approaches to Speech Recognition in the Presence of Multi-Channel Recordings

Method	Proposed Around	Characteristics
MVDR beamformer (Frost, 1972)	1972	Minimizes output power of the beamformer under a constraint in the look direction
Generalized sidelobe cancellor (Griffiths and Jim, 1982)	1982	Reformulates MVDR criterion as an unconstrained optimization
Speech recognition with microphone arrays (Compernolle et al., 1990; Omologo et al., 1997)	1990-1997	First publications demonstrating the benefits of beamforming for ASR
Multi-stream ASR (Bourlard et al., 1996; Janin et al., 1999)	1996	Multiple input streams to ASR with fusion at feature, score, or decoding stage
Concept of relative transfer functions (Cohen, 2004; Gannot et al., 2001; Warsitz and Haeb-Umbach, 2007)	2001	Different methods to identify relative transfer functions
Multi-channel Wiener filter (Doclo and Moonen, 2002)	2002	Time domain realization of (parametric) multi-channel Wiener filter
Likelihood maximizing beamforming (Seltzer et al., 2004)	2004	Determines beamformer coefficients as to maximize the likelihood of the correct word hypothesis in the recognizer
Beamforming criterion based on higher-order statistics (Kumatani et al., 2009)	2009	Maximum negentropy criterion for beamformer design
Convolutional neural networks for multi-channel input (Swietojanski et al., 2014)	2014	Convolutional neural networks for distant speech recognition

A tutorial style article on microphone arrays for distant speech recognition is available in Kumatani et al. (2012), while further material on distant speech recognition is in the book by Wölfel and McDonough (2009).

An acoustic beamforming tool widely used in the ASR community can be found in Anguera (2006).

10.6 SUMMARY

Table 10.1 summarizes key publications in the field of multi-channel processing. We mention only very few from the field of beamforming, being aware of the fact that there are many more influential publications in this field. Acoustic beamforming for automatic speech recognition was pioneered in the 1990s by Compernolle and others. Multi-Stream ASR became popular in the late 1990s with works by Bourlard and Morgan, among others. Seltzer et al. investigated a closer coupling between acoustic beamforming and speech recognition. They optimized the beamformer coefficients using criteria relevant to ASR. Recently, various approaches to combine multi-channel data and deep neural networks have been investigated.

REFERENCES

Adcock, J., Gotoh, Y, Mashao, D., Silverman, H., 1996. Microphone-array speech recognition via incremental map training. In: Proc. of IEEE International Conference on Acoustics, Speech and Signal Processing (ICASSP), vol. 2, pp. 897-900.

Anguera, X., 2006. Beamformit: Fast and robust acoustic beamformer. http://www.xavieranguera.com/beamformit/.

Bertrand, A., 2011. Applications and trends in wireless acoustic sensor networks: A signal processing perspective. In: IEEE Symposium on Communications and Vehicular Technology in the Benelux (SCVT), pp. 1-6.

Bourlard, H., Dupont, S., Valais Suisse, M., Ris, C., 1996. Multi stream speech recognition. IDIAP Research Report 96-07.

Brandstein, M., Ward, D. (Eds.), 2001. Microphone Arrays. Springer, New York.

Capon, J., 1969. High resolution frequency-wavenumber spectrum analysis. Proc. IEEE 57 (8), 1408-1418.

Carletta, J., 2007. Unleashing the killer corpus: experiences in creating the multi-everything AMI meeting corpus. Lang. Resources Evaluat. 41, 181-190.

Cohen, I., 2004. Relative transfer function identification using speech signals. IEEE Trans. Speech Audio Process. 12 (5), 451-459. ISSN 1063-6676. http://dx.doi.org/10.1109/TSA.2004.832975.

Compernolle, D.V., Ma, W., Xie, F., Diest, M.V., 1990. Speech recognition in noisy environments with the aid of microphone arrays. Speech Commun. 9 (5-6), 433-442. ISSN 0167-6393. http://dx.doi.org/http://dx.doi.org/10.1016/0167-6393(90)90019-6. URL http://www.sciencedirect.com/science/article/pii/0167639390900196.

Dahl, G., Yu, D., Deng, L., Acero, A., 2011. Large vocabulary continuous speech recognition with context-dependent DBN-HMMs. In: Proc. International Conference on Acoustics, Speech and Signal Processing (ICASSP).

Dahl, G., Yu, D., Deng, L., Acero, A., 2012. Context-dependent pre-trained deep neural networks for large-vocabulary speech recognition. IEEE Trans. Audio Speech Lang. Process. 20 (1), 30-42.

Delcroix, M., Yoshioka, T., Ogawa, A., Kubo, Y, Fujimoto, M., Ito, N., 2014. Linear prediction-based dereverberation with advanced speech enhancement and recognition technologies for the reverb challenge. In: REVERB Challenge Workshop, Florence, Italy.

Deng, L., Yu, D., 2014. Deep Learning: Methods and Applications. Now Publishers, Hanover, MA.

Doclo, S., Moonen, M., 2002. Gsvd-based optimal filtering for single and multimicrophone speech enhancement. IEEE Trans. Signal Process. 50 (9), 2230-2244. ISSN 1053-587X. http://dx.doi.org/10.1109/TSP.2002.801937.

Doclo, S., Spriet, A., Wouters, J., Moonen, M., 2005. Speech distortion weighted multichannel wiener filtering techniques for noise reduction. In: Benesty, J., Makino, S., Chen, D. (Eds.), Speech Enhancement. Springer, New York.

Feng, X., Kumantani, K., McDonough, J., 2014. The CMU-MIT REVERB channel 2014 system: Description and results. In: REVERB Challenge Workshop, Florence, Italy.

Frost, O.L., 1972. An algorithm for linearly constrained adaptive array processing. Proc. IEEE 60 (8), 926-935. ISSN 0018-9219. http://dx.doi.org/10.1109/PROC.1972.8817.

Gannot, S., Burshtein, D., Weinstein, E., 2001. Signal enhancement using beamforming and nonstationarity with applications to speech. IEEE Trans. Signal Process. 49 (8), 1614-1626. ISSN 1053-587X. http://dx.doi.org/10.1109/78.934132.

Gannot, S., Habets, E., 2013. Linear and parametric microphone array processing. http://www.audiolabs-erlangen.de/fau/professor/habets/lectures/ICASSP-2013.

Griffiths, L.J., Jim, C., 1982. An alternative approach to linearly constrained adaptive beamforming. IEEE Trans. Antennas Propagat. 30 (1), 27-34. ISSN 0018-926X. http://dx.doi.org/10.1109/TAP.1982.1142739.

Haykin, S., 2001. Adaptive Filter Theory. Prentice Hall, Upper Saddle River, NJ.

Hinton, G., Deng, L., Yu, D., Dahl, G.E., Mohamed, A., Jaitly, N., et al., 2012. Deep neural networks for acoustic modeling in speech recognition: The shared views of four research groups. IEEE Signal Process. Mag. 29 (6), 82-97.

Hoshuyama, O., Sugiyama, A., Hirano, A., 1999. A robust adaptive beamformer for microphone arrays with a blocking matrix using constrained adaptive filters. IEEE Trans. Signal Process. 47 (10), 2677-2684. ISSN 1053-587X. http://dx.doi.org/10.1109/78.790650.

Janin, A., Baron, D., Edwards, J., Ellis, D., Gelbart, D., Morgan, N., et al., 2003. The ICSI Meeting Corpus. In: Proc. of IEEE International Conference on Acoustics, Speech and Signal Processing (ICASSP), vol. 1, pp. I-364-I-367.

Janin, A., Ellis, D., Morgan, N., 1999. Multi-stream speech recognition: Ready for prime time?. In: Proc. Eurospeech, Budapest.

Jeub, M., Herglotz, C., Nelke, C., Beaugeant, C., Vary, P., 2012. Noise reduction for dual-microphone mobile phones exploiting power level differences. In: Proc. of IEEE International Conference on Acoustics, Speech and Signal Processing (ICASSP), pp. 1693-1696.

Johnson, D., Dudgeon, D., 1993. Array Signal Processing: Concepts and Techniques. Prentice Hall, Upper Saddle River, NJ.

Katsamanis, A., Rodomagoulakis, I., Potamianos, G., Maragos, P., Tsiami, A., 2014. Robust far-field spoken command recognition for home automation combining adaptation and multichannel processing. In: Proc. of IEEE International Conference on Acoustics, Speech and Signal Processing (ICASSP), pp. 5547-5551.

Kinoshita, K., Delcroix, M., Yoshioka, T., Nakatani, T., Habets, E., Haeb-Umbach, R., et al., 2013. The REVERB challenge: A common evaluation framework for dereverberation and recognition of reverberant speech. In: Proc. of IEEE ASSP Workshop on Applications of Signal Processing to Audio and Acoustics, pp. 1-4.

Knapp, C., Carter, G., 1976. The generalized correlation method for estimation of time delay. IEEE Trans. Acoustics Speech Signal Process. 24 (4), 320-327. ISSN 0096-3518. http://dx.doi.org/10.1109/TASSP.1976.1162830.

Krueger, A., Warsitz, E., Haeb-Umbach, R., 2011. Speech enhancement with a gsc-like structure employing eigenvector-based transfer function ratios estimation. IEEE Trans. Audio Speech Lang. Process. 19 (1), 206-219. ISSN 1558-7916. http://dx.doi.org/10.1109/TASL.2010.2047324.

Kumatani, K., McDonough, J., Raj, B., 2012. Microphone array processing for distant speech recognition: From close-talking microphones to far-field sensors. IEEE Signal Process. Mag. 29 (6), 127-140. ISSN 1053-5888. http://dx.doi.org/10.1109/MSP.2012.2205285.

Kumatani, K., McDonough, J., Rauch, B., Klakow, D., Garner, P., Li, W., 2009. Beamforming with a maximum negentropy criterion. IEEE Transa. Audio Speech Lang. Process. 17 (5), 994-1008. ISSN 1558-7916. http://dx.doi.org/10.1109/TASL.2009.2015090.

Lawin-Ore, T., Doclo, S., 2012. Reference microphone selection for mwf-based noise reduction using distributed microphone arrays. In: ITG Conference on Speech Communication, Braunschweig, Germany.

Leng, Y., Dennis, J., Terrence, Z., Tran, H., 2014. PBF-GSC beamforming for ASR and speech enhancement in reverberant environments. In: REVERB Challenge Workshop, Florence, Italy.

McCowan, I., Bourlard, H., 2003. Microphone array post-filter based on noise field coherence. IEEE Trans. Speech Audio Process. 11 (6), 709-716. ISSN 1063-6676. http://dx.doi.org/10.1109/TSA.2003.818212.

Nelke, C., Beaugeant, C., Vary, P., 2013. Dual microphone noise PSD estimation for mobile phones in hands-free position exploiting the coherence and speech presence probability. In: Proc. of IEEE International Conference on Acoustics, Speech and Signal Processing (ICASSP), pp. 7279-7283.

Omologo, M., Matassoni, M., Svaizer, P., Giuliani, D., 1997, Apr. Microphone array based speech recognition with different talker-array positions. In: Proc. of IEEE International Conference on Acoustics, Speech and Signal Processing (ICASSP), 1, pp. 227-230 vol.1.

Renals, S., Swietojanski, P., 2014. Neural networks for distant speech recognition. In: Workshop on Hands-free Speech Communication and Microphone Arrays (HSCMA), pp. 172-176.

Reuven, G., Gannot, S., Cohen, I., 2008. Dual-source transfer-function generalized sidelobe canceller. IEEE Transa. Audio Speech Lang. Process. 16 (4), 711-727. ISSN 1558-7916. http://dx.doi.org/10.1109/TASL.2008.917389.

Seltzer, M., Raj, B., Stern, R., 2004. Likelihood-maximizing beamforming for robust hands-free speech recognition. IEEE Trans. Speech Audio Process. 12 (5), 489-498. ISSN 1063-6676. http://dx.doi.org/10.1109/TSA.2004.832988.

Simmer, K., Bitzer, J., Marro, C., 2001. Postfiltering techniques. In: Brandstein, M., Ward, D. (Eds.), Microphone Arrays. Springer, New York.

Swietojanski, P., Ghoshal, A., Renals, S., 2013. Hybrid acoustic models for distant and multichannel large vocabulary speech recognition. In: Proc. of IEEE Workshop on Automatic Speech Recognition and Understanding (ASRU), pp. 285-290.

Swietojanski, P., Ghoshal, A., Renals, S., 2014. Convolutional neural networks for distant speech recognition. IEEE Signal Process. Lett. 21 (9), 1120-1124. ISSN 1070-9908. http://dx.doi.org/10.1109/LSP.2014.2325781.

Taseska, M., Habets, E., 2014. Informed spatial filtering for sound extraction using distributed microphone arrays. IEEE Trans. Audio Speech Lang. Process. 22 (7), 1195-1207. ISSN 2329-9290. http://dx.doi.org/10.1109/TASLP.2014.2327294.

Van Trees, H., 2002. Detection, Estimation, and Modulation Theory; Part IV: Optimum Array Processing. Wiley, West Sussex, UK.

Van Veen, B., Buckley, K., 1988. Beamforming: a versatile approach to spatial filtering. IEEE ASSP Mag. 5 (2), 4-24. ISSN 0740-7467. http://dx.doi.org/10.1109/53.665.

Warsitz, E., Haeb-Umbach, R., 2007. Blind acoustic beamforming based on generalized eigenvalue decomposition. IEEE Transa. Audio Speech Lang. Process. 15 (5), 1529-1539. http://dx.doi.org/10.1109/TASL.2007.898454. URL http://nt.uni-paderborn.de/public/pubs/2007/WaHa07.pdf.

Wolf, M., Nadeu, C., 2013. Channel selection using n-best hypothesis for multi-microphone ASR. In: Proc. INTERSPEECH, pp. 3507-3511.

Wolf, M., Nadeu, C., 2014. Channel selection measures for multi-microphone speech recognition. Speech Commun. 57, 170-180. ISSN 0167-6393. http://dx.doi.org/10.1016/j.specom.2013.09.015. URL http://dx.doi.org/10.1016/j.specom.2013.09.015.

Wölfel, M., Fügen, C., Ikbal, S., McDonough, J., 2006. Multi-source far-distance microphone selection and combination for automatic transcription of lectures. In: Proc. INTERSPEECH.

Wölfel, M., McDonough, J., 2009. Distant Speech Recognition. Wiley, West Sussex, UK.

Yu, D., Deng, L., 2011. Deep learning and its applications to signal and information processing. In: IEEE Signal Processing Magazine, vol. 28, pp. 145-154.

Yu, D., Deng, L., 2014. Automatic Speech Recognition—A Deep Learning Approach. Springer, New York.

Zelinski, R., 1988. A microphone array with adaptive post-filtering for noise reduction in reverberant rooms. In: Proc. of IEEE International Conference on Acoustics, Speech and Signal Processing (ICASSP), pp. 2578-2581 vol.5.

Zhang, Z., Liu, Z., M., S, Acero, A., Deng, L., J., D., et al., 2004. Multi-sensory microphones for robust speech detection, enhancement and recognition. In: Proc. International Conference on Acoustics, Speech and Signal Processing (ICASSP).

Zheng, Y., Liu, Z., Zhang, Z., Sinclair, M., Droppo, J., Deng, L., et al., 2003. Air- and bone-conductive integrated microphones for robust speech detection and enhancement. In: IEEE Workshop on Automatic Speech Recognition and Understanding.

Summary and future directions

11

CHAPTER OUTLINE

In this book, we first presented fundamental concepts, components, models, and methods of automatic speech recognition (ASR). This introductory material included Gaussian mixture models (GMM), hidden Markov models (HMM) and variants, and the more recent deep neural networks (DNN) and related deep learning methods. We then provided background information about the robust ASR problem based on the GMM-HMM and DNN speech models. One most important concept in the robust ASR problem is acoustic distortion of speech signals and features. The effects of such acoustic distortion on both GMM-HMM and DNN models of speech are elaborated in a mathematically rigorous manner. To offer insight into the distinct capabilities of a wide range of noise-robust ASR techniques combating the acoustic distortion, we use a taxonomy-oriented approach for single-microphone nonreverberant speech recognition. The taxonomy adopted is based on five key attributes—feature vs. model domain processing, explicit vs. implicit distortion modeling, use of prior knowledge about distortion or otherwise, deterministic vs. uncertain processing, and joint vs. disjoint training. These attributes are used to organize the literature on the topic spanning over more than three decades and to demonstrate the commonalities and differences among the plethora of robust ASR methods described in this book. Each of the five attributes constitutes a separate chapter in the book. Then after the noise-robust techniques for single-microphone nonreverberant ASR are comprehensively discussed, we include two chapters, covering reverberant ASR and multi-channel processing for noise-robust ASR, respectively.

In this final chapter, we provide a summary of the book from a holistic view based on two distinctive eras of research, and point out and analyze future directions for noise-robust ASR.

Robust Automatic Speech Recognition. http://dx.doi.org/10.1016/B978-0-12-802398-3.00011-8

11.1 ROBUST METHODS IN THE ERA OF GMM

We here summarize the single-microphone methods developed for GMMs as the speech model in Table 11.1 using the five distinct attributes discussed in this book. Note that the column with the heading "explicit modeling" refers to the use of an explicit model for the physical relation between clean and distorted speech. As such, signal processing methods, such as PLP and RASTA, while having auditory modeling inside, are not classified as explicit distortion modeling. CMN is considered to be an explicit modeling method because it can remove the convolutive channel effect while CMVN is considered to be a representative of implicit modeling because cepstral variance normalization does not explicitly address any distortion. We classify observation uncertainty and front-end uncertainty decoding as hybrid (domain) methods in Table 11.1 because the uncertainty is obtained in the feature space and is then passed to the back-end recognizer by modifying the model covariance with a bias.

Note that some methods were proposed a long time ago, and they were revived when more advanced technologies were adopted. As an example, VTS was first proposed in 1996 (Moreno, 1996) for both model adaptation and feature enhancement. But VTS has only quite recently demonstrated an advantage with the advanced online re-estimation for all the distortion parameters (Li et al., 2007, 2009). Another example is the famous Wiener filter, proposed as early as 1979 (Lim and Oppenheim, 1979) to improve the performance on noisy speech. Only after some 20 years, in 2002, two-stage Mel-warped Wiener filtering was proposed to boost the performance of Wiener filtering in several key aspects, and has become the main component of the ETSI advanced front-end. Furthermore, ANN-HMM hybrid systems (Bourlard and Morgan, 1994) were studied in the 1990s, and again only after 20 years, they have been reformulated into deep architectures with new learning algorithms to achieve much greater successes in ASR and in noise robustness in particular (Dahl et al., 2011, 2012; Deng and Yu, 2014; Hinton et al., 2012; Yu et al., 2010). Hence, understanding current well-established technologies is important for providing a foundation for further technology development.

In the early years of noise-robust ASR research, the focus was mostly on feature-domain methods due to their efficient runtime implementation. Runtime efficiency is always a factor when it comes to the deployment of noise-robustness technologies. A good example is CMLLR which can be effectively realized in the feature space with very small cost although it can also be implemented in the model space with a much larger cost by transforming all of the acoustic model parameters instead of very limited ones in the feature space (e.g., a single vector per frame). From PLP to RASTA, and to PNCC, the research on auditory features has been active to extract key insights from human speech perception and production knowledge in order to benefit robust ASR. However, since there is no universally accepted auditory theory for robust speech recognition, it is sometimes very hard to set the right parameter values in auditory methods. Some parameters can be learned from data (Chiu et al., 2010), but this may not always be the case. Although the auditory-based

Table 11.1 Representative Methods Originally Proposed for GMMs, Arranged Alphabetically in Terms of the Names of the Methods

Methods	Time of Publication	Model vs. Feature Domain	Explicit vs. Implicit Distortion Modeling	Use Prior Knowledge about Distortion or Not	Deterministic vs. Uncertainty Processing	Joint vs. Disjoint Training
Acoustic factorization for GMM (Gales, 2001; Seltzer and Acero, 2011)	2001	Model	Both	Use	deterministic	disjoint
ANN-HMM hybrid systems (Bourlard and Morgan, 1994)	1994	Feature	Implicit	Not use	Deterministic	Disjoint
Bayesian predictive classification (BPC) (Huo et al., 1997)	1997	Model	Implicit	Not use	Uncertainty	Disjoint
Bottle-neck feature (Grézl et al., 2007)	2007	Feature	Implicit	Not use	Deterministic	Disjoint
Cepstral mean normalization (CMN) (Atal, 1974)	1974	Feature	Explicit	Not use	Deterministic	Disjoint
Cepstral mean and variance normalization (CMVN) (Viikki et al., 1998)	1998	Feature	implicit	Not use	Deterministic	Disjoint
Constrained MLLR (CMLLR) (Gales, 1998)	1998	Both	Implicit	Not use	Deterministic	Disjoint
Discriminative classifiers with adaptive kernels (Gales and Flego, 2010)	2010	Model	Implicit	Not use	Deterministic	Disjoint
Empirical cepstral compensation (Acero and Stern, 1990; Stern et al., 1996)	1990	Feature	Implicit	Use	Deterministic	Disjoint
Exemplar-based reconstruction use non-negative matrix factorization (Gemmeke and Virtanen, 2010; Raj et al., 2010)	2010	Feature	Explicit	Use	Deterministic	Disjoint
Eigenvoice (Kuhn et al., 2000)	2000	Model	Implicit	Use	Deterministic	Disjoint
ETSI advanced front-end (AFE) (ETSI, 2002)	2002	Feature	Explicit	Not use	Deterministic	Disjoint
Feature space noise adaptive training (NAT) (Deng et al., 2000)	2000	Feature	Implicit	Both	Deterministic	Joint
Front-end uncertainty decoding (Droppo et al., 2002; Liao and Gales, 2005)	2002	Hybrid	Implicit	Use	Uncertainty	Disjoint

Continued

Table 11.1 Representative Methods Originally Proposed for GMMs, Arranged Alphabetically in Terms of the Names of the Methods—cont'd

Methods	Time of Publication	Model vs. Feature Domain	Explicit vs. Implicit Distortion Modeling	Use Prior Knowledge about Distortion or Not	Deterministic vs. Uncertainty Processing	Joint vs. Disjoint Training
Histogram equalization method (HEQ) (Molau et al., 2003)	2003	Feature	Implicit	Use	Deterministic	disjoint
Irrelevant variability normalization (Hu and Huo, 2007; Wu and Huo, 2002)	2002	Both	Both	Both	Deterministic	Joint
Joint adaptive training (JAT) (Liao and Gales, 2007)	2007	Hybrid	Explicit	Not use	Uncertainty	Joint
Joint uncertainty decoding (JUD) (Liao and Gales, 2005)	2005	Hybrid	Explicit	Not use	Uncertainty	Disjoint
Mel-warped Wiener filtering (Li et al., 2004; Macho et al., 2002)	2002	Feature	Explicit	Not use	Deterministic	Disjoint
Maximum likelihood linear regression (MLLR) (Leggetter and Woodland, 1995)	1995	Model	Implicit	Not use	Deterministic	Disjoint
Missing feature (Cooke et al., 1994, 2001; Raj et al., 2004)	1994	Feature	Implicit	Not use	Uncertainty	Disjoint
Multi-style training (Lippmann et al., 1987)	1987	Model	Implicit	Use	Deterministic	Disjoint
Observation uncertainty (Arrowood and Clements, 2002; Deng et al., 2002a; Stouten et al., 2006)	2002	Hybrid	Implicit	Both	Uncertainty	Disjoint
Parallel model combination (PMC) (Gales, 1995)	1995	Model	Explicit	Not use	Deterministic	Disjoint
Perceptual linear prediction (PLP) (Hermansky et al., 1985)	1985	Feature	Implicit	Not use	Deterministic	Disjoint

Method	Year	Feature/Model	Implicit/Explicit	Use/Not use		Disjoint/Joint
Power-normalized cepstral coefficients (PNCC) (Kim and Stern, 2010)	2010	Feature	Implicit	Not use	Deterministic	Disjoint
Relative spectral processing (RASTA) (Hermansky et al., 1991; Morgan and Hermansky, 1992)	1991	Feature	Implicit	Not use	Deterministic	Disjoint
Speaker adaptive training (SAT) (Anastasakos et al., 1996)	1996	Model	Implicit	Not use	Deterministic	Joint
Spectral subtraction (Boll, 1979)	1979	Feature	Explicit	Not use	Deterministic	Disjoint
Stereo piecewise linear compensation for environment (SPLICE) (Deng et al., 2001, 2000, 2003, 2002b)	2000	Feature	Implicit	Use	Deterministic	Disjoint
TANDEM (Hermansky et al., 2000)	2000	Feature	Implicit	Not use	Deterministic	Disjoint
TempoRAL Pattern (TRAP) processing (Hermansky and Sharma, 1998)	1998	Feature	Implicit	Not use	Deterministic	Disjoint
Unscented transform (UT) (Hu and Huo, 2006; Li et al., 2010)	2006	Model	Explicit	Not use	Deterministic	Disjoint
Variable-parameter HMM (VPHMM) (Cui and Gong, 2003)	2003	Model	Implicit	Use	Deterministic	Disjoint
Vector Taylor series (VTS) model adaptation (Moreno, 1996; Acero et al., 2000; Li et al., 2009)	1996	Model	Explicit	Not use	Deterministic	Disjoint
VTS feature enhancement (Moreno, 1996; Stouten et al., 2003; Li et al., 2012)	1996	Feature	Explicit	Not use	Deterministic	Disjoint
VTS-JUD (Xu et al., 2009, 2011)	2009	Model	Explicit	Not use	Deterministic	Disjoint
VTS-NAT (Kalinli et al., 2009, 2010)	2009	Model	Explicit	Not use	Deterministic	Joint
Wiener filter (Lim and Oppenheim, 1979)	1979	Feature	Explicit	Not use	Deterministic	Disjoint

features can usually achieve better performance than MFCCs, they have a much more complicated generation process which sometimes prevents them from widely being used together with some noise-robustness technologies. In order to further move this area forward, more insightful and relevant knowledge should be explored with the consideration of the aforementioned factors. Feature normalization methods come with very low cost and hence are widely used. But they address noise-robustness problems in an implicit way. In contrast, spectral subtraction, Wiener filtering, and VTS feature enhancement use explicit distortion modeling to remove noise, and are more effective. Note that most feature-domain methods are decoupled from the ASR objective function, hence they may not perform as well as model-domain methods. In contrast, while typically achieving higher accuracy than feature-domain methods, model-domain methods usually incur significantly larger computational costs. With increasing computational power, research on model-domain methods is expected to become increasingly active.

Many of the model and feature domain methods use explicit distortion models to describe the physical relationship between clean and distorted speech. Because the physical constraints are explicitly represented in the models, the explicit distortion modeling methods require only a relatively small number of distortion parameters to be estimated. In contrast to the general-purpose techniques, they also exhibit high performance due to the explicit exploitation of the distorted speech generation process. One of the most important explicit distortion modeling techniques is VTS model adaptation (Moreno, 1996; Acero et al., 2000; Li et al., 2009), which has been discussed in detail in this book. However, all of the model adaptation methods using such explicit distortion models rely on validity of the physical constraints expressed by the particular features used. Even with a simple feature normalization technology such as CMN, the distortion model such as in Equation 3.14 is no longer valid. As a result, model adaptation using explicit distortion modeling cannot be easily combined with all of the feature postprocessing technologies, such as CMN, HLDA, fMPE, etc. One solution is to use VTS feature enhancement, which still utilizes explicit distortion modeling (Moreno, 1996; Stouten et al., 2003; Li et al., 2012) and can also be combined with other feature postprocessing technologies, despite the small accuracy gap between VTS feature enhancement and VTS model adaptation (Li et al., 2012). Another advantage of VTS feature enhancement is that it can reduce the computational cost of VTS model adaptation, which is always an important concern in ASR system deployment. JUD (Liao and Gales, 2006), PCMLLR (Gales and van Dalen, 2007), and VTS-JUD (Xu et al., 2009) have been developed also addressing such concerns. As shown in Chapter 6, better distortion modeling results in better algorithm performance, but also incurs a large computation cost (e.g., UT; Hu and Huo 2006; Li et al. 2010). While most adaptation is a one-to-one mapping between the clean Gaussian and the distorted Gaussian due to easy implementation, recent work has appeared covering distributional mappings between the GMMs (van Dalen and Gales, 2011). In conclusion, research in explicit distortion modeling is expected to grow, and an important direction will be how to combine better modeling with runtime efficiency and how to make it capable of working with other feature processing methods.

Without explicit distortion modeling, it is very difficult to predict the impact of noise on clean speech during testing if the acoustic model is trained only from clean speech. The more prior knowledge of that impact we have, the better we can recognize the corrupted speech during testing. The methods utilizing prior knowledge about distortion discussed in this book are motivated by such reasoning. Methods like SPLICE learn the environment-dependent mapping from corrupted speech to clean speech. The extreme case is exemplar-based reconstruction with NMF, which restores cleaned speech by constructing the noisy speech with clean speech and noise exemplars and by keeping only the clean speech exemplars. However, the main challenge in the future direction is that methods utilizing prior knowledge need to perform also well for unseen environments. One solution is to use online model combination, but a better way is to use technologies such as VPHMM to extrapolate model parameters with learned functions. We expect more research in this direction in the future.

Since neither feature enhancement nor model adaptation is perfect, there always exists uncertainty in feature or model space, and uncertainty processing has been designed to address this issue. The initial study in the model space modified the decision rule to use either the minimax rule or the BPC rule. Although the mathematics is well grounded, the computational cost is very large and it is difficult to define the model neighborhood. Research was subsequently switched to the feature space, resulting in the methods exploiting observation uncertainty and JUD. The latter also serves as an excellent approximation to VTS with much lower computational cost. Future research in uncertainty processing is expected to focus on the feature space, and on combining the technique with advances in other areas such as explicit distortion modeling.

Joint training is a good way to obtain canonical acoustic models. Feature-space NAT is a common practice that is now widely used while model-space NAT is much harder to develop and deploy partly because of the difficulty in finding closed-form solutions in model learning and because of the computational complexity. Despite the difficulty, joint training is more promising in the long run because it removes the irrelevant variability to phonetic classification during training, and multi-style training data is easier to obtain than clean training data in real-world applications. Better integrated algorithm design, improved joint optimization of model and transform parameters, and clever use of metadata as labels for the otherwise hidden "distortion condition" variables are all promising future research directions.

By comparing the methods in Table 11.1, which provides a comprehensive summary listing all major methods in GMM-based noise-robust ASR up to this date, we clearly see the advantages of explicit distortion modeling, using prior knowledge, uncertainty, and joint training methods over their counterparts. When developing a noise-robust ASR method, their combinations should be explored. For real-world applications, there are also some other factors to consider. For example, there is always a trade-off between high accuracy and low computational cost. Special attention should also be paid to nonstationary noise. Some methods such as NMF can handle nonstationary noise very well because the noise exemplars are extracted from a large dictionary which can consist of different types of noise. The effectiveness of many frame-by-frame feature compensation methods (e.g., spectral subtraction and Wiener filtering) depends on whether the noise tracking module is good at tracking nonstationary noise. Some methods such as the standard VTS may not directly handle nonstationary noise well because they assume the noise is Gaussian distributed. This problem can be solved by relaxing the assumption using a time-dependent noise estimate (Yoshioka and Nakatani, 2013). As in ETSI AFE, it is also important to combine several methods together to reach the best system performance.

11.2 ROBUST METHODS IN THE ERA OF DNN

The recent acoustic modeling technology, CD-DNN-HMM (Yu and Deng, 2014), brings new challenges to conventional noise-robustness technologies. As shown in Section 3.4, the DNN provides a layer-by-layer feature extraction strategy that automatically derives powerful noise-resistant features from primitive raw data for senone classification, resulting in good noise-invariance property although there is a recent study showing that further noise-robustness can be achieved by imposing manifold-based locality preserving constraints on the outputs of the DNN (Tomar and Rose, 2014). The DNN enjoys much better noise-invariance property than the GMM. In Seltzer et al. (2013), the CD-DNN-HMM trained with multi-style data easily matches the state-of-the-art performance obtained with complicated conventional noise-robustness technology on GMM systems (Wang and Gales, 2012). The noise, channel, and speaker factors may already be well normalized by the complex nonlinear transform inside the DNN. However, this does not mean that the robustness technologies are not necessary when used together with CD-DNN-HMMs. There are still plenty of methods that can help to improve the robustness of DNNs as presented in this book and the representative methods are listed in Table 11.2. Given that DNNs have only been used for acoustic modeling a few years ago, the methods list in Table 11.2 is pretty short, compared to the long GMM methods list in Table 11.1.

It is interesting to see that almost all the DNN methods in Table 11.2 can find their corresponding GMM methods with similar concepts, as shown in Table 11.3. All these robustness methods for DNNs are not developed from scratch. Instead, they are inspired by and follow the design concepts of the previous successful GMM methods. Therefore, it is still very necessary to study the classic robustness methods in the GMM era which have been built on top of concepts critical to the final success of robustness. Then, we can move from GMMs to DNNs using those critical concepts, in a similar way as the methods in Table 11.3. We are expecting more and more new robustness methods developed in this way for DNNs as time elapses.

Most methods in Table 11.3 are model-based methods. This is because most feature processing methods have been developed without dependency on the underlying acoustic models. Therefore, some of these methods can still be used for the

Table 11.2 Representative Methods Originally Proposed for DNNs, Arranged Alphabetically

Methods	Time of Publication	Model vs. Feature Domain	Explicit vs. Implicit Distortion Modeling	Use Prior Knowledge about Distortion or Not	Deterministic vs. Uncertainty Processing	Joint vs. Disjoint Training
Acoustic factorization for DNN (Li et al., 2014b)	2014	Model	Explicit	Not use	Deterministic	Disjoint
Deep unfolding (Hershey et al., 2014)	2014	Feature	Explicit	Use	Deterministic	Disjoint
DNN for noise removal (Lu et al., 2013)	2013	Feature	Implicit	Use	Deterministic	Disjoint
Feature discriminative linear regression (fDLR) (Seide et al., 2011)	2011	Feature	Implicit	Not use	Deterministic	Disjoint
Hidden activation adaptation (Zhao et al., 2015)	2015	Model	Implicit	Not use	Deterministic	Disjoint
Joint adaptive training for DNN	2014	Model	Implicit	Not use	Deterministic	Joint
Joint front-end and DNN model training (Narayanan and Wang, 2014)	2014	Both	Implicit	Not use	Deterministic	Joint
Learning hidden unit contribution (LHUC) (Swietojanski and Renals, 2014)	2014	Model	Implicit	Not use	Deterministic	Disjoint
Long, deep, and wide neural network (Li et al., 2014a)	2014	Model	Implicit	Not use	Deterministic	Disjoint
Online DNN model combination	2014	Model	Implicit	Use	Deterministic	Disjoint
RNN for noise removal (Maas et al., 2012)	2012	Feature	Implicit	Use	Deterministic	Disjoint
SVD bottleneck adaptation (Xue et al., 2014)	2014	Model	Implicit	Not use	Deterministic	Disjoint
Transformed feature normalization for DNN (Li and Sim, 2013)	2013	Model	Explicit	Not use	Deterministic	Disjoint
Uncertainty propagation through multi-layer perceptrons (Astudillo and da Silva Neto, 2011)	2011	Hybrid	Implicit	Not use	Uncertainty	Disjoint
Variable-component DNN (Zhao et al., 2014)	2014	Model	Implicit	Use	Deterministic	Disjoint

Table 11.3 The Counterparts of GMM-based Robustness Methods for DNN-based Robustness Methods

DNN Methods	Corresponding GMM Methods
Acoustic factorization for DNN	Acoustic factorization for GMM
Deep unfolding	Discriminative NMF
DNN for noise removal	SPLICE
Feature discriminative linear regression fDLR	CMLLR
Hidden activation adaptation	Diagonal MLLR
Joint adaptive training for DNN	Noise adaptive training
Joint front-end and DNN model training	fMPE
Learning hidden unit contribution	Diagonal MLLR
Long, deep, and wide neural network	TRAP
Online DNN model combination	Online GMM model combination
RNN for noise removal	SPLICE
SVD bottleneck adaptation	MLLR
Transformed feature normalization for DNN	VTS
Variable-component DNN	VPHMM

front-end of DNNs (e.g., feature VTS in Li and Sim (2013)). The only exception is the DNN/RNN-based noise removal method, which uses stereo data to learn the mapping from the noisy feature to the clean feature. In the GMM era, piecewise linear methods such as empirical cepstral compensation and SPLICE are used to do the mapping. With its powerful layer-by-layer nonlinear processing, the DNN can provide a much better mapping function. While a DNN uses a fixed-length context window, an RNN naturally handles the temporal information of speech sequence with its recurrent units. Furthermore, with the introduction of LSTM and BLSTM units, the original gradient vanishing or explosion training issues of RNNs can be effectively solved. We expect RNN-based noise removal with LSTM and BLSTM units to become more popular as an advanced feature enhancement method.

The general model adaptation techniques developed in the GMM era can easily find their counterparts in the DNN era, with examples including SVD bottleneck adaptation, learning hidden unit contribution, hidden activation adaptation, and feature discriminative linear regression, as shown in Table 11.3. All these methods add a single or a set of transform matrices to the original DNN structure. Because DNNs usually have very large matrices, methods are designed to work on the small SVD layer as in the case of SVD bottleneck adaptation, or even further, on the use of a diagonal transform which works on hidden unit as in the case of learning hidden unit contribution and hidden activation adaptation. With a small number of parameters to adapt, these methods satisfy the requirement of fast adaptation of DNN with limited amounts of data.

Variable-component DNN (VCDNN) outlined in Table 11.3 is a natural extension of VPHMM by constructing polynomial functions for any component instead of only model parameters in a DNN. With the polynomial functions, the model can be instantiated even in unseen cases by extrapolation, thus enjoying good generalization abilities. Generalization to unseen test cases is a very important future research direction because as shown in Section 3.4 DNNs can have excellent distortion invariance on observed data but still fail to generate good invariance on unseen data. Prediction with polynomial functions in VCDNN provides one solution to this problem, and more powerful functions may be expected in future work by DNN and ASR researchers to further improve the generalization ability of DNN-based noise-robust ASR systems.

11.3 MULTI-CHANNEL INPUT AND ROBUSTNESS TO REVERBERATION

The discussion in this chapter has so far been concerned with robustness toward additive environmental noise. Reverberation is a convolutional distortion of the speech signal which is quite different from additive noise: it leads to a dispersion of speech energy over time, resulting in a highly nonstationary distortion. As a consequence, approaches to mitigate the effect of reverberation are quite different than those mitigating additive noise, yet the same taxonomy that was used throughout this book to categorize noise-robustness techniques can also be applied to categorize them: we can distinguish between signal, feature, and model domain-based approaches. There are compensation techniques which use prior knowledge, for example, by learning the mapping between clean and reverberant speech from stereo training data. There exist physically motivated explicit distortion models and even uncertainty processing has been proposed for reverberant ASR.

When discussing enhancement methods for reverberated speech and in particular when comparing a GMM to a DNN back-end, the following considerations should be made.

First, the physical process of reverberation is quite well understood. Techniques exist to estimate the nonreverberant signal or features from the observed reverberant ones using an explicit distortion model. Nevertheless, due to the volatile nature of acoustic impulse responses, the estimation of the reverberation and the clean-up of the signal or the features are not easy tasks. Explicit distortion models require only a relatively small number of parameters to be estimated, which can often be done alongside recognition. No extra training data is required. In contrast, data-driven techniques, which do not rely on an explicit distortion model, require the availability of clean and distorted training data to learn the mapping of the one onto the other. While this is a severe impediment to their usability in case of additive noise, where noise-free training data may not be available, it does not seem to be equally problematic in the case of reverberation. Most of the currently available databases have been collected with close talking microphones, thus exhibiting no or low reverberation only. If this comprises the "clean data," the reverberant ones can be easily generated by convolving the clean data either with artificially generated or measured acoustic impulse responses. Thus stereo data to learn the mapping from clean to distorted or vice versa in a data-driven manner is readily available. This makes data-driven enhancement techniques, such as the use of a denoising autoencoder for feature enhancement an attractive alternative.

Second, in the era of DNN, it is common practice to present to the neural network a large temporal window of data, covering several hundred milliseconds. With this wide window, the network is able to see and thus learn the temporal dispersion of speech energy caused by reverberation. In contrast, GMM-based systems operate on frames of the size of a few tens of milliseconds, and the smearing of speech energy is much harder to model.

The third consideration particularly pertains to multi-channel processing. The DNNs used for ASR today operate on speech representations which are devoid of phase information. However, the phase differences between the signals of a microphone array are the primary carrier of spatial information. If the source signal is to be separated from noise or reverberant components impinging on the microphone from other directions than the target signal, the signal processing must exploit the phase differences to extract the source signal. Unless DNNs are developed which are able to exploit such phase information, signal processing techniques, like beamforming will have its right to exist and will deliver a performance advantage.

And, finally, a fourth consideration: with the ubiquity of mobile devices equipped with microphones (smartphones, tablet computers, laptops, etc.) more and more acoustic sensors, which are spatially distributed, will be available to capture an acoustic scene. In such scenarios, multi-channel processing, which is able to exploit the spatial diversity of sound sources to extract the signal of interest, will become even more important.

11.4 EPILOGUE

In this chapter, we have made some key observations regarding the long history of noise-robust ASR research. At a high level, many noise-robust techniques for ASR can be divided into those developed mainly for generative GMM-HMM models of speech and those for discriminative DNNs. Interestingly, these two classes of techniques share commonalities; for example, the powerful concept of noise adaptive training developed originally for GMM-based ASR systems is applicable for DNN-based systems, and the idea of exploiting stereo clean-noisy speech data, which is the basis of the SPLICE technique, can also be effectively applied to noise-robust ASR for DNN or RNN-based systems.

From the historical perspective, when context-dependent DNN-HMM systems showed its effectiveness in 2010-2011, one of the concerns back then was the lack of effective adaptation techniques. This is especially so since DNN systems have much more parameters to train than any of the earlier ASR systems. Various studies since 2011 demonstrated that adaptation of CD-DNN-HMM systems is effective. Most of these studies are for the purpose of speaker adaption. It will be interesting to examine whether similar techniques are equally effective for noise adaptation. One recent trend in fast adaptation for noise-robust ASR is acoustic factorization as part of the DNN design, where different acoustic factors are realized as factor vectors which are then connected to the DNN. Because the "factor" vectors usually are in a low-dimensional space, the number of parameters related to acoustic factors is small. More research is needed and is expected in the near future in the area of fast adaptation for DNN with limited amounts of data.

It is much easier to perform joint training in DNNs than in GMMs because all components of a DNN can be considered as part of a big ensemble DNN. Therefore, the standard back propagation update can be applied to optimize those components in the big DNN. Compared to the GMM counterparts such as fMPE or noise adaptive training, the joint front-end training and joint adaptive training for the DNN presented in this book have a much simpler optimization process. We believe this will also be a trend for future technology development; that is, to embed all the components of interest into a big network and to optimize them using the same criterion. The long, deep, and wide neural network is another huge ensemble network which addresses the robustness problem in a divide and conquer way. As the computational machines are becoming more and more powerful, it is possible to afford building such a huge ensemble network.

While achieving high successes in the GMM era, explicit distortion modeling techniques discussed earlier in this book have demonstrated much lower successes in the DNN era. This can be attributed to no simple and analytic relationship between the values of weight matrices and acoustic feature vectors. This contrasts sharply to the clear relations between the GMM mean values and acoustic feature vectors. This difference is closely connected to the use of the GMM as a generative model for charactering speech's statistical properties and the use of the DNN as a discriminative model for directly classifying speech classes. As such, the relations among the GMM model parameters for clean and noisy speech, such as Equation 6.20, are easily derived and made useful for the GMM-based ASR systems but not for the DNN-based systems. However, VTS with explicit distortion modeling (Li and Sim, 2013) and DOLPHIN (dominance-based locational and power-spectral characteristics integration) (Delcroix et al., 2013) have shown to be effective in improving ASR robustness when the recognizer's back-end is a DNN. One possible reason for the improvement is that the nonlinear distortion model used in VTS and the spatial information used in DOLPHIN are not available to the DNN. Another example is shown in Swietojanski et al. (2014) where delicate pooling structures are used to incorporate inputs from multiple channels together with DNNs and CNNs, but it cannot outperform the model utilizing explicit beamforming knowledge. Therefore, one potential direction to improve robustness of DNNs to noise is to incorporate the information not explicitly exploited in DNN training. One related recent study is deep unfolding (Hershey et al., 2014), which takes advantage of model-based approaches by unfolding the inference iterations as layers in a DNN.

The most important lesson learned through our past several years of investigation into why the DNN-based approach for ASR is so much better than the GMM-based counterpart is the distributed representation inherent in the DNN model that is missing in the GMM grounded on the localist representation. Localist and distributed representations are important concepts in cognitive science as two distinct styles of data representation. For the former, each neuron represents a single concept on a stand-alone basis. That is, localist units have their own meaning and interpretation. The latter pertains to an internal representation of concepts in such a way that they are modeled as being explained by the interactions of many hidden factors. A particular factor learned from configurations of other factors can often generalize well to new configurations, not so in the localist representation. Distributed representations, based on vectors consisting of many elements or units, naturally occur in the "connectionist" DNN, where a concept is represented by a pattern of activity across a number of units and where at the same time a unit typically contributes to many concepts. One key advantage of such many-to-many correspondence is that they provide *robustness* in representing the internal structure of the data in terms of graceful degradation and damage resistance. Such robustness is enabled by redundant storage of information. Another advantage is that they facilitate automatic generalization of concepts and relations, thus enabling reasoning abilities. Further, the distributed representation allows similar vectors to be associated with similar concepts and it allows efficient use of representational resources.

The above strengths of the distributed representation in the DNN make ASR, especially noise-robust ASR, highly effective. For example, the acoustic distortion that baffles high-accuracy ASR has the compositional properties discussed above, naturally permitting one distortion factor to be learned from configurations of other distortion factors. This accounts for why the factorization method works so well for DNN-based system for noise-robust ASR as well as for speaker-adaptive ASR. However, the attractive properties of distributed representations discussed above come with a set of weaknesses. These include non-obviousness in interpreting the representations, difficulties with incorporating explicit knowledge, and inconvenience in representing variable-length sequences. On the other hand, local representation, which is typically adopted for probabilistic generative models, has advantages of explicitness and ease of use. That is, the explicit representation of the components of a task is simple and the design of representational schemes for structured objects is easy. The most striking example of the above contrast for noise-robust ASR is the straightforward application of acoustic distortion knowledge to GMM-based ASR systems, while the same knowledge is hard to be incorporated into DNN-based systems. How to effectively embed speech dynamic and distortion knowledge so naturally expressed in generative models into deep learning-based discriminative models for noise-robust ASR is a highly promising but challenging research direction.

REFERENCES

Acero, A., 1993. Acoustical and Environmental Robustness in Automatic Speech Recognition. Cambridge University Press, Cambridge, UK.

Acero, A., Deng, L., Kristjansson, T., Zhang, J., 2000. HMM adaptation using vector Taylor series for noisy speech recognition. In: Proc. International Conference on Spoken Language Processing (ICSLP), pp. 869-872.

Acero, A., Stern, R., 1990. Environmental robustness in automatic speech recognition. In: Proc. International Conference on Acoustics, Speech and Signal Processing (ICASSP), vol. 2, pp. 849-852.

Anastasakos, T., McDonough, J., Schwartz, R., Makhoul, J., 1996. A compact model for speaker-adaptive training. In: Proc. International Conference on Spoken Language Processing (ICSLP), vol. 2, pp. 1137-1140.

Arrowood, J.A., Clements, M.A., 2002. Using observation uncertainty in HMM decoding. In: Proc. Interspeech, pp. 1561-1564.

Astudillo, R.F., da Silva Neto, J.P., 2011. Propagation of uncertainty through multilayer perceptrons for robust automatic speech recognition. In: INTERSPEECH, pp. 461-464.

Atal, B., 1974. Effectiveness of linear prediction characteristics of the speech wave for automatic speaker identification and verification. J. Acoust. Soc. Amer. 55, 1304-1312.

Boll, S.F., 1979. Suppression of acoustic noise in speech using spectral subtraction. IEEE Trans. Acoust. Speech Signal Process. 27 (2), 113-120.

Bourlard, H., Morgan, N., 1994. Connectionist speech recognition—A Hybrid approach. Kluwer Academic Press, Boston, MA.

Chiu, Y.H., Raj, B., Stern, R.M., 2010. Learning-based auditory encoding for robust speech recognition. In: Proc. International Conference on Acoustics, Speech and Signal Processing (ICASSP), pp. 4278-4281.

Cooke, M., Green, P.D., Crawford, M., 1994. Handling missing data in speech recognition. In: Proc. International Conference on Spoken Language Processing (ICSLP), pp. 1555-1558.

Cooke, M., Green, P.D., Josifovski, L., Vizinho, A., 2001. Robust automatic speech recognition with missing and unreliable acoustic data. Speech Commun. 34 (3), 267-285.

Cui, X., Gong, Y., 2003. Variable parameter Gaussian mixture hidden Markov modeling for speech recognition. In: Proc. International Conference on Acoustics, Speech and Signal Processing (ICASSP), vol. 1, pp. 12-15.

Dahl, G., Yu, D., Deng, L., Acero, A., 2011. Large vocabulary continuous speech recognition with context-dependent DBN-HMMs. In: Proc. International Conference on Acoustics, Speech and Signal Processing (ICASSP).

Dahl, G.E., Yu, D., Deng, L., Acero, A., 2012. Context-dependent pre-trained deep neural networks for large-vocabulary speech recognition. IEEE Trans. Audio Speech Lang. Process. 20 (1), 30-42.

Delcroix, M., Kubo, Y., Nakatani, T., Nakamura, A., 2013. Is speech enhancement pre-processing still relevant when using deep neural networks for acoustic modeling. In: Proc. Interspeech, pp. 2992-2996.

Deng, L., Acero, A., Jiang, L., Droppo, J., Huang, X.D., 2001. High-performance robust speech recognition using stereo training data. In: Proc. International Conference on Acoustics, Speech and Signal Processing (ICASSP), pp. 301-304.

Deng, L., Acero, A., Plumpe, M., Huang, X., 2000. Large vocabulary speech recognition under adverse acoustic environment. In: Proc. International Conference on Spoken Language Processing (ICSLP), vol. 3, pp. 806-809.

Deng, L., Droppo, J., Acero, A., 2003. Recursive estimation of nonstationary noise using iterative stochastic approximation for robust speech recognition. IEEE Trans. Speech Audio Process. 11, 568-580.

Deng, L., Droppo, J., Acero, A., 2002a. Exploiting variances in robust feature extraction based on a parametric model of speech distortion. In: Proc. Interspeech, pp. 2449-2452.

Deng, L., Wang, K.A., Acero, H.H., Huang, X., 2002b. Distributed speech processing in MiPad's multimodal user interface. IEEE Trans. Audio Speech Lang. Process. 10 (8), 605-619.

Deng, L., Yu, D., 2014. Deep Learning: Methods and Applications. NOW Publishers, Hanover, MA.

Droppo, J., Deng, L., Acero, A., 2002. Uncertainty decoding with SPLICE for noise robust speech recognition. In: Proc. International Conference on Acoustics, Speech and Signal Processing (ICASSP), vol. 1, pp. 57-60.

ETSI., 2002. Speech processing, transmission and quality aspects (STQ); distributed speech recognition; advanced front-end feature extraction algorithm; compression algorithms. ETSI.

Gales, M.J.F., 1995. Model-based techniques for noise robust speech recognition. Ph.D. thesis, University of Cambridge.

Gales, M.J.F., 1998. Maximum likelihood linear transformations for HMM-based speech recognition. Comput. Speech Lang. 12, 75-98.

Gales, M.J.F., 2001. Acoustic factorisation. In: Proc. IEEE Workshop on Automatic Speech Recognition and Understanding (ASRU), pp. 77-80.

Gales, M.J.F., Flego, F., 2010. Discriminative classifiers with adaptive kernels for noise robust speech recognition. Comput. Speech Lang. 24 (4), 648-662.

Gales, M.J.F., van Dalen, R.C., 2007. Predictive linear transforms for noise robust speech recognition. In: Proc. IEEE Workshop on Automatic Speech Recognition and Understanding (ASRU), pp. 59-64.

Gemmeke, J.F., Virtanen, T., 2010. Noise robust exemplar-based connected digit recognition. In: Proc. International Conference on Acoustics, Speech and Signal Processing (ICASSP), pp. 4546-4549.

Grézl, F., Karafiát, M., Kontár, S., Cernocký, J., 2007. Probabilistic and bottle-neck features for LVCSR of meetings. In: Proc. International Conference on Acoustics, Speech and Signal Processing (ICASSP), vol. IV, pp. 757-760.

Hermansky, H., Ellis, D.P.W., Sharma, S., 2000. Tandem connectionist feature extraction for conventional HMM systems. In: Proc. International Conference on Acoustics, Speech and Signal Processing (ICASSP), vol. 3, pp. 1635-1638.

Hermansky, H., Hanson, B.A., Wakita, H., 1985. Perceptually based linear predictive analysis of speech. In: Proc. International Conference on Acoustics, Speech and Signal Processing (ICASSP), vol. I, pp. 509-512.

Hermansky, H., Morgan, N., Bayya, A., Kohn, P., 1991. Compensation for the effect of communication channel in auditory-like analysis of speech (RASTA-PLP). In: Proceedings of European Conference on Speech Technology, pp. 1367-1370.

Hermansky, H., Sharma, S., 1998. TRAPs—classifiers of temporal patterns. In: Proc. International Conference on Spoken Language Processing (ICSLP).

Hershey, J., Le Roux, J., Weninger, F., 2014. Deep unfolding: Model-based inspiration of novel deep architectures. arXiv preprint arXiv:1409.2574.

Hinton, G., Deng, L., Yu, D., Dahl, G.E., Mohamed, A., Jaitly, N., et al., 2012. Deep neural networks for acoustic modeling in speech recognition: The shared views of four research groups. IEEE Signal Process. Mag. 29 (6), 82-97.

Hu, Y., Huo, Q., 2006. An HMM compensation approach using unscented transformation for noisy speech recognition. In: ISCSLP.

Hu, Y., Huo, Q., 2007. Irrelevant variability normalization based HMM training using VTS approximation of an explicit model of environmental distortions. In: Proc. Interspeech, pp. 1042-1045.

Huo, Q., Jiang, H., Lee, C.H., 1997. A Bayesian predictive classification approach to robust speech recognition. In: Proc. International Conference on Acoustics, Speech and Signal Processing (ICASSP), pp. 1547-1550.

Kalinli, O., Seltzer, M.L, Acero, A., 2009. Noise adaptive training using a vector Taylor series approach for noise robust automatic speech recognition. In: Proc. International Conference on Acoustics, Speech and Signal Processing (ICASSP), pp. 3825-3828.

Kalinli, O., Seltzer, M.L., Droppo, J., Acero, A., 2010. Noise adaptive training for robust automatic speech recognition. IEEE Trans. Audio Speech Lang. Process. 18 (8), 1889-1901.

Kim, C., Stern, R.M., 2010. Feature extraction for robust speech recognition based on maximizing the sharpness of the power distribution and on power flooring. In: Proc. International Conference on Acoustics, Speech and Signal Processing (ICASSP), pp. 4574-4577.

Kuhn, R., Junqua, J.C., Nguyen, P., Niedzielski, N., 2000. Rapid speaker adaptation in eigenvoice space. IEEE Trans. Speech Audio Process. 8 (6), 695-707.

Leggetter, C., Woodland, P., 1995. Maximum likelihood linear regression for speaker adaptation of continuous density hidden Markov models. Comput. Speech Lang. 9 (2), 171-185.

Li, B., Sim, K.C., 2013. Noise adaptive front-end normalization based on vector Taylor series for deep neural networks in robust speech recognition. In: Proc. International Conference on Acoustics, Speech and Signal Processing (ICASSP), pp. 7408-7412.

Li, F., Nidadavolu, P., Hermansky, H., 2014a. A long, deep and wide artificial neural net for robust speech recognition in unknown noise. In: Proc. Interspeech.

Li, J., Deng, L., Yu, D., Gong, Y., Acero, A., 2007. High-performance HMM adaptation with joint compensation of additive and convolutive distortions via vector Taylor series. In: Proc. IEEE Workshop on Automatic Speech Recognition and Understanding (ASRU), pp. 65-70.

Li, J., Deng, L., Yu, D., Gong, Y., Acero, A., 2009. A unified framework of HMM adaptation with joint compensation of additive and convolutive distortions. Comput. Speech Lang. 23 (3), 389-405.

Li, J., Huang, J.T., Gong, Y., 2014b. Factorized adaptation for deep neural network. In: Proc. International Conference on Acoustics, Speech and Signal Processing (ICASSP).

Li, J., Liu, B., Wang, R.H., Dai, L., 2004. A complexity reduction of ETSI advanced front-end for DSR. In: Proc. International Conference on Acoustics, Speech and Signal Processing (ICASSP), vol. 1, pp. 61-64.

Li, J., Seltzer, M.L., Gong, Y., 2012. Improvements to VTS feature enhancement. In: Proc. International Conference on Acoustics, Speech and Signal Processing (ICASSP), pp. 4677-4680.

Li, J., Yu, D., Gong, Y., Deng, L., 2010. Unscented transform with online distortion estimation for HMM adaptation. In: Proc. Interspeech, pp. 1660-1663.

Liao, H., Gales, M.J.F., 2005. Joint uncertainty decoding for noise robust speech recognition. In: Proc. Interspeech, pp. 3129-3132.

Liao, H., Gales, M.J.F., 2006. Joint uncertainty decoding for robust large vocabulary speech recognition. University of Cambridge.

Liao, H., Gales, M.J.F., 2007. Adaptive training with joint uncertainty decoding for robust recognition of noisy data. In: Proc. International Conference on Acoustics, Speech and Signal Processing (ICASSP), vol. 4, pp. 389-392.

Lim, J.S., Oppenheim, A.V., 1979. Enhancement and bandwidth compression of noisy speech. Proc. IEEE 67 (12), 1586-1604.

Lippmann, R., Martin, E., Paul, D., 1987. Multi-style training for robust isolated-word speech recognition. In: Proc. International Conference on Acoustics, Speech and Signal Processing (ICASSP), pp. 705-708.

Lu, X., Tsao, Y., Matsuda, S., Hori, C., 2013. Speech enhancement based on deep denoising autoencoder. In: Proc. Interspeech, pp. 436-440.

Maas, A.L., Le, Q.V., O'Neil, T.M., Vinyals, O., Nguyen, P., Ng, A.Y., 2012. Recurrent neural networks for noise reduction in robust ASR. In: Proc. Interspeech, pp. 22-25.

Macho, D., Mauuary, L., Noé, B., Cheng, Y.M., Ealey, D., Jouvet, D, et al., 2002. Evaluation of a noise-robust DSR front-end on Aurora databases. In: Proc. International Conference on Spoken Language Processing (ICSLP), pp. 17-20.

Molau, S., Hilger, F., Ney, H., 2003. Feature space normalization in adverse acoustic conditions. In: Proc. International Conference on Acoustics, Speech and Signal Processing (ICASSP), vol. 1, pp. 656-659.

Moreno, P.J., 1996. Speech recognition in noisy environments. Ph.D. thesis, Carnegie Mellon University.

Morgan, N., Hermansky, H., 1992. RASTA extensions: Robustness to additive and convolutional noise. In: ESCA Workshop Proceedings of Speech Processing in Adverse Conditions, pp. 115-118.

Narayanan, A., Wang, D., 2014. Joint noise adaptive training for robust automatic speech recognition. In: Proc. International Conference on Acoustics, Speech and Signal Processing (ICASSP).

Raj, B., Seltzer, M.L., Stern, R.M., 2004. Reconstruction of missing features for robust speech recognition. Speech Commun. 43 (4), 275-296.

Raj, B., Virtanen, T., Chaudhuri, S., Singh, R., 2010. Non-negative matrix factorization based compensation of music for automatic speech recognition. In: Proc. Interspeech, pp. 717-720.

Seide, F., Li, G., Chen, X., Yu, D., 2011. Feature engineering in context-dependent deep neural networks for conversational speech transcription. In: Proc. IEEE Workshop on Automatic Speech Recognition and Understanding (ASRU), pp. 24-29.

Seltzer, M.L., Acero, A., 2011. Separating speaker and environmental variability using factored transforms. In: Proc. Interspeech, pp. 1097-1100.

Seltzer, M.L., Yu, D., Wang, Y., 2013. An investigation of deep neural networks for noise robust speech recognition. In: Proc. International Conference on Acoustics, Speech and Signal Processing (ICASSP), pp. 7398-7402.

Stern, R., Acero, A., Liu, F.H., Ohshima, Y., 1996. Signal processing for robust speech recognition. In: Lee, C.H., Soong, F.K. Paliwal, K.K. (Eds.), Automatic Speech and

Speaker Recognition: Advanced Topics. Kluwer Academic Publishers, Boston, MA, pp. 357-384.

Stouten, V., Hamme, H.V., Demuynck, K., Wambacq, P., 2003. Robust speech recognition using model-based feature enhancement. In: Proc. European Conference on Speech Communication and Technology (EUROSPEECH), pp. 17-20.

Stouten, V., Hamme, H.V., Wambacq, P., 2006. Model-based feature enhancement with uncertainty decoding for noise robust ASR. Speech Commun. 48 (11), 1502-1514.

Swietojanski, P., Ghoshal, A., Renals, S., 2014. Convolutional neural networks for distant speech recognition. IEEE Signal Process. Lett. 21 (9), 1120-1124. ISSN 1070-9908. http://dx.doi.org/10.1109/LSP.2014.2325781.

Swietojanski, P., Renals, S., 2014. Learning hidden unit contributions for unsupervised speaker adaptation of neural network acoustic models. In: Proc. IEEE Spoken Language Technology Workshop.

Tomar, V., Rose, R.C., 2014. Manifold regularized deep neural networks. In: Proc. Interspeech.

van Dalen, R.C., Gales, M.J.F., 2011. A variational perspective on noise-robust speech recognition. In: Proc. IEEE Workshop on Automatic Speech Recognition and Understanding (ASRU), pp. 125-130.

Viikki, O., Bye, D., Laurila, K., 1998. A recursive feature vector normalization approach for robust speech recognition in noise. In: Proc. International Conference on Acoustics, Speech and Signal Processing (ICASSP), pp. 733-736.

Wang, Y., Gales, M.J.F., 2011. Speaker and noise factorisation on aurora4 task. In: Proc. International Conference on Acoustics, Speech and Signal Processing (ICASSP), pp. 4584-4587.

Wang, Y., Gales, M.J.F., 2012. Speaker and noise factorisation for robust speech recognition. IEEE Trans. Audio Speech Lang. Process. 20 (7), 2149-2158.

Wu, J., Huo, Q., 2002. An environment compensated minimum classification error training approach and its evaluation on Aurora2 database. In: Proc. Interspeech, pp. 453-456.

Xu, H., Gales, M.J.F., Chin, K.K., 2009. Improving joint uncertainty decoding performance by predictive methods for noise robust speech recognition. In: Proc. IEEE Workshop on Automatic Speech Recognition and Understanding (ASRU), pp. 222-227.

Xu, H., Gales, M.J.F., Chin, K.K., 2011. Joint uncertainty decoding with predictive methods for noise robust speech recognition. IEEE Trans. Audio Speech Lang. Process. 19 (6), 1665-1676.

Xue, J., Li, J., Yu, D., Seltzer, M., Gong, Y., 2014. Singular value decomposition based low-footprint speaker adaptation and personalization for deep neural network. In: Proc. International Conference on Acoustics, Speech and Signal Processing (ICASSP), pp. 6359-6363.

Yoshioka, T., Nakatani, T., 2013. Noise model transfer: Novel approach to robustness against nonstationary noise. IEEE Trans. Audio Speech Lang. Process. 21 (10), 2182-2192.

Yu, D., Deng, L., 2014. Automatic Speech Recognition—A Deep Learning Approach. Springer, New York.

Yu, D., Deng, L., Dahl, G., 2010. Roles of pretraining and fine-tuning in context-dependent DBN-HMMs for real-world speech recognition. In: Proc. NIPS Workshop on Deep Learning and Unsupervised Feature Learning.

Zhao, R., Li, J., Gong, Y., 2014. Variable-component deep neural network for robust speech recognition. In: Proc. Interspeech.

Zhao, Y., Li, J., Xue, J., Gong, Y., 2015. Investigating online low-footprint speaker adaptation using generalized linear regression and click-through data. In: Proc. International Conference on Acoustics, Speech and Signal Processing (ICASSP).

Index

Note: Page numbers followed by *f* indicate figures and *t* indicate Tables.

Principal component analysis (PCA), 72-73
Prior knowledge, 58, 261, 267
  deep neural network (DNN), compensation
    with, 130t
  Gaussian mixture models (GMMs),
    compensation with, 129t
  multi-environment data, 116-128
  stereo data, 108-115
Processing domain, 57

R

Real-recording data set (RealData), 229
Rectified linear units (ReLU), 27
Recurrent neural network (RNN), 224
Relative spectral processing (RASTA), 69-70
Relative transfer functions (RTF), 250-253
REVERB. *See* Reverberant voice enhancement
    and recognition benchmark (REVERB)
Reverberant speech recognition
  acoustic impulse response (AIR), 206-211,
    207f
  acoustic model domain approaches, 225-228
  on automatic speech recognition (ASR)
    performance, 213-214, 213f
  CHiME-2 challenge, 231
  chronological order, 232t
  data-driven enhancement, 221-225, 223f
  in different domains, 211-213
  feature domain approaches, 218-225
  feature normalization, 219
  linear filtering approaches, 214-217
  magnitude or power spectrum enhancement,
    217-218
  model based feature enhancement, 219-221
  REVERB challenge, 228-231
  reverberation robust features, 218-219
Reverberant voice enhancement and recognition
    benchmark (REVERB), 228-231
Robust automatic speech recognition,
    compensation
  acoustic distortion, 58
  deterministic *vs.* uncertainty processing, 59
  disjoint *vs.* joint model training, 60
  explicit *vs.* implicit distortion modeling, 59
  feature domain *vs.* model domain, 57-58, 57f
Robust methods
  deep neural network (DNN), 268-271, 269t,
    270t
  Gaussian mixture models (GMM), 262-268,
    263t, 270t
Robustness, 274

to noisy environments, 2, 3f
  software algorithmic processing, 2
Robust speech recognition, 2
  acoustic environments, 43-46, 47f
  automatic speech recognition (ASR), 57-60
  DNN modeling, 50-55, 51f, 52f, 54f
  framework for, 55-57
  Gaussian modeling, 46-50, 48f, 49f
  standard evaluation databases, 41-43
Room reverberation time, 208, 209, 210

S

Sampling-based methods
  data-driven parallel model combination
    (DPMC), 154
  Gaussian assumption, 156
  unscented transform, 154-156
Short-time discrete Fourier transform (STDFT),
    43-44, 211
Signal-to-noise ratios (SNRs), 42-43, 204,
    239-240
Simulation data set (SimData), 229
Singular value decomposition (SVD), 89-90
SLDM. *See* Switching linear dynamic model
    (SLDM)
SNR-dependent cepstral normalization (SDCN),
    108-109
SNR-dependent PDCN (SPDCN), 109
Soft margin estimation (SME), 91
Soft-mask-based MMSE estimation, 180
Source normalization training (SNT), 189-190
Sparse auditory reproducing kernel (SPARK),
    67
Sparse classification (SC) method, 121
Speaker adaptive training (SAT), 189-190, 190f
Spectral subtraction (SS), 188
Speech distortion weighted multi-channel
    Wiener filter (SDW-MWF), 247
Speech in noisy environments (SPINE), 41-42
Speech recognition. *See also* Hidden Markov
    models (HMM)
  components of, 9-11
  deep learning and deep neural networks,
    21-31
  Gaussian mixture models, 11-13
  hidden Markov models and variants, 13-21
  history of, 21
  optimal word sequence, 10-11
  spoken speech signal, 10-11
SPINE. *See* Speech in noisy environments
    (SPINE)

Printed in the United States
By Bookmasters